Elements of the Behavioral Code

Elements of the Behavioral Code

FRANCIS V. DeFEUDIS
and
PATRICIA A. F. DeFEUDIS
Autonomous University Medical School,
Madrid, Spain

1977

ACADEMIC PRESS
LONDON NEW YORK SAN FRANCISCO
A Subsidiary of Harcourt Brace Jovanovich, Publishers

ACADEMIC PRESS INC. (LONDON) LTD.
24/28 Oval Road,
London NW1

United States Edition published by
ACADEMIC PRESS INC.
111 Fifth Avenue
New York, New York 10003

Library of Congress Catalog Card Number: 77–74362
ISBN: 0–12–208760–7

Text set in 11/12pt Monotype Baskerville, printed by letterpress, and bound in Great Britain at The Pitman Press, Bath

Prologue

We are grateful to Academic Press for providing us with the opportunity to write this monograph. Data obtained by investigators working in a variety of scientific disciplines have been analyzed and integrated in an attempt to provide "Elements of the Behavioral Code". A theory for the "Behavioral Code" is provided which is based on findings which have revealed that *environment alters neuromorphology*. Emphasis has been placed on the changes in synaptic morphology and chemistry that accompany informational input to the CNS and environmentally induced behavioral modifications. Attempts have been made to answer several basic psychological questions using data provided by biochemical and morphological experiments.

It is proposed that, in vertebrates, various environmental inputs serve as "non-specific" stressors that increase the levels of "arousal" and limbic system response which then cause general metabolic changes in the whole CNS (and in some peripheral tissues). However, enduring changes in behavior, characterized by learning and memory, appear to occur only in the synaptic elements of those specific regions of the CNS that subserve the functions required for the adaptive response to particular environmental changes. These enduring changes that govern behavior appear to be mediated by alterations in the *amount of synaptic protein synthesized*, which is reflected by changes in the number or size of synaptic boutons but which may not be reflected as changes in the total protein content of the CNS or of CNS regions. This theory is in accord with both the "aggregate-field" and "cellular-connectionist" concepts, since it includes events that affect the total CNS as well as those that persist as memories and which may be specific to synaptic connections of certain functional pathways.

We are also grateful to the many investigators who have supplied us with reprints and unpublished manuscripts of their work and to the investigators and publishers who have granted us permission to reproduce the data which are presented in the tables and figures of this monograph. Special thanks are due to Professor W. Buño, Professor E. Garcia-Austt, Dr B. Kulig, Professor E. Kvamme, Dr A. Parreño,

Professor J. H. Quastel, Professor G. Ramirez, Professor J. Segundo, Dr R. N. Walsh, and Professor B. L. Welch for their expert criticism of various chapters, to Miss Gaynor Fox and Miss Marcia Danby for typing several drafts of this manuscript, to Señorita Conchita Muela for help with the photographic work, and to Señora Mª Angustias Pérez-de-Tudela of the Instituto Cajal for assistance with the bibliography.

Madrid F. V. DeFeudis, B.S., M.A., Ph.D.
June 1976 P. A. F. DeFeudis, B.A., M.A., Ph.D.

Foreword

Attempts have been made for many years to understand memory in terms of some sort of physical change in the brain. Socrates thought that there existed in the mind something like a block of wax that received an impression and when this impression disappeared all recollection of whatever gave rise to the impression disappeared. Descartes, in the seventeenth century, conceived that parts in the brain, on receiving an impression, became distended and that this distension, while it lasted, constituted a memory record. Huxley (T. H. Huxley, On the hypothesis that animals are automata. In "Methods and Results" (1874), pp. 199–250, Essays, Appleton, New York, 1917) during the last century had ideas that were rather more in keeping with modern thought. He proposed that "each sensory impression leaves behind a record in the structure of the brain—an ideagenous molecule, so to speak, which is competent to reproduce in a familiar condition the state of consciousness which corresponded with the original sensory impression. It is these ideagenous molecules which are the physical basis of memory". Magoun has recently briefly reviewed a considerable body of information and hypotheses that has accumulated in modern times. (Neural plasticity and the memory process, In "Biological Treatment of Mental Illness", 1966 (Ed. M. Rinkel) pp. 154–193, L. C. Page and Co.)

Drs Francis and Patricia DeFeudis have, in the present informative book, analyzed and integrated the results of many modern investigations, neurological, neurochemical, neurophysiological, in an attempt to give a rational understanding, in terms current at the present day, of memory and behavioral phenomena. They suggest that the major event occurring during memory is the formation of new protein, both structural and enzymatic, that is required for synaptic structures and which regulates behavioral changes. "This chemo-morphological change, which is triggered by nervous impulses and which affects membrane events, and subsequently, the regulation of gene expression, may persist in localized neural pathways by a form of specificity that could depend mainly upon the nature of the environmental stimulus"

(p. 207). The collation of recent experimental data and present day hypotheses is a valuable feature of this book which should be of great service to all workers in this complex and important field of investigation.

September, 1976

J. H. Quastel, Ph.D., D.Sc., F.R.S., F.R.S.C., C.C.
Professor of Neurochemistry,
University of British Columbia,
Vancouver, Canada

Contents

Dedicated to the memory of
Professor Santiago Ramón y Cajal

1. Introduction

A. Concept of the Synapse; Origin of the "Behavioral Code"

In 1891, Waldeyer gathered the known facts about nerve cells and devised what is known today as the "neurone doctrine" (i.e. that the nervous system is composed of discrete elements rather than of a syncytial net, and that the neurone represents the genetic, anatomic, functional, and trophic unit of the nervous system). However, it remained for Ramón y Cajal to demonstrate nerve cells in their entirety. Using the knowledge that nerve fibers possess an affinity for silver salts, which they reduce to a deep brown silver proteinate, Ramón y Cajal made preparations which revealed nerve cells and their axons and dendrites, all stained in various shades of brown.

> Qu'il s'agisse de ramifications finales de cylindres-axes, de collatérales de cylindres-axes ou de prolongements protoplasmiques, dans tous ces cas, ce sont toujours des branchilles *libres* qui forment les terminaisons. La cellule nerveuse avec tout l'ensemble de ses divisions et subdivisions constitue donc une individualité absolument indépendante et, pour employer l'expression de Waldeyer, nous l'appellerons *neurone*. . . . Libres, les ramifications nerveuses terminales ne s'unissent donc point à d'autres ramifications nerveuses. Non, elles s'appliquent purement et simplement, *par contact*, sur le corps ou sur les expansions protoplasmiques d'autres éléments cellulaires.
>
> (Ramón y Cajal, 1952a, p. 31)

(What concerns us is the terminal branches of axis cylinders, of collaterals of axis cylinders or protoplasmic extensions, in all these cases, *free* branchlets always form the terminals. The nerve cell with all its divisions and subdivisions thus forms a completely independent unit and, to use Waldeyer's expression, we shall call it *neurone*. . . . Being free, the terminal nerve branches therefore never join other nerve branches. No, they function purely and simply, *by contact*, upon the body or protoplasmic expansions of other cellular elements.)

Ramón y Cajal also was among the first to postulate that learning and memory processes ("cerebral exercise") might involve alterations in synaptic structure (Ramón y Cajal, 1895, 1911; see also Tanzi, 1893). In his conclusions regarding histological studies of the mammalian cerebral cortex, Ramón y Cajal (1895) stated:

> Dans le premier cas, on pourrait supposer que la gymnastique cérébrale, puisqu'elle ne peut produire des cellules nouvelles (les cellules nerveuses ne se multiplient pas comme les cellules musculaires) porte un peu plus loin que l'ordinaire le développement des expansions protoplasmiques et des collatérales neuveuses, en forçant l'établissement de connexions intercorticales nouvelles et plus étendues. Dans ce processus, pour expliquer le maintien du même volume cérébral, on peut imaginer une diminution corrélative du corps des éléments nerveux ou une réduction de la trame nevroglique.
>
> (In the first case, one could assume that cerebral exercise, since it cannot produce new cells (nerve cells do not multiply like muscle cells) carries further than usual the development of protoplasmic expansions and nerve collaterals, forcing the establishment of new and more extended intercortical connections. In order to explain the maintenance of a constant cerebral volume in this process, one could imagine a correlative diminution of the body of nervous elements or a reduction of the neuroglial web.)

His contributions toward establishing the neurone as the structural unit of the nervous system (e.g. Ramón y Cajal, 1888, 1891, 1895, 1911) along with those of Sherrington in developing the concepts of the "synapse" and "reciprocal inhibition" (Foster and Sherrington, 1897; Sherrington, 1906), stimulated searches for causal relationships between changes in neural structure and the processing of environmental information by the organism. Such studies are at the core of a major intellectual issue of our time—the nature of the *Behavioral Code*.

B. Environment Causes Behavior

In order to approach this issue, it should be understood that behavior, which includes the organism's outward expression of learning (acquisition of information) and memory (retention of information) is *caused by the environment* (e.g. Skinner, 1938, 1953), though its limits appear to be set by heredity and temporarily adjusted by the organism's levels of "arousal" (Lindsley, 1951; Schlosberg, 1954; Hebb, 1955; Jasper *et al.*, 1958) and limbic system activation (Papez, 1937; MacLean, 1958; Delgado, 1967; DiCara, 1974; see Fig. 1). An organism's perceptual

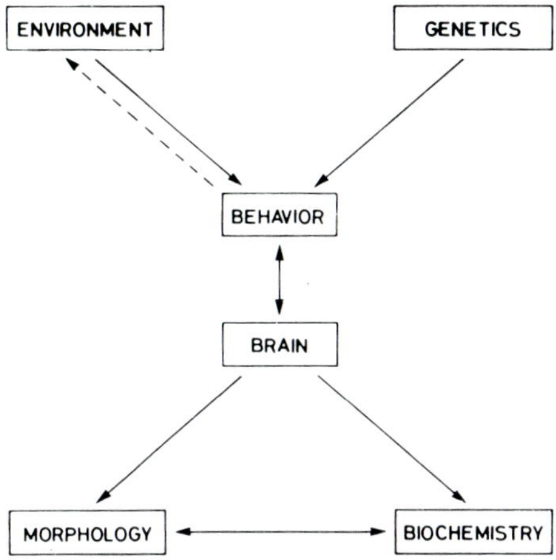

FIG. 1. General interrelationships involved in studies concerning the morphological and chemical bases of behavior.

experience and knowledge of its *external* (physical) environment constitute its behavior. In Skinner's (1938) words: "It is more to the point to say that behavior is that part of the functioning of an organism which is engaged in acting upon or having commerce with the outside world." Modifications of an organism's perception of its *external* environment, produced by changes in its *internal* environment (e.g. by organic lesions, infectious microorganisms, drug administration, or by other insults which affect homeostasis), also play major roles in behavioral regulation.

Claude Bernard (1859, 1878) was one of the first to distinguish *internal* from *external* environment, the "milieu intérieur" consisting mainly of body fluids which bathe the cells and the "milieu extérieur" being the organism's external environment (but, see Haldane, 1917 and Chapter 5).

> La *fixité du milieu intérieur est la condition de la vie libre, indépendente*; le mécanisme qui la permet est celui qui assure dans le *milieu intérieur* le maintien de toutes les conditions nécessaires à la vie des éléments.
>
> (Bernard, 1878, p. 113)

> (The stability of the internal environment is the condition for free independent life; the mechanism which allows this is that which secures, in the internal environment, the maintenance of all conditions necessary for the life of the elements.)

After Bernard had provided this concept of internal stability, Cannon (1932) coined the term *homeostasis* to describe the relative stability and constancy of an organism's *internal* environment (physicochemical state):

> The constant conditions which are maintained in the body might be termed *equilibria*. . . . The coordinated physiological processes which maintain most of the steady states in the organism are so complex and so peculiar to living beings—involving, as they may, the brain and nerves, the heart, lungs, kidneys and spleen, all working cooperatively—that I have suggested a special designation for these states, *homeostasis*. It means a condition—a condition which may vary, but which is relatively constant.

The potency of homeostatic mechanisms is readily appreciated when it is considered that both circadian and ultradian rhythms exist in the behaviors of many animals, including man (e.g. Richter, 1927; Sollberger, 1971; Kripke, 1974; Delgado-Garcia *et al.*, 1976).

Experiments by Richter (1942) broadened this concept by showing that when disturbances in homeostasis occur, these are compensated for by behavior which serves to return the organism to its physicochemical steady state. It is easy to conceive that motivation, emotion, drives, needs and arousal,* then, may all act upon the organism by causing disturbances in homeostasis. Instinctive and autonomic mechanisms are among the important regulators of homeostasis.

---

* *Motivation* refers (in a general sense) to those conditions, *both external and internal*, which lead to organized behavior. It is a mechanism which impels an organism to act. Particular motivations (e.g. those of hunger and sex) refer to tendencies which appear intermittently and which vary in strength. Motivational processes account for the initiation and direction, as well as the observed variability, of behavior.

Emotion is generally considered to be a component of motivation and refers to the affective (or subjective) characteristics of bodily reactions or states resulting from *internal or external changes*.

Drives refer to *internal changes* (usually induced by deprivation) which predispose the organism to execute behaviors which oppose these internal changes. Drives are based on the notion of homeostasis; i.e. when there is an imbalance in bodily processes, the organism behaves to restore its equilibrium. A drive may be regarded as a specific conception about the way in which motivational processes function. Drive may also be considered in a more generalized sense as an "energizer" (see "arousal").

Arousal refers to altered activation levels of the organism which influence behavior; too low an arousal level in a particular situation may inhibit responding, whereas too high an arousal level may disrupt responding. Arousal may also be viewed as a generalized drive state having the function of maintaining cortical activity at a level sufficient for appropriate responding (see Hebb, 1955).

Needs refer to basic organismic conditions that must be satisfied for survival of the organism or the species (e.g. hunger, thirst, sex and avoidance of pain).

Learning and Memory may be considered to be instruments which serve motivational processes.

However, homeostatic theory cannot explain all behaviors, although changes in drive (*internal changes*) appear to be sufficient for the occurrence of some behaviors (e.g. consummatory acts, such as eating, drinking, and sexual behavior). In organized "goal-oriented" behaviors *external* environmental factors are decisive determinants of the responses. Motivational factors arising from external environmental stimuli ("incentive stimuli") interact with ongoing internal conditions (drives) to determine such behaviors (see Bindra, 1968). Hence, human motivation is governed to a great extent by non-homeostatic determinants (see also Hebb, 1949; Harlow, 1953). According to Harlow (1953): "Most, if not all, of man's complex learning is motivated by non-emotional or mildly pleasurable stimuli and is disrupted or inhibited by intense affective states."

Thus, the probability that an organism will emit a given behavior can be predicted by studying its *internal* and *external* environments. It appears likely that exposure of animals, including man, to various environmental stresses, especially during periods of rapid cerebral development, can be etiological factors in the development of behavioral disturbances. It should be noted further that although a particular environmental change should present similar problems of adaptation to any animal, the animal's behavioral response is expected to depend upon its central synaptic organization and upon its interpretation of the environmental change.

Environmental inputs are continuously filtered, integrated, stored and retrieved by the organism in its efforts to emit behavioral responses. Thus, the central nervous system (CNS) transforms environmental signals into neuronal signals that form the basis of the learning and memory code, or the *Behavioral Code.* Evidence for such a coding mechanism stems from the "law of specific energies" (Müller, 1826) which states that the "quality of the information is independent of the nature of the stimulus but is determined entirely by the central connections of the pathway stimulated". However, it should be noted that subsequent experimentation has revealed that particular afferent pathways of various mammals are not exclusively involved with a particular sensory modality (e.g. recordings from sensory neurones have indicated that fibers which are sensitive to tactile stimulation at discrete points on the skin can also be sensitive to temperature or to visual stimuli; see, e.g. refs in Stein *et al.*, 1976). However, there does not seem to be a great deal of overlap of modalities within a given sensory pathway although interaction among neighboring pathways is frequent (see, e.g. P. M. Milner, 1970).

Sinclair (1955) and Wall (1960) have suggested an alternative to the idea of specificity of sensory cells and have proposed that all sensations are the result of specific spatio-temporal patterns of input. Moreover, experiments by Mountcastle and Powell (1959) have indicated that information about the locus of stimulation is represented in the cortex by the activity of many cells having overlapping peripheral fields. Hence, the presence of a specific sensory neurone, implied by Müller's model, is not necessarily required in order for the nervous system to organize a particular response to a given stimulus. Rather, a response appears to be produced by a *pattern* of neural activity, and if one considers only the response that a stimulus may evoke, then the nervous system can convert directly from a sensory pattern to a motor pattern (see P. M. Milner, 1970).

C. Behavior, Learning and Memory

Behavior is a dynamic process of extreme complexity, but its dynamics and associated neural events must be dealt with while attempting to formulate the *Behavioral Code*. Consistent behavioral and neural changes must be discovered and subjected to a scientific analysis. It seems essential to derive "molar conceptions of neural action" (Hebb, 1949), which necessarily require an integration of data acquired by diverse scientific approaches. The organism should be studied from both its "inside" and its "outside", and it should be assumed that experientially induced neural changes that constitute memories are in some way *structural* and *static* (Hebb, 1949). In the words of Young (1965): "Memory is not something subjective. It is a physical system, by means of which there is made a record or representation of certain past events." One might even consider that perhaps the only "energy reservoir" inside the organism may be its structural memories (see Jelliffe, 1918). The crucial role of memory in behavioral regulation was brought out clearly by Sechenov (1952) who stated that memory

> "obviously serves the animal as a guide to purposeful action—without which it could not differentiate wood from food, would confuse a tree with an enemy and in general would be unable to orient itself among surrounding objects for one minute."

At the experimental level, one needs only to realize that retrograde amnesia can be produced by electroconvulsive shock (ECS; see, e.g. McGaugh and Herz, 1972; Essman, 1973), local electrical stimulation (e.g. McGaugh and Gold, 1974; Wilburn and Kesner, 1974), antibiotics (e.g. puromycin, cycloheximide; see, e.g. Agranoff, 1973), hypoxia (e.g. Thompson and Pryor, 1956; Sara, 1973), hypothermia (e.g. Beitel and Porter, 1968), hypercapnia (e.g. Quinton, 1966), spreading depression (e.g. Bureš *et al.*, 1958; Bureš and Burešová, 1960, 1963; Avis and Carlton, 1968; Hughes, 1969; Schneider, 1973;

Bureš *et al.*, 1974), cholinolytic agents and cholinesterase inhibitors (e.g. Deutsch *et al.*, 1966; Ilyutchenok, 1969; Deutsch, 1971, 1972; Ilyutchenok *et al.*, 1973; DeFeudis, 1974b; Biederman, 1974; Puerto *et al.*, 1976) and other drugs (e.g. Mark and Watts, 1971), and that effective sites for eliciting such amnestic responses include the hippocampus (e.g. Lidsky and Slotnick, 1970; Vardaris and Schwartz, 1971; Zornetzer *et al.*, 1973), the caudate-putamen region (e.g. Gold and King, 1972; Wilburn and Kesner, 1972, 1974) and amygdala (e.g. Kesner and Doty, 1968; Bresnahan and Routtenberg, 1972) to understand that memory processes are associated with diverse cerebral events. Also, memory can be facilitated by certain drugs (e.g. McGaugh, 1973; Dawson and McGaugh, 1973; Deutsch, 1973), by stimulation of the brain stem reticular formation (e.g. Bloch, 1970; Denti *et al.*, 1970) and perhaps by stimulation of some other CNS regions (e.g. Erickson and Patel, 1969; for detailed coverage of the effects of brain stimulation on learning and memory, see Kesner and Wilburn (1974) and Routtenberg (1975). These diverse environmental insults that affect learning and memory may share some common elements (e.g. changes in cerebral oxidative metabolism and in protein synthesis; the production of surface-negative gradients across the cerebral cortex), which could lead to short- or long-lasting changes in synaptic transmission. Consequently, it seems unlikely that memory will be explained in terms of a single postulated neurotransmitter system, even though limited regions of the nervous system could be required for the learning or retention of particular activities (see also Chapter 2, Part A and Chapter 8, Part E).

D. Role of Genetics

The nervous systems of higher vertebrates are "fully" developed at birth, or shortly thereafter, and young animals possess many innate ("pre-coded") neural pathways which enable them to emit reflexes and complex instinctive behaviors. These innate processes develop rather independently of experience, whereas the postnatal development of other neurobehavioral processes depends upon an adequate amount of environmental input. However, it should be noted that certain genetically transmitted behavioral tendencies may remain latent until triggered by the essential environmental stimuli.

Studies on the neurochemical and morphological correlates of innate behaviors, especially in genetic mutants (see, e.g. refs in Sotelo, 1975a, b), appear to be of great value in elucidating elements of the *Behavioral Code*, especially since such behaviors may depend on mechanisms which are similar to, or identical to, some behaviors that are triggered by individual experience. Environmental conditions must be monitored in such studies, since genetically based differences in environmental interactions could modify enzyme activities and membrane structures involved in neurotransmission, and since instinctive behaviors can be modified to some extent by environmental influences.

The analogy between chemical coding of genetic information and memory processes first became obvious to Katz and Halstead (1950). Initially, this idea led investigators to consider RNA as the memory molecule, but today emphasis is being placed on both nucleic acids and proteins (see Chapter 4). As examples of relationships between genetics and possible CNS neurotransmitters, it has been shown: (1) that norepinephrine (NE) concentrations are lower in the hypothalami and brain stems of genetically hypertensive rats than in control strains of rats (Yamori *et al.*, 1970); (2) that correlations exist between cerebral NE and serotonin (5-HT) concentrations and susceptibility to audiogenic seizures in certain inbred strains of mice (Schlesinger *et al.*, 1965, 1970); (3) that there appears to be a genetically linked difference in the affinity of NE uptake by synaptosomal fractions of mouse brain (Moisset *et al.*, 1975; Fig. 2). Other studies with genetically different strains of mice involved the phenotyping of these animals for both behavior and cerebral neurotransmitter metabolism. Using eight strains of rat known to differ in some behavioral parameters, Rick *et al.* (1971) showed that GABA production in the somatosensory cortex was inversely related to behavioral measures of learning and activity. After measuring the activities of five enzymes (glutamate-α-decarboxylase (GAD); choline acetylase (ChAc); acetylcholinesterase; monoamine oxidase; catechol-O-methyl transferase) in whole brain, Tunnicliff *et al.* (1973) suggested that: (1) an inverse relationship exists between the central cholinergic system, maintenance of activity and latency of initial responses to shock; (2) a positive association exists between monoamine systems and maintenance of activity and shock escape latency; (3) the GABA system showed relationships similar to, but less significant than those of the cholinergic system. Most recently, Gaitonde and Festing (1976) have shown, using ten strains of mice, that an inverse correlation existed between whole forebrain GAD and exploratory activity (see Fig. 3), a finding which indicated that less GAD-containing synapses might exist in the brains of mice which display higher exploratory activity. (Recently, an extensive bibliography about studies with genetically defined mice has been published; see Sprott and Staats, 1975.)

Although our main problem is to discover the mechanisms by which *new* patterns of behavior are learned, stored, and used by the organism for adaptation to its ever-changing environment, it must be appreciated that the mechanisms involved in new behavioral patterns reside on a baseline that includes innate components.

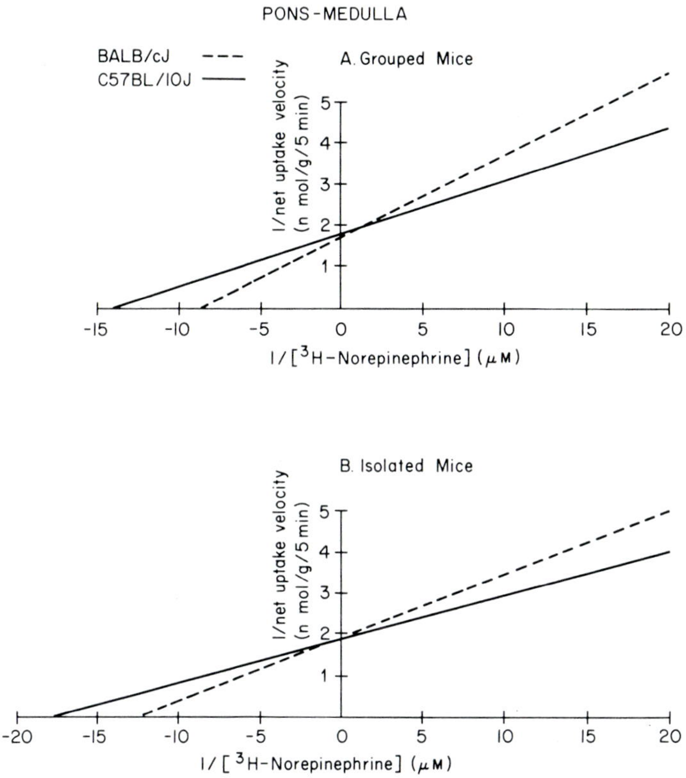

FIG. 2. Lineweaver-Burk analysis of strain differences in ^{3}H-norepinephrine uptake kinetics in homogenates from pons-medulla of male BALB/cJ and C57BL/10J mice. Net uptake at 37°C minus uptake at 0°C was determined in 5-min homogenates. Plotted values are averages derived from least squares regressions calculated for individual mice. (Reproduced with permission from Moisset *et al.*, 1975.)

E. Role of Maturation

It is well known that an animal's behavior changes during its development and that such changes are accompanied by changes in cerebral energy metabolism, in neurotransmitter systems, and in many other cerebral and peripheral chemical substances, as well as by changes in synaptic morphology (see, e.g. Waelsch, 1955; Purpura and Schadé, 1964; Davison and Dobbing, 1968; Dobbing, 1968; Jacobson, 1970a, b; W. A. Himwich, 1970; Benjamins and McKhann, 1972; Brazier, 1975; Chapter 8, Parts B and G), and in cerebral electrical activity (e.g. Crain, 1952; Bradley *et al.*, 1960; Deza and Eidelberg, 1967). Many of our present concepts regarding neuronal ontogeny, as

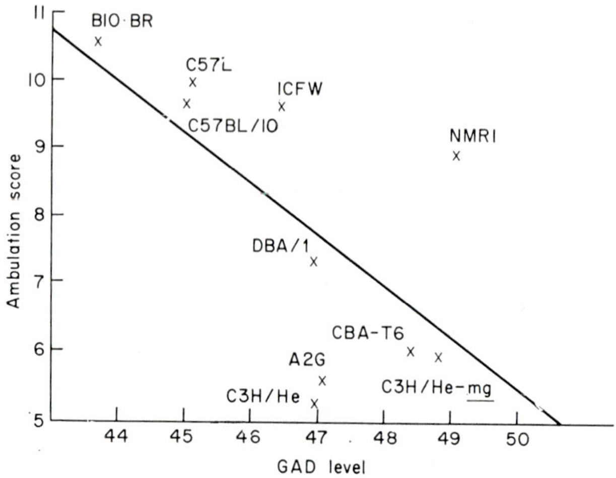

FIG. 3. The inverse correlation between brain glutamic acid decarboxylase level (μmole glutamate decarboxylated/g wet wt./h) and the exploratory activity of 10 inbred strains of mice. The data fitted the equation $y = 42{\cdot}805 - 0{\cdot}747\times$ with a coefficient of correlation $r = -0{\cdot}64$ ($\rho < 0{\cdot}05$). Ambulation score is the square root of the number of squares entered in a 2 min period. (Reproduced with permission from Gaitonde and Festing, 1976.)

well as degeneration and regeneration, have been established mainly through the classic histological work of Ramón y Cajal (1929), who used modified Golgi (1873) methods. It is also well known from these early studies that the rate of growth of the brain and its constituent elements decreases during development.

Therefore, it is expected that the level of maturation of an animal will influence markedly its responses to various environmental insults and hence its mechanisms of learning, memory and adaptation. As appears to be the case with nutritional insults, the brain is perhaps more vulnerable to environmental insults during its period of most rapid growth (see, e.g. Dobbing, 1968, 1976).

F. Role of Nutrition

An important role of nutrition (a component of *internal* environment) in governing memory formation and behavior is evident from the findings that certain dietary treatments can reverse or prevent the symptoms of some types of brain injury and genetic biochemical impairment (e.g. phenylketonuria). Profound influences of under-nutrition on the neural (e.g. reduction in cell number and myelination;

EEG changes) and behavioral (e.g. hyper-reactivity; inferior motor performance) development of animals, including man, have also been reported, especially when undernutrition had its onset during the suckling period (see, e.g. Stoch and Smithe, 1967; Culley and Lineberger, 1968; Howard and Granoff, 1968; Dobbing, 1968, 1976; Chase *et al.*, 1969; Altman *et al.*, 1970, 1971a, b; Dobbing and Sands, 1973; Dobbing and Smart, 1973; Leathwood *et al.*, 1974; Zimmerman *et al.*, 1974, 1975; Gramsbergen, 1976).

Dietary modifications also influence putative neurotransmitter systems of the brain. For example, with regard to the cerebral mechanism of phenylketonuria, in which the principal metabolic lesion is a deficiency of hepatic phenylalanine hydroxylase activity (Kaufman, 1958), it has been shown that phenylalanine-rich diets caused decreases in cerebral 5-HT (e.g. Yuwiler and Louttit, 1961). This change in brain 5-HT was accompanied by behavioral impairment in some learning tasks (Yuwiler and Louttit, 1961). Further studies revealed that a reduction of brain 5-HT in phenylalanine-fed rats was caused by a decrease in 5-HT formation and that inhibition of both tryptophan hydroxylation and precursor transport were involved (Yuwiler *et al.*, 1965; Yuwiler and Geller, 1969). As in phenylketonuria, mental retardation is a prominent clinical feature in "maple syrup disease" (branched-chain ketonuria). Yuwiler and Geller (1965) showed that dietary leucine decreased brain 5-HT in a fashion similar to that caused by dietary phenylalanine. These studies have indicated that the mental deficiency characterizing phenylketonuria and maple syrup disease might have a common basis; i.e. a decrease in cerebral 5-HT.

In accord with this notion, it has been shown that increases or decreases in dietary tryptophan over periods of several weeks were reflected by parallel changes in cerebral 5-HT (e.g. Culley *et al.*, 1963). More recently, it has been shown that the increased concentration of plasma tryptophan produced in rats by ingestion of carbohydrate was accompanied by increases in cerebral tryptophan and 5-HT contents (Fernstrom and Wurtman, 1971). These workers suggested that 5-HT neurones might be involved in systems with which the rat's brain integrates information about its metabolic rate in relation to its control of homeostasis and behavior. Further studies have revealed that the concentrations of tryptophan, 5-HT and 5-hydroxyindoleacetic acid (5-HIAA) were all increased in the spinal cord and in several brain regions two hours after fasted rats had begun to consume a carbohydrate-fat meal (Colmenares *et al.*, 1975). These results indicated that such changes had occurred in both cell bodies and terminals of 5-HT neurones. Such modified diets may be of use for reversing the mental deficiency of phenylketonuria and maple syrup disease.

The acetylcholine (ACh) concentrations of whole brain or brain regions of rats have been shown to increase after ingestion of choline-rich diets (Cohen and Wurtman, 1976). Such results indicated that certain functional disorders in humans that are associated with deficient cerebral cholinergic activity (e.g. tardive dyskinesia) might be treated successfully with large doses of dietary choline (see also Davis *et al.*, 1975). In another study, 25% less norepinephrine (NE) was found in the brains of rats that had been undernourished from mid-gestation and killed at weaning than in those of adequately fed litter-mates; tyrosine hydroxylase activity was significantly increased in the brains of these animals (Shoemaker and Wurtman, 1971). Even five days of water deprivation can produce a selective decrease in the 5-HT concentrations of the neural lobe and pars distalis of the rat's pituitary gland (Piezzi and Wurtman, 1970). However, the central cholinergic system of developing brains of protein-energy undernourished rats and pigs appeared to exhibit increased activities in both pseudo-cholinesterase (ChE) and true acetylcholinesterase (AChE) (Im *et al.*, 1971, 1973; Adlard and Dobbing, 1972), whereas ChAc activity was decreased in the brain stems of undernourished rats (Eckhert *et al.*, 1976).

A decreased threshold for "arousal" existed in male mice that were undernourished by feeding their mothers restricted diets during pregnancy and lactation (Smart, 1975). Also, some enduring cerebral and behavioral effects of early growth restriction in rats (Dobbing and Smart, 1974) and in mice (Castellano and Oliverio, 1976), whose only handicap was to be suckled in families larger than normal during the brain "growth spurt", have been reported.

These findings indicated that changes occur in the biochemical (Eckhert *et al.*, 1976) and/or ultrastructural (Gambetti *et al.*, 1974) maturation of the brains of animals that are undernourished during the suckling period. Early malnutrition (like environmental impoverishment; see, e.g. Welch, 1965; Schultz, 1965; DeFeudis, 1975a) delays brain maturation and behavioral development (see also Chapter 8, Part G, and Castellano and Oliverio, 1976). Furthermore, such untoward effects of early malnutrition in rats have been reversed by providing the animals with an "enriched" environment (Levitsky and Barnes, 1972). Evidence also exists at the human level for the interaction of nutrition and social environment (see also Chapter 8, Parts A and H). Malnutrition and environmental impoverishment might act synergistically in retarding mental development (Levitsky and Barnes, 1972), and exposure of previously malnourished children to stimulating environments can improve mental development (see Winick *et al.*, 1975).

G. Neural Plasticity

Plasticity is concerned with the nature and scope of recovery of function after neural *damage* and this capacity appears to decline during development of the organism. Studies on neural plasticity have provided useful clues about the natures and mechanisms of neuromorphological changes associated with learning, memory and behavior. Biochemical and morphological changes which characterize *developmental plasticity* may be similar to those associated with learning and memory, the differences being mainly in "scale, address and connectivity" (see Rose *et al.*, 1976).

The most basic evidence for neural plasticity derives from the effects of injury after section of axons and their subsequent regeneration (see, e.g. Guth and Windle, 1970, 1973; Guth, 1974, 1975; Bjorklund *et al.*, 1974; Kerr, 1975; Grafstein, 1975). Regeneration of the CNS has been shown to occur in several "cold-blooded" animals (see, e.g. Gaze, 1970), and recently a recovery of function and a recruitment of alternative pathways involved in feeding behavior have been shown in the marine polyclad flatworm, *Notoplana acticola*, after severing the nerve plexus on one side of its body (Koopowitz *et al.*, 1975). Although such effective regeneration is not easily shown in the mammalian CNS, the "collateral sprouting" that occurs in both the peripheral nervous system and CNS of mammals (that involves formation of new sprouts from the intact axons) has provided evidence for synaptic regeneration and plasticity (see, e.g. McCouch *et al.*, 1958; Kandel and Spencer, 1968; Raisman, 1973; Guth, 1974).

Recent studies have indicated that many non-myelinated axon systems of the adult mammalian CNS possess a high capacity for regenerative "sprouting". Using fluorescence histochemistry, a regrowth of central serotonergic neurones was shown after intraventricular 5,6-dihydroxytryptamine-induced axotomy of the rat hypothalamus, brain stem, and spinal cord; extensively damaged neurones displayed a partial ability for axonal re-growth along original paths by "sprouting", especially in the inferior olive (Bjorklund *et al.*, 1973, 1974; Nobin *et al.*, 1973; Wiklund and Bjorklund, 1975). "Sprouting" in the lower brain stem of rats following intraventricular injections of another strongly reducing congener of 5-HT, 5,7-dihydroxytryptamine, has also been described (Baumgarten, and Lachenmayer, 1972). Earlier, a vigorous "sprouting" from severed axons was observed following mechanical or electrolytic lesions of cerebral catecholaminergic and indoleaminergic neurones of adult rats (Bjorklund *et al.*, 1971; Katzman *et al.*, 1971). Other fluorescence histochemical studies have revealed that ingrowths of noradrenergic, cholinergic, dopaminergic and serotonergic fibers can occur into irides transplanted into rat CNS (Bjorklund and Stenevi, 1971; Svendgaard *et al.*, 1976). This ability of the iris to support regenerative ingrowth by fibers led to a further investigation with GABA. Using GAD as a marker for GABA terminals, it was shown that non-myelinated axon systems of the adult rat hippocampus and caudate nucleus also have a high capacity for regenerative "sprouting" into transplanted irides, thus providing the first evidence for regeneration of "GABA-ergic" neurones in the adult mammalian CNS (Emson *et al.*, 1976). Although these responses, and those mentioned above (see, e.g. Raisman, 1973), were provoked by destructive lesions, it should be noted that they could contribute to the plasticity of the normal mammalian CNS (see Guth, 1974).

The ultimate aim of these studies is to correlate structural and functional changes in the nervous system and to discover how such changes can be produced by excess use or by disuse. Excess use (or *increased normal use*) could enhance function and lead to memory storage by increasing the synthesis and availability of transmitter substances (e.g. by changing the activities of transmitter-synthesizing enzymes), by causing an enlargement of synaptic boutons, or by initiating the "sprouting" of new synaptic boutons. Since increases in the size or number of synaptic boutons could secure increased synaptic action for the lifetime of the organism (i.e. for as long as the memory persists), such changes could constitute the basis of "long-term" memory. The concept that synaptic growth is caused by increased function is supported further by the findings that the syntheses of RNA and protein increase during enhanced activity and during learning and memory (see, e.g. Hydén, 1973a, b, c; Chapter 4). An adaptive neuronal model, based on studies of synaptic regulatory function in relation both to genetic differences and to changes in the *internal* and *external* environments, and its relation to affective disorders, has been provided by Mandell (1975b); see Fig. 4. This model emphasizes the idea that changes in neural plasticity which govern behavior are mediated by small molecules associated with mechanisms of neurotransmitter synthesis, uptake and release.

H. Social Isolation

Studies on environmental improverishment, or "social isolation", which causes pronounced cerebral and behavioral effects in many vertebrates species, including man (e.g. Spitz, 1945; Beach and Jaynes 1954; Thompson and Heron, 1954; Bexton *et al.*, 1954; Hebb, 1958; Kuo, 1960; Burns and Kimura, 1963; Mason, 1963; Scott, 1964; Harlow *et al.*, 1964, 1965; Welch, 1965; Lorenz, 1966; Baenninger, 1967; Miller, 1968; Zubek, 1969; Prescott, 1971; DeFeudis, 1971b, 1974c, 1975a; Sackett, 1972, 1974; Gendreau *et al.*, 1972; Valzelli, 1973a, b; Harlow and Suomi, 1974; Konrad and Melzack, 1975; Leigh and Hofer, 1975; Hoffman *et al.*, 1975), will be particularly

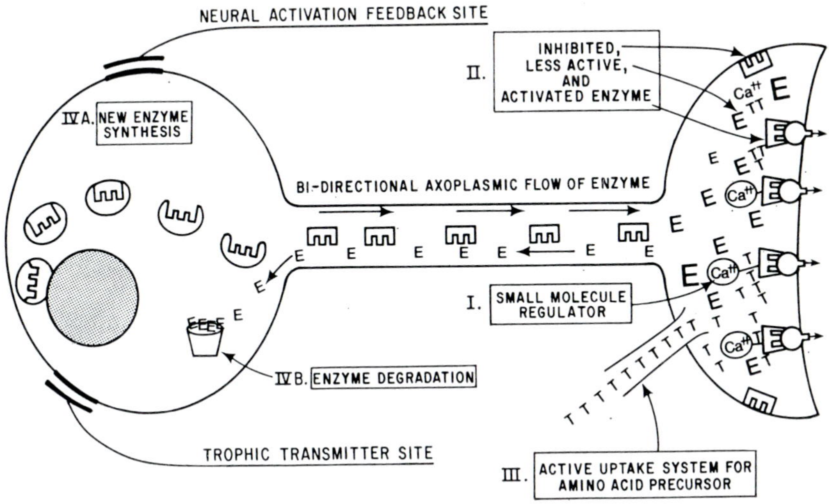

FIG. 4. A view in aggregate of the neurobiological adaptive model. Studies of synaptic regulatory function in relation to drugs, genetic strains, and environmental manipulations have focused on: (I) Small-molecule regulators of nerve ending neurotransmitter biosynthetic enzymes, exemplified by calcium and heparin, which activate tryptophan hydroxylase and tyrosine hydroxylase, respectively. Ca^{2+} activation of tryptophan hydroxylase is pictured here in juxtaposition to nerve ending structures. (II) Macromolecular alteration of neurotransmitter biosynthetic activity by membrane binding. In model synaptosomal systems *in vitro*, the activity of the soluble form of tyrosine hydroxylase is intermediate between that of membrane-occluded (inhibited) and membrane-activated forms. (III) Synaptosomal preparations evidence low-affinity ($K_m \simeq 1 \times 10^{-3}$ M) and high-affinity ($K_m \simeq 1 \times 10^{-5}$ M) uptake systems for tryptophan. High-affinity transport, which responds to changes in pH, temperature, energy, drugs, and ion and substrate concentrations, is highly correlated with the capacity of synaptosomal systems to convert tryptophan to 5-HT. (IV A and B) Via either receptor-mediated feedback affecting discharge rate or trophic transmitter action, the state of synaptic activity influences the synthesis and/or the degradation of neurotransmitter biosynthetic enzyme in the cell body. The manifestation of these changes in nerve endings is seen as a complex function of synthesis, release from cell body protein synthesizing systems, and axoplasmic flow. These adaptive changes tend to move in the direction necessary to oppose the influence of the acute or chronic treatments used as the independent variable. (Reproduced with permission from Mandell, 1975b.)

emphasized herein. Social isolation produces behavioral changes (e.g. increased aggressivity; decreased sexual activity and problem-solving behavior) in both young and adult animals, some of which are similar to those produced by undernutrition (see above). These findings and their applicability to man were revealed clearly by Hebb (1958) who stated:

> Without physical pain, without drugs, the personality can be badly deformed simply by modifying the perceptual environment. It becomes

evident that the adult is still a function of his sensory environment in a very general sense, as the child is.

Inherent to these words of Hebb are the concepts that the nervous system functions to receive information from its *external* (social) and *internal* environments by means of sensory receptors (exteroceptors and proprioceptors, respectively) which govern an organism's level of motivation. One cannot overestimate the effect which an individual's social environment has on his level of motivation and associated performance. Such environmentally induced behavioral changes have now been correlated with chemical and morphological alterations in the CNS (see following Chapters), with changes in the brain's responses to centrally-active agents (see, e.g. Baumel *et al.*, 1969; refs in DeFeudis, 1975a), and with changes in endocrine function (e.g. Sigg, 1969; Brain and Nowell, 1971a, b; see Chapter 4, Part E).

The factors causing these alterations appear to be related to the adaptation of the organism to abnormally low levels of sensory stimulation (e.g. Melzack, 1959; Welch, 1965, 1967) and to the stress promoted by the absence of con-specifics (e.g. Moore, 1968; Welch and Welch, 1968a; see also Chapters 7 and 8). It is noteworthy that many effects of environmental impoverishment can be reversed by restoring the animals to group housing or by providing them with increased conspecific interaction (e.g. Bronson and Chapman, 1968; Plaut and Grota, 1971; Harlow *et al.*, 1971; Suomi *et al.*, 1972, 1974; DeFeudis, 1973, 1975a; Bennett *et al.*, 1974; Walsh and Cummins, 1975; see Chapter 2, Part B, and Chapter 8, Part A). It has been shown also that "self-disturbance" behaviors in monkeys, caused by keeping them under conditions of partial social isolation during their first 11 months of life, were decreased significantly by treating the animals with the major tranquilizer, chlorpromazine (McKinney *et al.*, 1973). Other environmentally induced effects, especially those associated with pre- or post-weaning isolation, may be quite resistant to change (e.g. Harlow, 1962; Mason, 1968; Prescott, 1971; Leigh and Hofer, 1973, 1975). Some neurochemical and behavioral effects caused by isolation-rearing appear to occur primarily in males (e.g. Marcucci *et al.*, 1968; Sackett, 1972, 1974), and appear to undergo diurnal variation (e.g. DeFeudis and Black, 1972; Ziesenis *et al.*, 1975).

Interestingly, studies by Harlow *et al.* (1969) indicated that, at least in monkeys, social isolation had no effect on learning performance, and thus the induced behavioral changes might be attributed to a greater extent to emotional factors than to intellectual deficiency. However, the earlier studies of Bexton *et al.* (1954) revealed that sensory deprivation in human volunteers led to inferior performance in intellectual problems, perhaps due to changes in cognitive functions or motivational state.

I. Major Questions to be Answered

Several major questions related to the transformation of "mind" to "brain" (and *vice-versa*) must be answered in order to elucidate the *Behavioral Code*: (1) Are learning and memory achieved by mechanisms of anatomical and chemical plasticity, i.e. by increases in the number of, or size of synaptic connections (or boutons), or by changes in the resistance of synapses, or by changes in the synthesis or availability of some neurotransmitters, or by changes in the number of receptor sites for transmitters? (2) If such changes occur, do these represent altered rates of synthesis of pre-existing molecules, or are new species of molecules synthesized? (3) Is there a simple unit of memory, or "mnemon", as suggested by Young (1965)? (4) If so, is this unit a neurone, a synapse, or a macromolecule? (5) Is a general change in intermediary metabolism associated with learning and memory? (6) Are specific memories localized within specific regions of the brain? With regard to this last question, it should be noted that evidence exists for successful operant learning in rats subjected to bilateral removal of neocortex, striatum, hippocampus and amygdala (Huston and Borbély, 1973, 1974; Huston, 1975). Also, the early works of Lashley (1929, 1938) should be recalled (see Beach *et al.*, 1960).

Since the human brain contains about 10^{10} neurones (each of which can be connected to about 10^4 other neurones) and about 10^{11} glial cells, it is obvious that a simplifying approach is required. From the discussion to be presented herein, it seems likely that the *Behavioral Code* involves quantitative, and possibly qualitative, changes in the structure of synaptic membranes, or in the number of synaptic boutons; changes that can be produced in large populations of cells by altering the *amount* of environmental stimulation (experience) received by the organism. It should be noted further, as emphasized by Dunn (1975), that if the formation of a memory does involve the formation of a new synapse, all of the molecules of that synapse could be turned over, but provided that the synapse remained, the memory could persist.

It should be possible now to derive quantitative molecular and morphological expressions for behavior and for its associated environmental stimuli. But, how can one show what is stored or modified in the CNS, for a given environmental change? Alterations in the efficacy of synaptic transmission or in the connectivity of large numbers of neurones that may be reflected as macromolecular (structural) changes, appear to be involved. Certainly, the behaving animal should be able to adapt its neural structural elements to various environmental insults during efforts to ensure its survival.

J. Major Approaches to the Problem

Two major approaches for studying the *Behavioral Code* have evolved: (1) It is believed that synaptic connections are strengthened by repeated use, so that learning and memory can be explained as being due to "growth through experience". It seems noteworthy that these synaptic connections must necessarily be strengthened by an influence on presynaptic boutons, since "disuse" can also promote increases in the number of postsynaptic receptors by mechanisms related to "denervation supersensitivity" (see Cannon and Rosenblueth, 1949). However, growth by "denervation supersensitivity", which does not infer development of neural mechanisms, but, in contrast, is concerned with their destruction and responses to injury, could be related to the mechanism by which "social isolation" of animals affects neural structure (see, e.g. Prescott, 1971). (2) Macromolecular mechanisms may be involved in information storage, either indirectly via the synthesis of enzymes regulating neural metabolic or transport systems, or directly as receptors or as effectors acting on receptors (e.g. Glassman, 1969; Glassman and Wilson, 1972; Hydén, 1973a, b, c; Mandel *et al.*, 1973; Agranoff, 1974; Entingh *et al.*, 1975; Dunn, 1975). These two approaches are actually components of a single approach since changes in the syntheses or conformations of membrane macromolecules appear to constitute the mechanism by which experience causes synaptic growth.

The notion that neural constituents display increased growth through experience (i.e. the "cellular-connectionist" hypothesis; that activation of synapses increases their efficacy by causing enduring changes in their fine structure) was advanced long ago (Spurzheim, 1815; Darwin, 1874; Tanzi, 1893; Ramón y Cajal, 1895) and gained support from theorists of the mid 1900s (e.g. Konorski, 1948; Hebb, 1949; Eccles, 1953). Hebb (1949) attributed learning to a positive growth of synaptic knobs, and Eccles (1964) suggested that the strengthening of existing connections could be the key to learning and memory. However, other suggested mechanisms for learning and memory have involved a weakening or loss of synaptic connections (e.g. Rosenzweig *et al.*, 1972c and refs therein; see also Chapter 7).

Another postulate which implied that learning is a dynamic process involving continuously circulating patterns of impulses in closed neural chains is based on the assumptions that information is stored in a "bioelectric field" generated by the combined activity of many neurones rather than in single cells or synapses (the "aggregate-field" hypothesis; see Lashley, 1929, 1938, 1942), and that the nervous system's reactions to given sensory inputs can be changed in a unique way so long as these impulses continue to circulate (Young, 1938;

Rachevsky, 1938; Hilgard and Marquis, 1940; Householder and Landahl, 1945). This hypothesis emphasizes the notion that global aspects of neural activity should be considered when attempting to elucidate the *Behavioral Code.*

The anatomical localizations of learning, memory and behavior cannot be defined without considering both of these approaches. Although Lashley's (1929) statement that, "the learning process and the retention of habit are not dependent upon any finely localized structural changes within the cerebral cortex", is in a sense, against the hypothesis to be presented herein and has also been criticized by other workers (e.g. Hunter, 1930; Zangwill, 1961; see Kandel and Spencer, 1968), it will emerge as an integral part of the present theory.

The essentiality of applying an interdisciplinary approach cannot be overestimated. It has already led to the expression of various behavioral states (e.g. convulsions, self-stimulation behavior, sleep and wakefulness; see, e.g. Jouvet, 1972; DeFeudis, 1974b; Morgane and Stern, 1974) in terms of the associated chemical anatomies of brain circuits. Descartes emphasized the importance of such an approach long ago:

> If therefore, anyone wishes to search out the truth of things in serious earnest, he ought not to select one special science; for all the sciences are conjoined with each other and interdependent. . . .

The search must now be continued for those biological events which characterize a broad range of environmentally induced behavioral modifications; it is their integration that is necessary for expression of the *Behavioral Code*—the subject of this monograph.

2. Changes in Gross Neuromorphology

A. Introduction and Rationale for Studies

Charles Darwin (1874) first observed that the brains of feral rabbits were significantly heavier than those of their domesticated counterparts, thereby relating a change in an animal's cerebral gross morphology to its experience.

> I have shown that the brains of domestic rabbits are considerably reduced in bulk, in comparison with those of the wild rabbit or hare; and this may be attributed to their having been closely confined during many generations, so that they have exerted their intellect, instincts, senses and voluntary movements but little. (Darwin, 1874, p. 53.)

Nearly eight decades later, Hebb (1947, 1949) showed that rats reared in "enriched" environments displayed better problem-solving abilities than rats kept under "impoverished" conditions (see also Bingham and Griffiths, 1952; Forgays and Forgays, 1952). Since these original investigations, a large body of evidence has accumulated which supports the concept that the level of environmental complexity is a major determinant of the performances of experimental animals at various learning and motivational tasks (see, e.g. reviews by Bennett and Rosenzweig, 1971; DeFeudis, 1975a; Greenough, 1976). About a decade after Hebb's work, it was shown that environmental "enrichment" produced increases in the weight and thickness of the rat's cerebral cortex that were correlated with a small decrease in acetylcholinesterase (AChE) activity and a small increase in pseudo-ChE (ChE) activity (when expressed on a unit weight basis). These changes in AChE and ChE indicated that cortical growth caused by "enrichment" could have occurred mainly in elements possessing low AChE activity, such as glia and non-cholinergic neurones (Krech *et al.*, 1960,

1962a, b, 1963, 1966; Altman and Das, 1964; Diamond *et al.*, 1964, 1966; Das and Altman, 1966). In contrast, further studies on differentially-housed rats revealed that neither the magnitude nor the pattern of differences in two predominantly glial enzymes (pseudo-ChE and carbonic anhydrase) supported the contention of Diamond *et al.* (1966) (see also Chapter 3, Part C).

However, a new era for this type of study had been initiated, and recently many other studies have revealed that increases in CNS function or environmental "enrichment" can produce increases in cerebral "growth" (see, e.g. refs in Gaito, 1972; Rosenzweig and Bennett, 1972; Ansell and Bradley, 1973; Globus, 1975; Hudgens *et al.*, 1975).

By merely considering the general characteristics of the nervous system (e.g. its organizational complexity; its redundancy; its susceptibility to modification by environmental and ontogenetic factors), one can readily understand that gross morphological changes are associated with behavior. For example, its redundancy exists in the extensive chemical and functional overlap of its neural pathways, most of which contain some long neurones (see, e.g. refs in DeFeudis, 1974b; Morgane and Stern, 1974) and has been revealed clearly by the early ablation studies of Lashley (1929, 1930a, b, 1931, 1938) in which large portions of the brain were removed without significantly altering some behaviors (see also Penfield and Jasper, 1954; Huston, 1975; Berlucchi and Buchtel, 1975). These early findings of Lashley (1929, 1938) led him to postulate several integrative concepts about learning and memory which, though speculative, are of value in the present discussion. Such concepts include: (1) ". . . all of the cells of the brain are constantly active and are participating, by a sort of algebraic summation, in every activity. There are no special cells reserved for special memories" (Lashley, 1950); (2) "The engram of a new association, far from consisting of a single bond or neuron connexion, is probably a reorganization of a vast system of associations involving the interrelations of hundreds of thousands or millions of neurons" (Lashley, 1950); (3) "The trace of any activity is not an isolated connexion between sensory and motor elements. It is tied in with the whole complex of spatial and temporal axes of nervous activity which forms a constant substratum of behavior." (Lashley, 1950); (4) "There is no great excess of cells which can be reserved as the seat of special memories." (Lashley, 1950); (5) "The billions of neurons in the cerebral cortex are organized into a large number of systems. Each system consists of the traces of a number of habits or memories. . . . The systems are not anatomically separate, and the same neurons, in different permutations, may participate in many systems." (Lashley, 1958); (6) "Reduction in the tonic level of the system, as in sleep or under anesthetics, or violent invasion, as in a convulsive seizure, would destroy the organization necessary for memory and the continuity of the conscious state." (Lashley, 1958). These integrative concepts about the CNS, along with its redundancy, based on its possible constitution of large sets of parallel information processors (see, e.g. Deutsch, 1975), further support the existence of gross morphological correlates of behavior.

It remains difficult indeed to designate certain brain regions as being involved in learning on the basis of ablation studies. Upon ablation of a given region, one can only say, if learning persists, that another region was initially involved or has taken over the functions of the ablated region. Also, it seems likely that the regions remaining after ablation perform different functions since their physiology and metabolism would be altered by the lack of afferent (or other) input normally provided by the region ablated (see also Berlucchi and Buchtel, 1975). This argument also applies to nervous diseases characterized by brain lesions. For example, Parkinsonian symptoms could be caused by areas other than those bearing lesions, and L-DOPA could act on these other areas (see also Chapter 8, Part D); these other areas would have abnormal functions and drug responses due to the decreased afferentation caused by the areas bearing lesions. Perhaps the difficulty in relating CNS structure to higher functions is due mainly to the contention that even very simple behavioral acts result from many overlapping "mental" activities, so that alleviation of one of these activities may not cause any detectable effect. This, of course, would vary with the CNS region considered. As examples, primary input and output pathways (e.g. those responsible for simple conditioning) can be easily localized using lesions or drugs, but higher processes (i.e. complex behaviors that depend upon learning and memory) are more difficult to localize, especially since these involve the organism's levels of motivation and "arousal".

Furthermore, the ageing process, which appears to produce a general effect on the brain and its blood supply (see, e.g. Chapter 8, Part B), has been associated with a deficiency in recall that can extend to much, if not all, of remote memory (e.g. Welford, 1958; Squire, 1974). Also, cortical atrophy, narrowing of the gyri and widening of the sulci are common pathological findings in presenile or senile dementia (Alzheimer's disease) which involves memory loss. These findings have served to strengthen the associations between memory, behavior and gross neuromorphology.

Thus, many elements of the *Behavioral Code*; all complex behaviors, learning and memory, on the one hand, and the gross regulators of behavior, learning and memory (e.g. cerebral blood flow; stress and its associated pituitary-adrenal system; "arousal" level), on the other, all involve gross regions of the nervous system.* One can understand further the relationship between behavior and cerebral gross morphology by drawing analogies with neurological disorders that affect learning, memory and behavior. In Chapter 8 (Part D) such a model has been constructed for Parkinsonism, a nervous disease characterized by pronounced behavioral symptoms which involve gross regions of the CNS (see, e.g. Munkvad *et al.*, 1968; Stern and Potkonjak, 1970; Larochelle *et al.*, 1971; DeFeudis, 1974b; Poirier *et al.*, 1975; Cools and Van Rossum, 1976); Alzheimer's disease is mentioned above and in Chapter 8 (Part B). Another good example would be the loss of "recent memory" which occurs in Korsakoff's syndrome (Korsakoff, 1889; Milner and Teuber, 1968) which is associated with organic brain lesions (e.g. Adams *et al.*, 1962).

Other findings have indicated that specific sensory stimulation can affect gross regions of the brain. For example, visual attention affects cerebral blood flow (CBF), a finding that could be significant in the interpretation of studies on learning and memory or environmental "enrichment" (see Bondy *et al.*, 1974; for reviews on mechanisms underlying "attention", in relation to the processing and storage of information in the CNS, see Worden (1966), Broadbent (1971), Regan (1972), Kahneman (1973) and Posner (1975). Such visually induced alterations in CBF could alter cerebral metabolism and gross anatomy. Bondy and Morelos (1971) have shown further that visual deprivation can decrease the complexity of cerebral afferentation in the young chick, an effect which could impair adequate vascularity and thus retard maturation. Visual deprivation can also alter dendritic morphology in cerebral areas other than those subserving the visual system (Vrensen and DeGroot, 1975; see also Chapter 3, Part B2), adding further support to this view. These findings also lend themselves to "dissection" by the newly devised methods based on 2-deoxy-D-glucose (see, e.g. Sokoloff, 1975) and regional CBF (see, e.g. Ingvar and Gustafson, 1970).

When one considers specific projection areas of the cerebral cortex in higher animals, suppression of specific behaviors may be produced by their ablation. For example, Area 13 of the cerebral cortex, which is the projection area for the vagus nerve (Bailey and Bremer, 1938) and which is involved in the parasympathetic reflexes associated with salivation and digestion (Patton and Amassian, 1950), upon extirpation

* The reader is encouraged to review works which emphasize the gross and cellular morphologies of the nervous system in relation to functional systems and behavior in order to gain a thorough understanding of the discussion which follows in the remaining chapters of this monograph (see, e.g. Jasper *et al.*, 1958; Campbell, 1965; Woodburne, 1967; Gardner, 1968; Cole and Maltzman, 1969; Arnold, 1970; Curtis *et al.*, 1972; Vierck, 1974; DeFeudis, 1974b; Wall, 1975; articles in Gazzaniga and Blakemore, 1975).

in the monkey causes a profound change in behavior (Ruch and Shenkin, 1943) characterized by increased restlessness and hyper-motility (Livingston *et al.*, 1947). Many other functions have been linked to specific regions of the brain, but such findings have shed little light on mechanisms of learning and memory storage.

After examining the effects of cortical lesions on maze learning in rats, Lashley (1929) concluded that retention of the task was dependent upon the size of the lesion ("law of mass action") but not upon its localization in the cortex ("law of equipotentiality"). Hence, the cortex appeared to be equipotential for learning and memory, but not necessarily for behaviors such as motor movements, vision, and speech, which are known to be differentially affected by the locus of the cortical lesion (see P. M. Milner, 1970). Since it is known that after a habit is learned it is not essential that particular receptor cells receive the stimulus (Lashley, 1929) or that particular motor cells carry out the act (DeFeudis, 1968), learning or memory are not likely to be functions of individual cells, but rather, these appear to be functions of the total mass of cerebral tissue. Therefore, though the organization of learned responses seems to be a function of the cerebrum, particular associations are most likely not represented by particular neurones. For example, even though the visual cortex as a whole serves visual functions, removal of its component parts does not produce a loss of discrimination of particular visual figures (Lashley, 1930b, 1935). As long as some tissue is left (regardless of which bit), function can continue. In addition, a non-visual role in maze-learning has been proposed for the visual cortex. Lashley (1929) showed that peripherally blinded rats performed better on a maze-learning task than did such blinded rats with visual cortex lesions. Hence, it may be concluded that learning and memory are similar functions since they both appear to require the total mass of tissue rather than individual cells (see Dunbar, 1954; see also Chapter 8, Parts B and C).

B. Environmental Effects on Gross Neuromorphology

The brains of rats subjected to environmental complexity (EC) are heavier than those of their stimulus-deprived (IC) counterparts (e.g. Bennett *et al.*, 1964; Geller *et al.*, 1965; Rosenzweig *et al.*, 1968, 1969; Ferchmin *et al.*, 1970; Geller, 1971; Walsh *et al.*, 1971; Bennett, 1976). Gross morphological changes between EC and IC rats appear to be more pronounced after about 30 days of differential housing when differential housing is begun at post-weaning ages (Diamond *et al.*, 1972); such treatments cause even greater effects in pre-weaning animals (Malkasian and Diamond, 1971). These effects were maximal in occipital cortex. Other studies have indicated that 30 days of environmental "enrichment" in rats caused a differential development (increase in tissue weight) of the cerebral cortex with respect to three other brain areas that were examined (Massey *et al.*, 1973). EC–IC differences in female rats were slightly smaller than in male rats (Diamond *et al.*, 1971). In other experiments, Altman *et al.*, (1968) showed that the normal length of the cerebrum of rats was not attained under certain rearing conditions and that it could be surpassed by increasing the complexity of the environment. Behavioral manipulations were thus shown to affect the antero-posterior dimension of the cerebrum (see also, Walsh *et al.*, 1971, 1973). However, an extensive analysis of mouse brain (rostral to the inferior colliculi; excluding cerebellum) has revealed that individually housed animals possess slightly heavier brains than their group-reared littermates though the brain weight/body weight ratio was greater for the "aggregated" mice (DeFeudis, 1974a; Table 1). Bell *et al.* (1971) had shown previously that mouse brain weight decreased with increasing group size when housing was carried out in a fixed space, but that this measure increased with group size at a constant density. Regardless of the direction of change, these results have indicated that environment can alter gross cerebral morphology. The histological and ultrastructural studies discussed in Chapter 3 provide further details about this change.

Another index of gross morphological change is cerebral water content. The brains of "aggregated" mice have been shown to contain a greater percentage of water than those of their individually housed littermates (DeFeudis, 1972a). In an earlier study, Geller (1971) found that water content, nitrogen content, and cholesterol were about the same in the brains of "enriched" and "impoverished" rats, but that lipid content was slightly greater in "enriched" brains and that RNA and DNA contents were somewhat greater in "impoverished "brains. These latter results are not in accord with the notion that the increase in brain weight caused by environmental complexity is accompanied by increased glial proliferation, as suggested by Diamond (1967); see also Chapter 3, Part C. Geller (1971) suggested that such differences indicated that the neurones of "enriched" rats might increase in size and degree of myelination.

Rosenzweig *et al.* (1972a) have suggested that maturation is not a major factor contributing to the effects of differential housing since they found that there is no critical period for producing the EC–IC effect on brain weight (see also Rosenzweig and Bennett, 1976a). However, this should not be taken to mean that maturation (i.e. *biological maturation*, rather than chronological maturation) is not involved in this change. The extent of biological maturation, which is

TABLE 1. Brain and body weights of "isolated" and "aggregated" male, C-57 black mice.

	Brain weight (g)	Body weight (g)	$\left(\frac{\text{Brain weight}}{\text{Body weight}}\right)$ (100)
"Isolated" mice	0·317 ± 0·0014 (245)	22·88 ± 0·20 (296)	1·386 ± 0·013
"Aggregated" mice	0·308 ± 0·0013 (251)	20·03 ± 0·18 (303)	1·538 ± 0·016
% Difference	2·9	14·2	11·0
p-value	≪0·001	≪0·001	≪0·001

All p-values were determined using a two-tailed Student's *t*-test; means ± standard errors; number of mice in parentheses. Combined standard errors for brain weight/body weight ratios were calculated with the following equation:

$$\text{CSE} = \pm \sqrt{\frac{\left(\frac{\text{Ba}}{\text{A}}\right)^2 + \text{b}^2}{\text{A}}},$$

in which B = brain weight, A = body weight and ± a and b were the respective standard errors of A and B. (Reproduced with permission from DeFeudis, 1974a.)

governed by the environment, could continue for as long as the organism has the capacities for "arousal", learning and memory (see Chapter 8, Part G). Therefore, the effects of "enriched" environments could be explained by postulating an increase in *biological maturation.*

Another significant concept which has evolved from recent studies involves the reversibility, or persistence of environmental effects on brain and behavior. Such reversibility adds impetus to the notion that therapeutic measures for behavioral disorders in man could be based on environmental modification (see also Chapter 5 and Chapter 8, Part A and articles in Walsh and Greenough, 1976). Rosenzweig *et al.* (1972a) reported that differences in cortical weight and thickness between EC and IC rats persisted for as long as these animals were differentially housed (see also Geller *et al.*, 1965; Walsh *et al.*, 1969). However, more recent studies have indicated that the magnitudes of these differences decrease over time (Walsh *et al.*, 1974; cited by Greenough, 1976). Bennett *et al.* (1974) studied the persistence of changes in brain weight produced by "enriched" experience by subjecting rats to the IC condition after they had been kept for 30 or 80 days in the EC condition. They showed that the EC effect could be reversed to a great extent but not abolished by this treatment (see Fig. 5). Earlier results by Geller (1971) revealed that many of the effects of both EC and IC on cerebral anatomy and chemistry could be largely

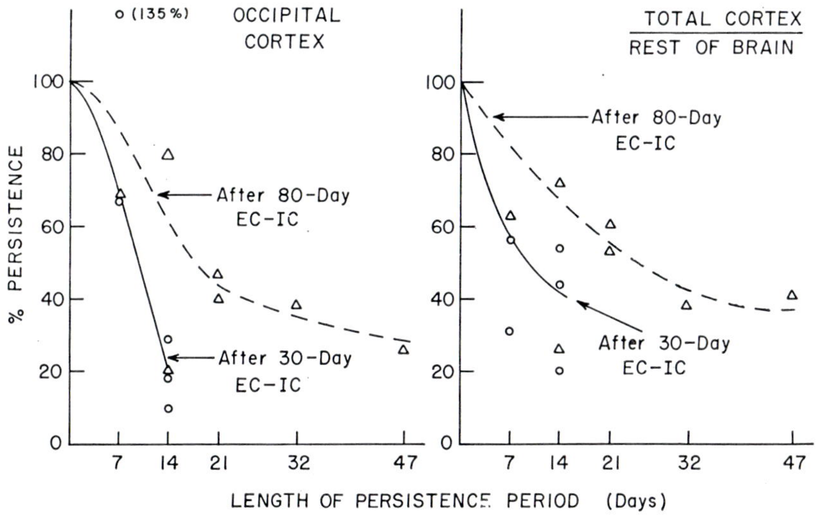

FIG. 5. Persistence of brain weight for occipital cortex and for the ratio Total Cortex/ Rest of Brain of rats transferred to IC after 30 days in EC (○———○) and after 80 days in EC (△– – – – – –△). (Reproduced with permission from Bennett *et al.*, 1974.)

TABLE 2. Values obtained from animals maintained for 30 days in EC or IC conditions, then rehoused in pairs for an additional 30 days.[a]

Parameter	EC (mean ± S.E.M.)		IC (mean ± S.E.M.)	p
Original body wt. (g)	41·1 ± 1·91		41·5 ± 1·76	
Final body wt. (g)	409·7 ± 17·4	(14)	401·4 ± 15·8	
Liver wt. (g)	16·862 ± 0·809	(15)	16·549 ± 0·723	
Brain wt. (g)	1·9607 ± 0·0400	(15)	1·9613 ± 0·0399	
Adrenal wt. (mg)	47·7 ± 1·9	(13)	46·1 ± 1·7	
ACS (μg/g)	9·08 ± 1·49	(6)	17·22 ± 2·56	<0·001
SCS (μg/100 ml)	8·78 ± 1·04	(10)	10·35 ± 1·42	
AAA (mg/100 g)	481·0 ± 19·6	(10)	505·3 ± 14·2	<0·05
Serotonin (μg/g)	0·682 ± 0·046	(11)	0·705 ± 0·085	
Serotonin (μg/brain)	1·315 ± 0·079	(12)	1·352 ± 0·124	
Dopamine (μg)	0·683 ± 0·066	(7)	0·960 ± 0·154	<0·05
Dopamine (μg/brain)	1·336 ± 0·131	(6)	1·784 ± 0·241	<0·05
Norepinephrine (μg/g)	0·585 ± 0·078	(9)	0·602 ± 0·092	<0·05
Norepinephrine (μg/brain)	1·089 ± 0·123	(10)	1·014 ± 0·138	<0·05

Parameter	EC (mean ± S.E.M.)	IC (mean ± S.E.M.)	p
Acetylcholinesterase (μmol/h/g Pr.)	0·136 ± 0·004 (13)	0·131 ± 0·004	
5-HTP decarboxylase (μmol/h/g Pr.)	30·45 ± 8·43 (11)	41·22 ± 9·78	
Glutamic decarboxylase (μmol/h/g Pr.)	21·83 ± 2·59 (12)	23·99 ± 4·11	
Transaminase (μmol/min/g Pr.)			
Tyrosine	1·677 ± 0·273 (12)	1·858 ± 0·362	
Tyrosine + pp	12·082 ± 0·774 (13)	12·255 ± 0·830	
Tryptophan × 10^2	11·426 ± 1·034 (13)	11·735 ± 1·030	
Tryptophan + pp × 10^2	14·042 ± 0·947 (17)	13·631 ± 0·940	$<0·05$
Phenylalanine hydroxylase (μmol/h/g Pr.)	62·53 ± 3·20 (12)	63·57 ± 3·72	
Tryptophan hydroxylase (μmol/h/g Pr.)	3·511 ± 0·579 (15)	3·251 ± 0·393	
Pyrrolase (μmol/h/g Pr.)	5·395 ± 0·499 (11)	5·949 ± 0·821	
Pyrrolase + hematin	16·455 ± 1·808 (13)	15·548 ± 1·431	

[a] Numbers in parentheses are numbers of pairs. The following abbreviations are used: ACS, adrenal corticosteroids; SCS, serum corticosteroids; AAA, adrenal ascorbic acid; 5-HTP, 5-hydroxytryptophan; pp, pyridoxal phosphate. (Reproduced with permission from Geller, 1971.)

TABLE 3. Effects of environment and methamphetamine on brain and body weights of rats.

Strain	Brain measure	Per cent differences of brain and body wts from Saline-HC group			between Meth-EC and Saline-EC
		Meth-HC	Saline-EC	Meth-EC	
S_1	Cortex	N = 10	N = 43	N = 48	N = 43
	Occipital	2·5	9·7‡	13·3‡	4·3†
	Total	1·3	4·6‡	6·4‡	1·9*
	Rest of brain	2·2*	1·4	1·9*	0·4
	Cortex/Rest	−0·9	3·3‡	4·4‡	1·4†
Fischer	Cortex	N = 9	N = 26	N = 26	N = 26
	Occipital	−1·4	4·1*	8·0‡	3·7
	Total	1·2	4·9‡	6·5‡	1·5
	Rest of brain	−1·2	2·5*	1·7	−0·9
	Cortex/Rest	2·4*	2·2‡	4·8‡	2·5‡
S_3	Cortex	—	N = 11	N = 11	N = 11
	Occipital	—	11·8‡	17·5‡	5·1*
	Total	—	4·3*	6·7‡	2·3
	Rest of brain	—	−0·9	0·6	1·5
	Cortex/Rest	—	5·1‡	6·2‡	1·0
Overall	Cortex	N = 19	N = 80	N = 86	N = 80
	Occipital	0·6	8·0‡	12·1‡	4·2‡
	Somesthetic	0·5	4·7‡	6·5‡	2·0*
	Rem. dorsal	2·9*	5·7‡	6·8‡	1·1
	Ventral	−0·1	2·4†	4·4‡	2·0*
	Total	1·2	4·5‡	6·3‡	1·8†
	Rest of Brain	0·6	1·4*	1·6*	0·2
	Total brain	0·8	2·8‡	3·7‡	0·9
	Cortex/Rest	0·7	3·1‡	4·6‡	1·7‡
	Terminal body weight	−3·2	0·2	0·9	0·4

* $p < 0{\cdot}05$; † $p < 0{\cdot}01$; ‡ $p < 0{\cdot}001$. HC = home cage; EC = "enriched" condition. (Reproduced with permission from Bennett *et al.*, 1973.)

reversed by re-housing 30-day-exposed animals in pairs for 30 days (see Table 2). With regard to the therapeutic significance of these studies, Will *et al.* (1976a, b) have shown that exposure of rats to "enriched" environments (i.e. both immediate and delayed treatments) enhanced their recoveries from bilateral lesions of the occipital cortex, as tested by using Hebb-Williams problems. Such environmental treatments appeared to reverse both the anatomical and behavioral effects of brain lesions.

Rosenzweig and Bennett (1972) have shown also that the increased environmental interaction produced by administration of stimulant drugs (e.g. methamphetamine) to rats is an adequate substitute for "enriched" experience with respect to its effect of increasing brain weight. Bennett *et al.* (1973) confirmed this effect of amphetamines (see Table 3), and showed further that pentobarbital decreased brain weights; corresponding motor activities of the drug-treated rats are shown in Fig. 6. Hence, the most significant variable involved in the

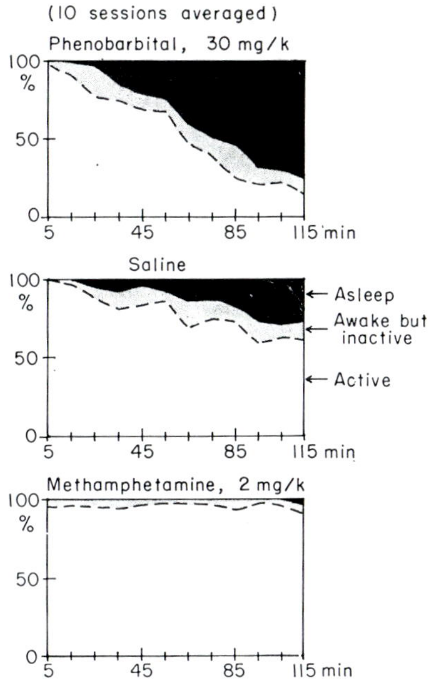

FIG. 6. Activity of S_1 rats during 2 h periods in the "enriched" condition. Littermates were given saline, methamphetamine or phenobarbital and were observed simultaneously over 10 sessions. (Reproduced from Bennett *et al.*, 1973.)

modification of brain weight appears to be the degree of interaction of the animal with a rather complex environment. This interaction can be primed by stimulant drugs which increase the animal's "arousal" level and which could be addictive (see also Chapter 8, Part F). Further experiments have indicated that *active contact* with an "enriched" environment appears to be essential for the development of EC effects on brain weights (Ferchmin *et al.*, 1975). Rats placed in small "observer" cages inside a large "enriched" condition cage (Fig. 7) possessed brains that were not significantly heavier than those of IC rats. Hence, specific sensory (e.g. auditory, olfactory, visual) inputs

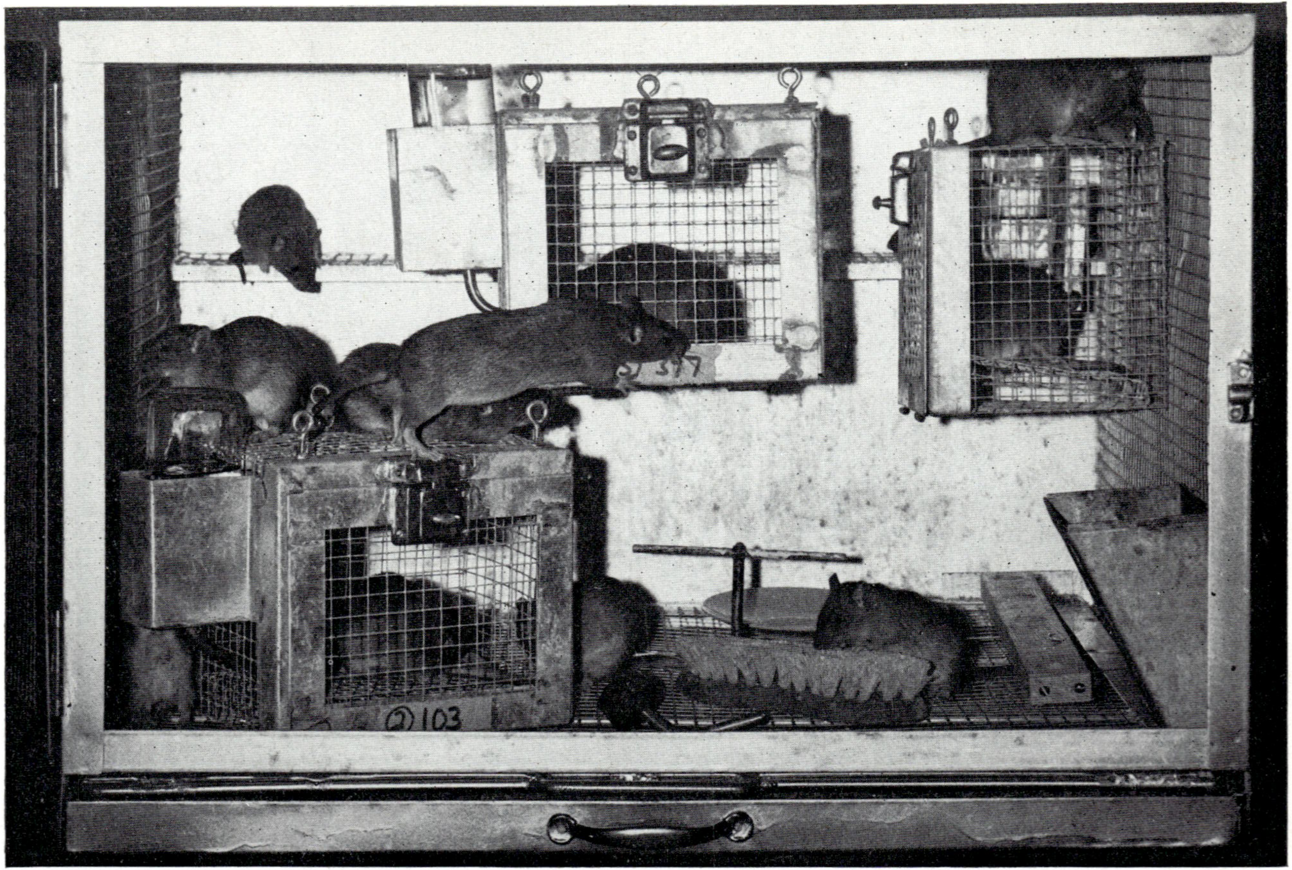

FIG. 7. Three small observer condition cages inside a large condition cage. (Reproduced with permission from Ferchmin *et al.* 1975.)

do not appear to contribute appreciably to changes in *brain growth*, measured as brain weight, although such specific sensory stimulation can produce changes in gross regions of the brain (e.g. in metabolism and in CBF; see p. 23 in Section A, above). The main effect appears to be produced by an interaction of the animal, either with objects or with con-specifics. These latter studies support the contention that somatosensory input is a major determinant of normal behavioral responses (Prescott, 1971; DeFeudis, 1974c; see also Chapter 1, Part H).

3. Changes in Cellular and Subcellular Neuromorphology

A. Neurones and Neuronal Nuclei

Changes in cellular and subcellular morphology can be produced by modifying an animal's neural function. For example, Hamberger and Hydén (1945) showed clearly that the dimensions and constituents of neural perikarya could adjust to alterations in functional load. More recent studies have revealed that physical exercise can promote shrinkage of pyramidal neurones of the rat cerebral cortex (Altman and Das, 1966; Table 4) and that cells of the locus coeruleus of the pontine tectum varied in size during sleep and "arousal" produced by stimulation of the thalamus and reticular formation (Bebenik and Monnier, 1972).

Spinal motoneurones of the guinea pig increased in size after activity of short duration (Edstrom, 1957), whereas the sizes of both their nuclei and perikarya were decreased by prolonged intense activity (Geinsmann *et al.*, 1971). Also, partial denervation of afferent inputs

TABLE 4. Size distribution (in percentage) of pyramidal cells of the cerebral cortex in unexercised and exercised rats.

Cell size	10–14 μ	14–18 μ	18–22 μ	>22 μ
Unexercised rats	35·0	32·0	21·0	12·0
Exercised rats	43·5	34·5	18·5	3·5

Note the systematic shift in the frequency distribution in the sizes of pyramidal cells, with a 5% decrease in the number of larger cells, and a similar increase in the number of smaller cells, in the exercised group of animals. (Reproduced with permission from Altman and Das, 1966.)

to the mammalian CNS has been shown to cause synaptic reorganization of some central neurones in both adult and immature animals (Raisman, 1969; Nakamura *et al.*, 1974; Tsukahara *et al.*, 1975). Gerchman *et al.*, (1975) have shown that rats subjected to physical exercise displayed changes in neuronal morphology and associated acid phosphatase, glucose-6-phosphate dehydrogenase and malate dehydrogenase activities.

It has also been shown that neuronal nuclei of epileptic foci were smaller than those of normal tissue (Westrum *et al.*, 1964). Nuclear and perikaryal abnormalities also characterized cells of cortical mirror foci in rabbits (see Morrell, 1969). Nuclei of neurones of the lateral geniculate body were reduced in size by subjecting kittens to dark-rearing (Hubel and Wiesel, 1963). Cell nuclei of visual receiving areas of the cerebral cortex were smaller in mice possessing inherited retinal dystrophy (Gyllensten and Lindberg, 1964) and in dark-reared mice (Gyllensten, 1959; Gyllensten *et al.*, 1965) than in control mice. In pre-weaned (Malkasian and Diamond, 1971) and post-weaned (Diamond, 1967) rats, the dimensions of neuronal nuclei and perikarya were altered by environmental changes. Diamond *et al.* (1975) showed further that the neuronal nuclear area of "enriched" rats was significantly greater (11·7%) than that of "impoverished" rats with 25 or 105 days of exposure; for 60 to 90 day exposures, the perikarya of "enriched" animals were slightly greater (4·4%) than those of "impoverished" animals, whereas no difference existed in neuronal nuclear area. Other studies showed that smaller neuronal nuclear and nucleolar cross-sectional areas of several subcortical brain regions characterize female and neonatally castrated male rats, as opposed to normal male rats (Pfaff, 1966); similar differences have been shown in the medial amygdaloid nucleus of monkeys (Bubenik and Brown, 1973).

B. Dendrites, Dendritic Spines and Pre-synaptic Boutons

1. *Introduction*

Growth and establishment of the pattern of dendritic arborization of neurones occur both pre- and post-natally in mammals, the final dendritic characteristics appearing to be controlled by both genetic and extrinsic factors (see, e.g. Rakic, 1974, 1975). Extrinsic factors are of major importance with regard to the present discussion since these can be modified by environmentally induced alterations of neighboring cells.

Ramón y Cajal (1891, 1896, 1911, 1934) first recognized that dendritic structure appears to depend upon interaction with other cells, and that dendritic spines represent normal postsynaptic structures which reflect the functional maturation of the nervous system (see also Conel, 1939–1963; Marin-Padilla, 1967; Part B5 below). Hence, the number of dendritic spines might be expected to provide a measure of synaptic boutons or of postsynaptic receptor regions of neurones, and has been useful for showing morphological correlates of CNS function and behavior (Globus and Scheibel, 1966, 1967a, b; Valverde, 1967, 1968; Valverde and Esteban, 1968). Changes in dendritic branching and in dendritic spine number in the cerebral cortex occur during the ageing process (e.g. Schadé *et al.*, 1964; Valverde and Ruiz-Marcos, 1969; Scheibel *et al.*, 1975; see above), and can be caused by subjecting animals to "long-term" environmental modification (e.g. Holloway, 1966; Globus and Scheibel, 1967a, b; Globus *et al.*, 1972, 1973b; Volkmar and Greenough, 1972; see also Greenough, 1975, 1976).

Increases in the number of dendritic spines per neurone and in the dendritic branching of neurones of the occipital cortices of rats caused by environmental "enrichment" have been correlated with an increase in the size of synaptic regions, but not with a change in synaptic density (Møllgaard *et al.*, 1971; Rosenzweig *et al.*, 1972c). However, Cummins *et al.* (1975) have claimed that a greater numerical density of synapses occurred in the occipital cortices of "enriched" rats, as compared with their "isolated" counterparts. They concluded that a *de novo* growth of synapses of small size had been caused by environmental enrichment. An increase in synaptic boutons of rat hippocampus has also been correlated with ontogenetic and environmental changes (Chronister *et al.*, 1973). Such studies have indicated that animals subjected to increased sensory stimulation possess more synapses per neurone.

Other studies have shown that electrically induced fatigue of neurotransmission at medullary axo-axonic synapses of the hatchetfish is paralleled by a depletion of synaptic vesicles and by an accumulation of irregular membranous structures in the presynaptic region; such changes were reversible (Fig. 8; Model *et al.*, 1975). Most recently, such changes in synaptic morphology have been correlated with synaptic

---

FIG. 8. Depletion of vesicles correlated with fatigue of transmission at a vertebrate central synapse. A, B and C are electron micrographs of portions of stimulated medullary axo-axonic synapses from Mauthner to giant fibers of the hatchetfish (*Gasteropelecus*). In the synaptic regions of the presynaptic terminals (Mf), synaptic vesicles are largely replaced by membranous structures of various shapes and sizes. Some of these structures are in the form of flattened cisternae (B and C; *arrows*), but most are irregular in outline. An occasional apparently unattached coated vesicle is present (A; *arrowhead*) and coated vesicles are fused with the membranes delimiting some of the intra-terminal compartments (C and D; *arrowheads*). Some synaptic vesicles of normal appearance remain; mitochondria (m) are swollen. ×64 125 (Reproduced with permission from Model *et al.*, 1975.)

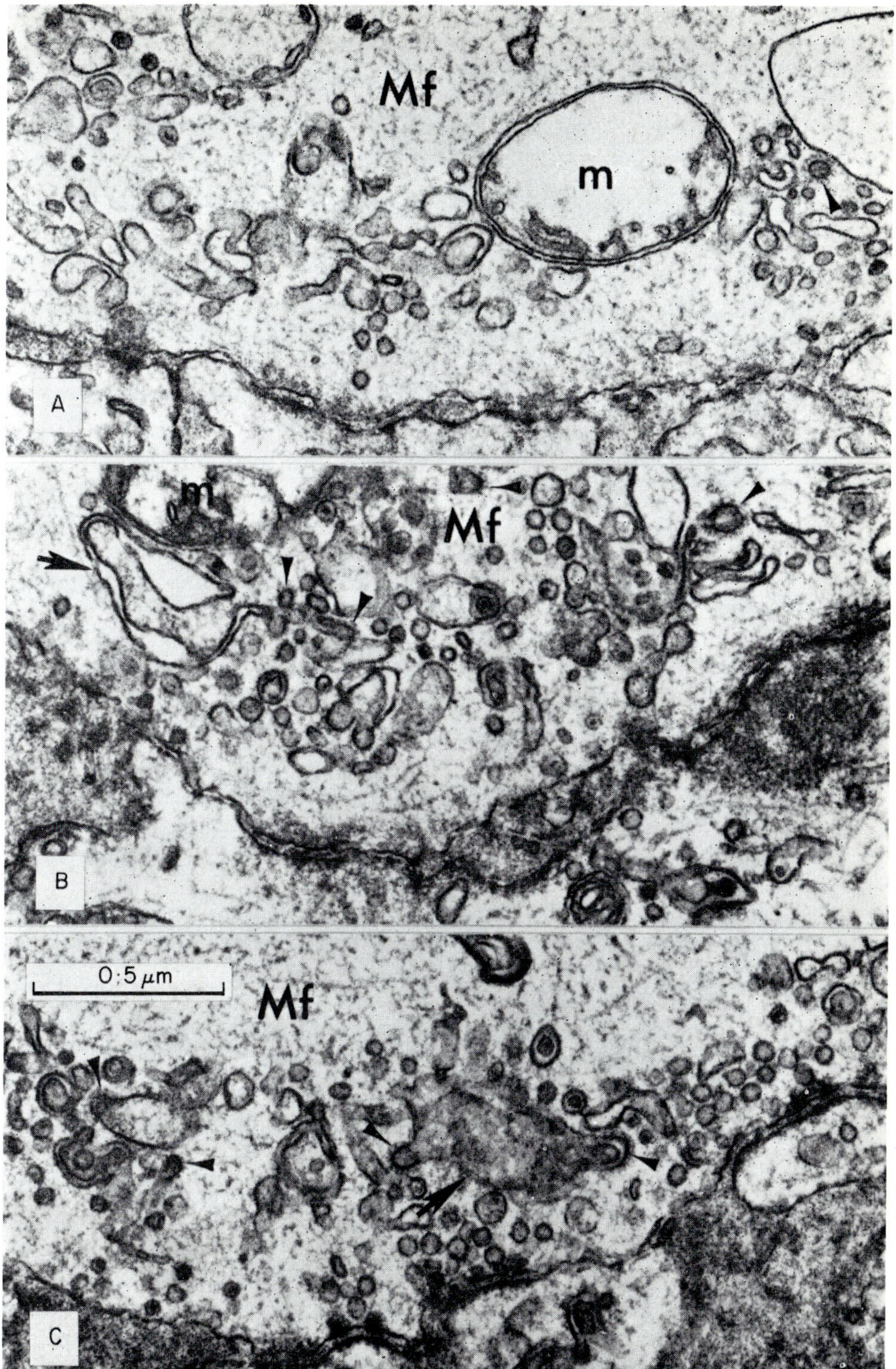
Mf
m
A
m
Mf
B
0.5 μm
Mf
C

transmission in the lamprey spinal cord (Christensen, 1976). Also, decreases in the number of synaptic connections per neurone (Cragg, 1972a, b) and in the size and density of presynaptic endings of rat cerebral cortex (Gambetti *et al.*, 1974) have been caused by malnutrition. It is noteworthy that changes in synaptic vesicle number in the anterior and posterior hypothalamus and caudate nucleus of rats exhibited a circadian rhythmicity (Walker *et al.*, 1971; Table 5).

Neuronal changes produced by chronic experimental epilepsy also involved a loss of dendritic spines (Westrum *et al.*, 1964), as well as neuronal depopulation and increased glial function (e.g. Harris, 1972, 1975). Convulsions induced in rats by methionine sulphoximine have been shown to cause swelling of nerve endings and a loss of synaptic vesicles of the cerebral cortex (DeRobertis *et al.*, 1967; Fig. 9). Furthermore, five days after entorhinal cortical lesions in adult rats, there occurred a 30% decrease in dendritic spines in the outer molecular layer of the dorsal leaf of the dentate gyrus of the ipsilateral hippocampus, but spine density was back to normal 60 days after the lesion (Lynch *et al.*, 1974). This study revealed that, at least in this area of the mammalian CNS, such dendritic changes may be reversible. In the most exaggerated case, the central effects of nerve section are known to include the transneuronal effect of withdrawal of synaptic boutons from the perikaryal surface (Blinzinger and Kreutzberg, 1968; Hamberger *et al.*, 1970).

Certain neurological mutant strains of mice (e.g. Rakic, 1975; Sotelo, 1975a, b, and refs therein) are also characterized by abnormalities of dendrites and their spines. As in the "weaver" mouse, cerebellar granule cells are virtually absent in mice treated with the toxin, cycasin, at the perinatal stage, (see Hirano and Dembitzer, 1975) or in animals inoculated with feline panleukopenia virus (Herndon *et al.*, 1971). The major effect appears to be that the Purkinje cells contain unattached dendritic spines due to their lack of presynaptic elements (i.e. granule cells). Other methods of destroying granule cell precursors (external germinal cells) in the developing cerebellum include: (1) administration of cyclophosphamide (Nathanson *et al.*, 1969) and floxuridine (Langeman *et al.*, 1972); (2) surgical extirpation of the rhombic lip (Forstronen, 1963); (3) intramuscular administration of thiophene (thiofuran), a noxious industrial solvent (Christomanos and Scholtz, 1975; see review by Herndon and Oster-Granite, 1975).

Woodward *et al.* (1975) showed that several types of alterations in Purkinje cell dendrites were produced by injecting neonatal rats with the anti-mitotic drug, methyl-azoxymethanol, which selectively destroys proliferating cells of the external granular layer of the cerebellum. Two of these alterations, the "disoriented dendrite" and the "S-shaped dendrite" (Figs 10 and 11), as well as the frequency of occurrence of

TABLE 5. Synaptic vesicle counts in the anterior hypothalamus, posterior hypothalamus and caudate nucleus.

Tissue	Time of day (h)			
	03:00	15:00	03:00	15:00
	Granular vesicles (per 15·5 μ^2)		Agranular vesicles (per 0·176 μ^2)	
Control				
Anterior hypothalamus	13·6 ± 3·5 (4)	6·6 ± 3·6 (26)†	78·9 ± 7·0 (4)	76·5 ± 14·0 (24)
Posterior hypothalamus	2·9 ± 1·4 (6)	3·7 ± 2·1 (28)	83·3 ± 3·4 (6)	76·0 ± 9·4 (26)
Caudate nucleus	6·4 ± 1·6 (5)	3·2 ± 1·6 (34)‡	78·5 ± 4·1 (4)	75·3 ± 16·0 (32)
Reserpine				
Anterior hypothalamus	8·0 ± 4·4 (23)[a]	7·2 ± 2·9 (11)	48·6 ± 7·6 (22)[c]	54·3 ± 5·0 (11)*,[c]
Posterior hypothalamus	7·0 ± 3·0 (20)[b]	9·5 ± 4·0 (20)*,[c]	41·6 ± 5·9 (20)[c]	49·8 ± 9·4 (20)†,[c]
Caudate nucleus	3·8 ± 1·8 (19)[a]	2·6 ± 1·7 (22)*	47·6 ± 6·1 (17)[c]	65·3 ± 7·4 (22)‡,[c]
dl-DOPA-MAOI				
Anterior hypothalamus	7·3 ± 3·6 (23)[b]	9·3 ± 3·5 (19)[a]	35·2 ± 4·4 (21)[c]	65·2 ± 9·9 (19)‡,[b]
Posterior hypothalamus	9·2 ± 4·5 (20)[b]	6·2 ± 2·4 (20)*,[c]	49·9 ± 8·8 (18)[c]	69·0 ± 6·9 (20)‡,[b]
Caudate nucleus	6·6 ± 2·4 (20)	5·2 ± 1·0 (20)*,[c]	70·0 ± 11·9 (20)	87·8 ± 6·4 (19)‡,[c]

Means ± standard deviations; numbers of determinations in parentheses; *, † and ‡ indicate, respectively, $p < 0{\cdot}05$, $p < 0{\cdot}01$ and $p < 0{\cdot}001$ for differences between 03:00 and 15:00 h; [a], [b] and [c] indicate, respectively, $p < 0{\cdot}05$, $p < 0{\cdot}01$ and $p < 0{\cdot}001$ for differences between values from control and treated rats. (Reproduced with permission from Walker *et al.*, 1971.)

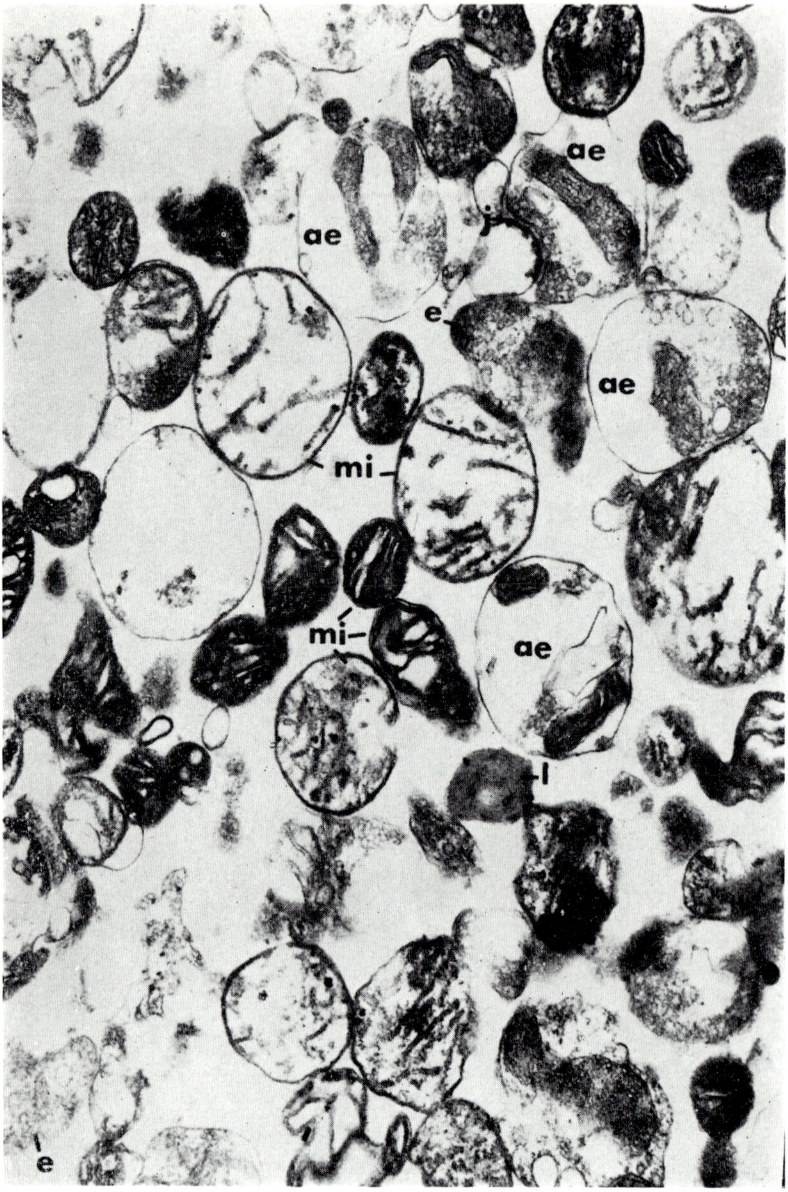

FIG. 9. Ultrastructural changes in the crude mitochondrial fraction of a rat convulsed by intraperitoneal injection of methionine sulphoximine (400 mg/kg). Notice that several nerve endings are altered (ae); they appear vacuolized and contain a decreased amount of synaptic vesicles. mi, free mitochondria; l, lysosome-like body; e, normal nerve ending; j, synaptic junction. ×40 500. (Reproduced with permission from DeRobertis *et al.*, 1967.)

dendritic abnormalities are shown (Fig. 12). Similar changes in dendritic morphology have been produced by subjecting neonatal rats to X-irradiation (see Altman, 1973).

2. *Visual System as a Model*

It has been long known that sustained modifications of visual input cause significant changes in retinal nerve cells (Mann, 1894; Carlson, 1902–1903), and such findings have been recently confirmed with respect to light deprivation (Chow *et al.*, 1957; Weiskrantz, 1958; Riesen, 1960). Other studies have revealed that an increase in the percentage of amacrine-amacrine connections of the inner plexiform layer of the albino rat retina occurred after eyelid suturing (Fifkova, 1972), which added support to the previous finding that light deprivation of hooded rats increased the total number of synapses in the inner plexiform layer (Sosula and Glow, 1971). However, studies of the possible occurrence of such changes in brain cells have been only recently initiated.

Binocular visual deprivation (dark-rearing) reduced the number of dendritic spines in visual receiving areas of mouse and rabbit cerebral cortices (Valverde, 1967, 1968, 1971; Valverde and Ruiz-Marcos, 1967, 1969; Globus and Scheibel, 1967a, b; Valverde and Esteban, 1968), as well as diminishing dendritic length and reducing the number of dendrites per cell (Coleman, 1965; Coleman and Riesen, 1968). After studying apical dendritic spines of the visual cortices of mice subjected to visual deprivation, Valverde (1967) concluded:

> The visual sensory deprivation does affect the fine structure of the central nervous system. Our results give some support to the assumption that some structural changes in the nerve cells occur as a result of experience.

Valverde (1971) concluded from his further studies on the visual cortices of mice, dark-reared from birth, that some spines would not grow without normal visual inputs and that some spine development can occur in some pyramidal cells even when the animals are not exposed to light until more than 20 days after birth (see Fig. 13).

On the other hand, visual overstimulation (i.e. exposure of rats to 35 days of continuous illumination, from birth) caused an increase in the spine density of the visual cortex which indicated an enhancement of postsynaptic growth of pyramidal cells (Parnavelas *et al.*, 1973b; Parnavelas and Uylings, 1975). Also, increases in dendritic branching and in the mean total dendritic length of the dorsal lateral geniculate nucleus occurred when rats were reared under continuous illumination, as compared with rats exposed to a light-dark cycle (Parnavelas *et al.*,

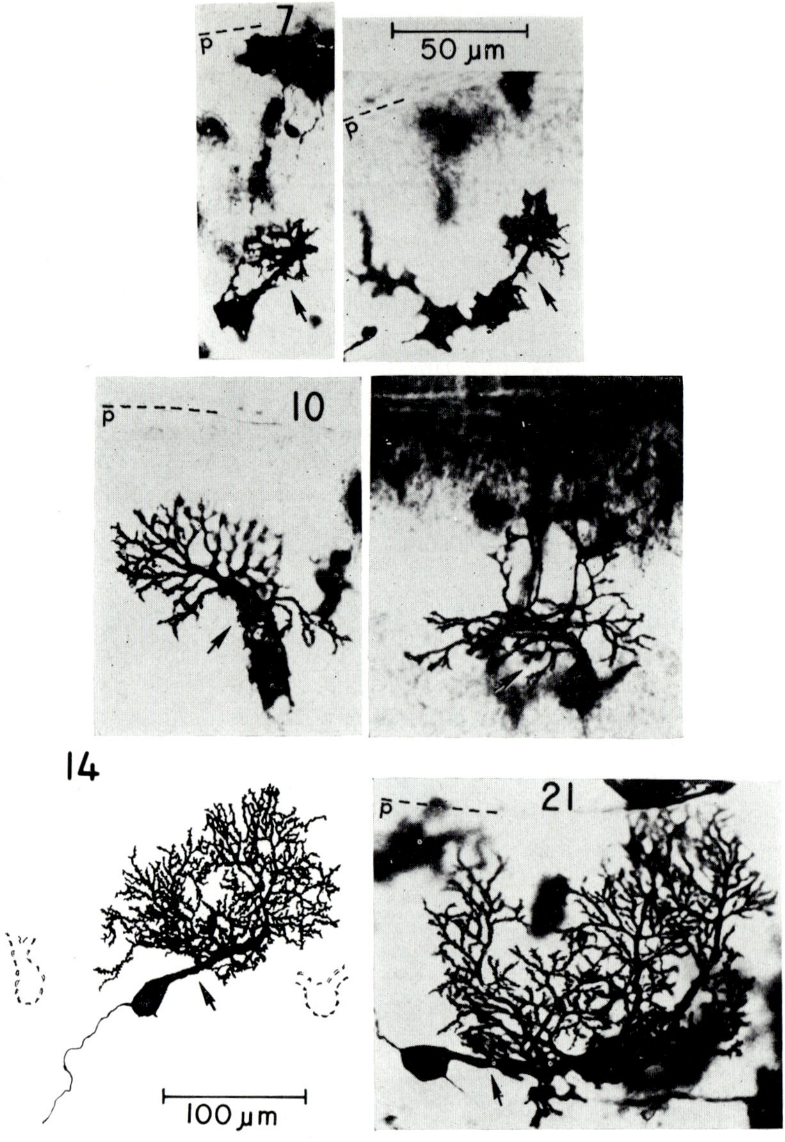

FIG. 10. Purkinje cell dendritic changes after transient injury to the external granular layer during development. Disoriented dendrite. Examples are shown of primary dendrites progressing markedly away from the vertical orientation. Arrows denote the "disoriented dendrites". The apical dendritic system for day 10 (on the left) also exhibits a phenomenon in which several processes extend vertically and penetrate into the external granular layer. (Reproduced with permission from Woodward *et al.*, 1975.)

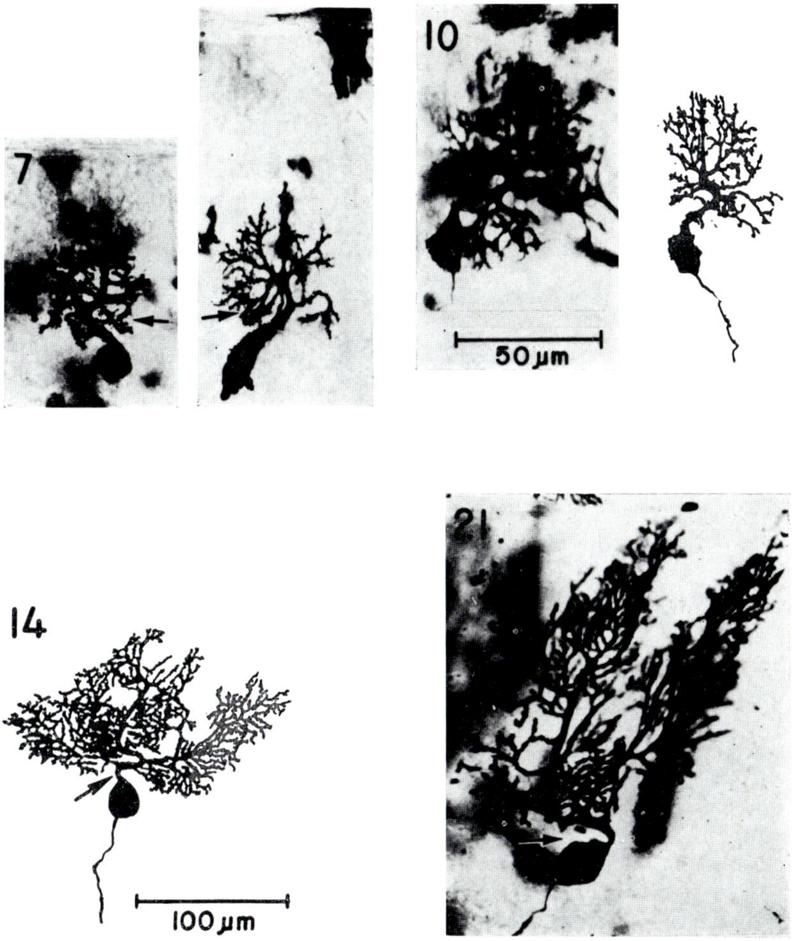

FIG. 11. Purkinje cell dendritic alterations after transient developmental injury of the external granular layer. "S-shaped" dendrites are illustrated at days 7, 10, 14 and 21. The primary dendrite is given off at various initial directions, then traverses in a rough S-shape while ascending in the molecular layer. (Reproduced with permission from Woodward *et al.*, 1975.)

1973a). These studies indicated that visual stimulation increases the synaptic complexity of cerebral visual areas.

Other studies have revealed that exposure of animals to patterned light produced increases in synaptic size in some areas of the visual system (Cragg, 1967; Fifkova, 1970). In addition, a recent ultrastructural study has revealed that monocular deprivation of rabbits caused a significant increase in synaptic density in both visual and motor areas

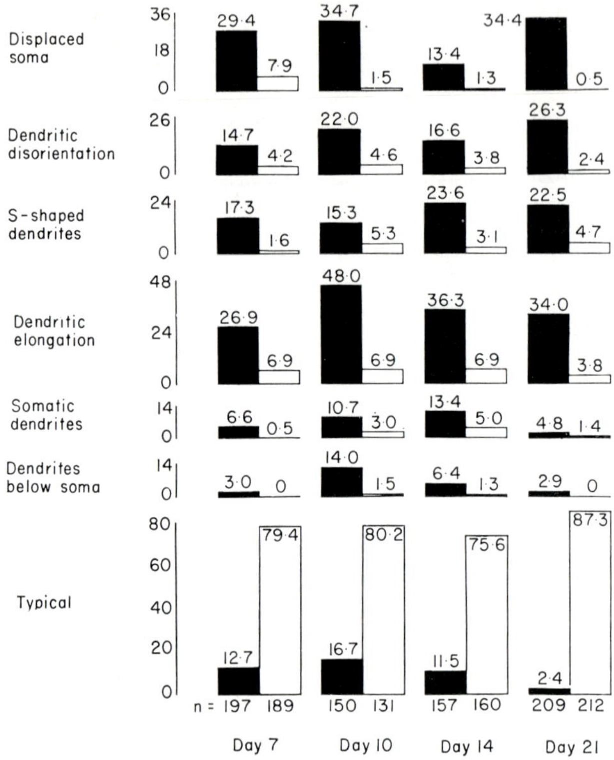

FIG. 12. Frequency of occurrence of dendritic abnormalities in control rats and in rats subjected to transient developmental injury of the external granular layer of the cerebellum. N at the bottom of columns denotes number of Purkinje cells evaluated in each group. *Black bars* correspond to per cent in height of all cells evaluated exhibiting a given characteristic. Percent of cells with a given characteristic is also indicated by numbers on top. *White bars* are for control. Scales at left represent percentage. (Reproduced with permission from Woodward *et al.*, 1975.)

of the deprived cortex and a significant decrease in the synaptic vesicle content of the deprived visual cortex (Vrensen and DeGroot, 1975; Table 6, Fig. 14; see also Vrensen and DeGroot, 1974a). Such studies further substantiated the findings of Wiesel and Hubel (1965a, b) which revealed that monocular deprivation in cats during a critical neonatal period caused amblyopia (reduction or loss of visual ability) which was correlated with marked changes in the functional organization of their visual systems (see also Barlow, 1975). Such amblyopia results in a loss of binocular input to visual cortical neurones which is caused by the asymmetrical visual experience. Recent studies have shown that intravenously administered bicuculline (a GABA antagonist) can make

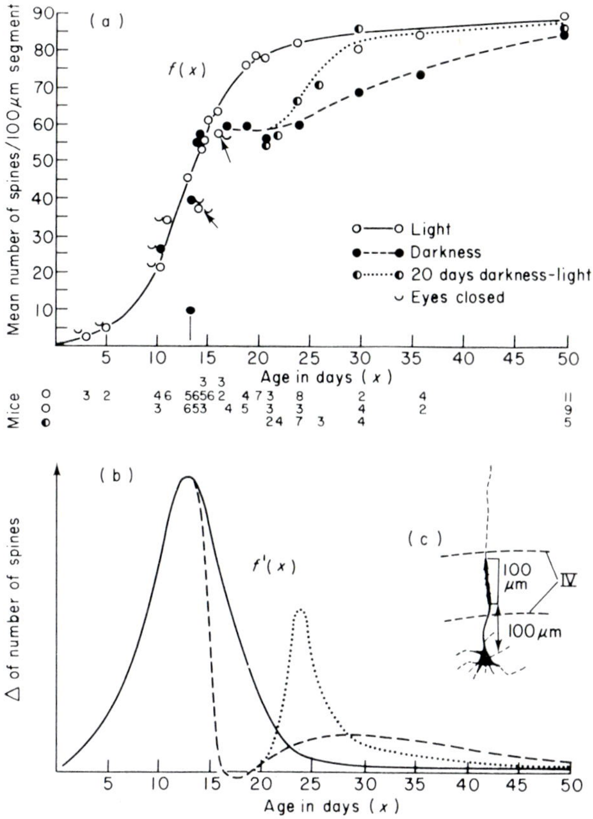

FIG. 13. (a) Mean number of dendritic spines per segment in apical dendrites of layer V pyramidal cells of the visual cortex as a function of age. The number of animals from which the mean values were obtained for each age group is indicated below the horizontal axis. (b) First derivative function indicating the rate of increase of the number of spines in an arbitrary scale as a function of age. Geometrical reconstructions from fitted curves of a. (c) The location of the standard segment in which the number of spines were counted on each apical dendrite. (Reproduced with permission from Valverde, 1971.)

visual cortical neurones accessible to stimulation of the deprived eye in cats with monocular deprivation-induced amblyopia, and it was postulated that this restoration of binocular input by bicuculline occurred by reduction of "GABA-ergic" inhibitory mechanisms (Duffy *et al.*, 1976). Such a finding indicates that feline amblyopia may not involve a permanent loss of visual function, i.e. that either associated morphological changes are reversible or that these may be overcome by GABA blockade.

TABLE 6. The effect of monocular deprivation (right eye) on the number of synaptic vesicles (N_{ves}) in the visual and motor cortices of rabbits.

	N_{ves} per terminal[a]		
	Left	Right	
Visual cortex, medial	49·9 ± 2·8 (8)[b]	57·8 ± 2·1 (8)	$t_7 = 1{\cdot}89$; $p = 0{\cdot}05$[c]
Visual cortex, lateral	57·1 ± 2·9 (9)	70·1 ± 5·0 (9)	$t_8 = 2{\cdot}39$; $0{\cdot}025 > p > 0{\cdot}01$
or cortex	55·7 ± 4·0 (9)	54·2 ± 4·0 (9)	N.S.

[a] Mean number of synaptic vesicles per terminal in 3 superficial layers (0, 100 and 200 μm). [b] inter-individual mean ± S.E.M.; numbers of animals in parentheses. [c] Statistical analysis with paired Student's *t*-test. (Reproduced with permission from Vrensen and DeGroot, 1975.)

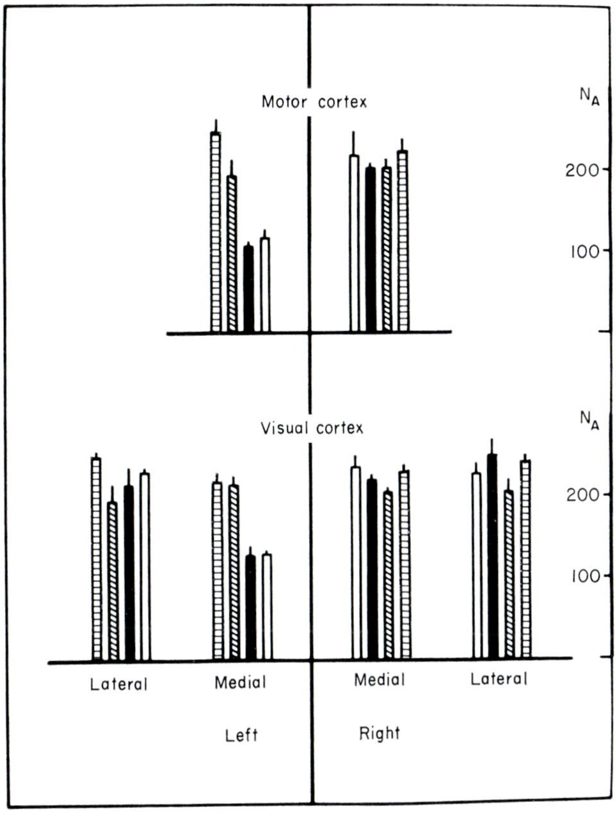

FIG. 14. Number of synaptic contact zones $N_A/1000\ \mu m^2$ in the superficial layers (0, 100 and 200 μm) of the visual and motor cortices of the left and right hemispheres of control (*open bars*), dark reared (*black bars*), dark reared-recovered (*obliquely hatched bars*) and monocularly deprived (*right eye; horizontally hatched bars*) rabbits. Note the difference in synaptic density between the medial aspects of the visual cortex in the left and right hemispheres of control and dark reared animals. A similar difference occurred in the motor cortex. (Reproduced with permission from Vrensen and DeGroot, 1975.)

Other studies have shown that binocular visual deprivation of rabbits for 7 months caused a 40% irreversible decrease in the number of synaptic vesicles of visual cortical cells while not altering the number of synaptic contact zones (Vrensen and Van Hof, 1975). Also at the ultrastructural level, Garey and Pettigrew (1974) showed an increase in synaptic vesicle density in small "type 1" terminals of layers III and IV of the cerebral cortex after exposing deprived kittens to moving stripes. These could be the terminals of the axons of spiny stellate cells that provided the "conditioning signal" in the model of Blakemore and Van Sluyters (1975); see below. Furthermore, such spiny stellate cells of layer IV, which form the principal target of the visual afferents in kittens, were susceptible to ribosomal disaggregation by visual (especially binocular) deprivation, whereas spinefree cells were not affected (LeVay, 1975). Hence, at the beginning of the second month in kittens, ribosomal aggregation (to polysomes, for protein synthesis) in spiny stellate cells appeared to become dependent on normal visual input (LeVay, 1975). Also noteworthy in this context is the recent finding that dark-rearing suppressed the incorporation of ^{3}H-lysine into a rapidly labelling neuronal fraction of rat's visual cortex (Rose and Sinha, 1974).

The above mentioned findings provide a morphological basis for electrophysiological studies which have shown that most cells of the visual cortices of kittens raised under conditions in which their eyes were exposed selectively to lines of a given orientation reacted to that particular orientation (Blakemore and Cooper, 1970; Hirsch and Spinelli, 1970, 1971). Studies by Blakemore and Mitchell (1973) have revealed further that visual cortical cells in kittens, reared in darkness until six weeks of age, adapted their responses in only one hour when exposed to a vertically striped cylinder (Fig. 15). Further studies of the roles of innate and environmental factors on the development of the kitten visual cortex have allowed the classification of the receptive fields of all cortical neurones of kittens of any age or rearing history into five distinct, mutually exclusive groups (see Blakemore and Van Sluyters, 1974): (1) visually unresponsive; (2) non-oriented (Fig. 16); (3) pure direction sensitive (Fig. 17); (4) orientational bias; (5) orientation selective. Also, an environment of randomly scattered spots led to a specific selectivity for spots of the same size, and exposure to high-contrast stripes or to a pattern of large rectangles produced a visual cortex in which almost all neurones were orientation selective, the preferred orientations matching those seen early in life (see also Hirsch and Spinelli, 1970; Blakemore and Cooper, 1970; Pettigrew and Garey, 1974).

Electrophysiological studies have also revealed that the amplitudes of visual evoked responses, recorded from the occipital cortices of rabbits, were markedly decreased in animals deprived of light for 7 months after birth (Kobayashi and Van Hof, 1972; see also Bonaventure *et al.*, 1971). Although the electroretinogram (ERG) of rabbits was shown to be relatively resistant to light deprivation (Reuter *et al.*, 1971), the "β-wave" of the ERG of guinea pigs born and raised in darkness for 3 months was considerably smaller than that of animals of the same age reared under normal conditions (Legein and Van Hof, 1970); exposure to normal illumination for 10 days greatly restored the ERG. Prolonged exposure to darkness also reduced "β-wave" amplitude in cats (Baxter and Riesen, 1961; Cornwell and Sharpless, 1967).

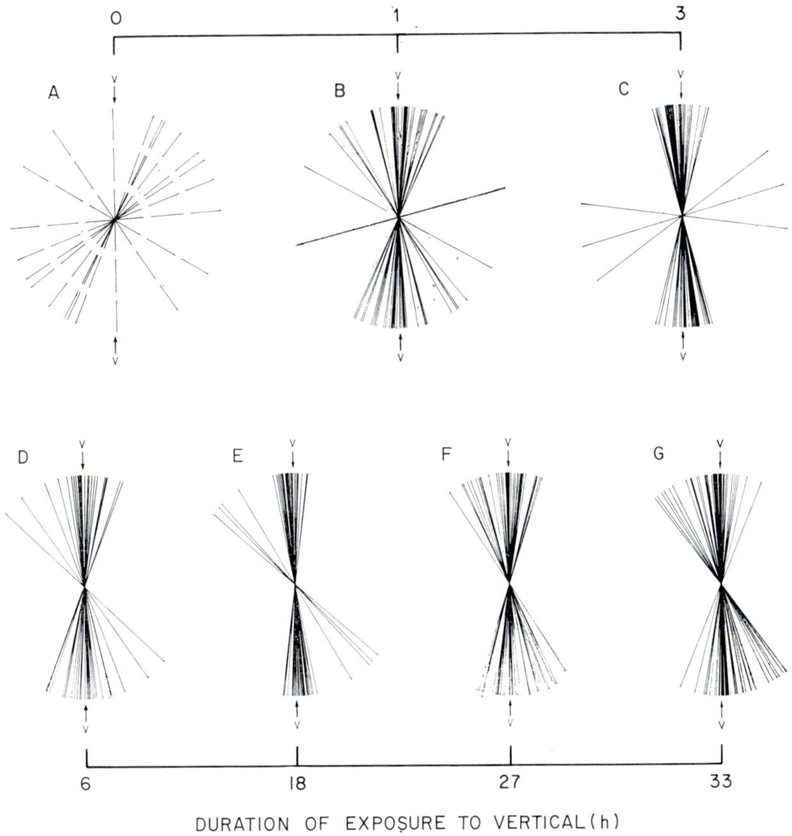

FIG. 15. Diagrams summarizing the effects on the visual cortex of different lengths of time spent in an environment of vertical stripes. Each line represents the optimal orientation for one cortical neurone: for the binocular deprived animal (A) the lines are interrupted to indicate that these neurones showed only the slightest preference for the angle shown and also responded to all other orientations. The number of hours of exposure for each kitten is indicated on the abscissae. (Reproduced with permission from Blakemore and Mitchell, 1973.)

However, it should be noted that even though rabbits raised in darkness for 7 months after birth behave abnormally after leaving the dark room (e.g. Goodman, 1931), this abnormal behavior disappears within a few weeks. Also, vertical *vs* horizontal and 45° *vs* 135° visual dise criminations, both with striations and rows of dots, were learned at about the same rate by normal and 7-month light-deprived rabbits (Van Hof and Kobayashi, 1972), whereas visual acuity, determined by changing the widths of white and black bars, was considerably diminished in light-deprived animals (see also Van Hof, 1967; Fig. 18).

Further studies on the alteration of visuomotor behavior which underlies behavioral changes caused by light deprivation have been conducted with cats (Van Hof-Van Duin, 1976a, b). In general, the

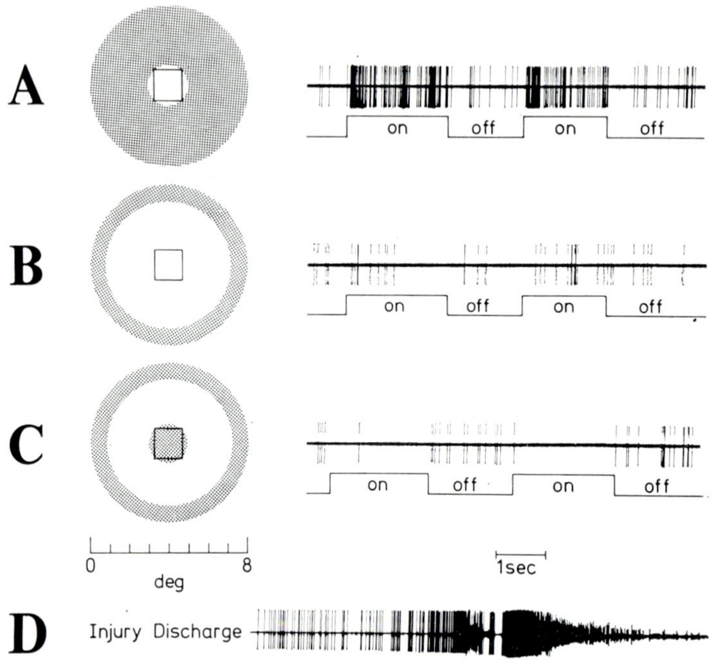

FIG. 16. Responses of a non-oriented cell from a kitten exposed to small spots and recorded at the age of 10·5 weeks. A, responses to a 2 deg diameter light spot centered over the receptive field and flashed on and off. B, responses to a 6 deg diameter light spot. C, responses to a 6 external ×2 deg internal diameter light annulus. The approximate times when the stimuli were turned on and off are indicated below the responses. D, the prolonged injury discharge obtained from the cell when the electrode was slightly advanced. Responses are displayed (positive upwards) next to the stimulus which evoked them. Luminance of the dark part of the screen was approximately 5 cd m^{-2} and of the light part of the screen approximately 20 cd m^{-2}. (Reproduced with permission from Blakemore and Van Sluyters, 1975.)

results indicated that behavioral defects after monocular deprivation are due to deficiencies in visuomotor control rather than in pattern identification (Van Hof-Van Duin, 1976b). Also, using two groups of light-deprived cats, one of which had been dark-reared for 4 months and the other for 7 months, it was shown that all light-deprived cats completely recovered their visuomotor behavior; 7-month deprived cats recovered within 10 weeks and 4-month deprived cats within 7 weeks. With respect to maturation, the time during which various responses in light-deprived cats developed after dark-rearing was found to agree roughly with that of normal visually inexperienced kittens after birth; i.e. the properties of visual cortical neurones of dark-reared cats

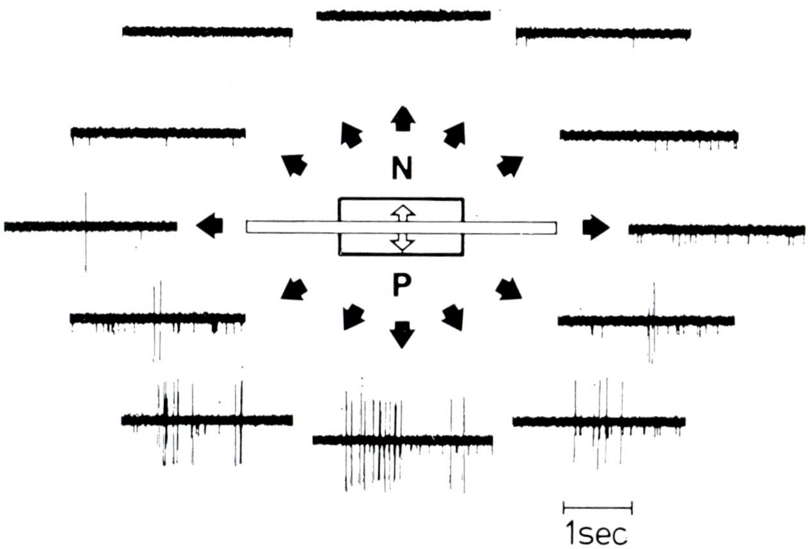

FIG. 17. Responses of a pure direction-selective cell recorded from a kitten exposed to an environment of stripes (K31). A $\frac{1}{3} \times 9$ deg light bar was moved across the receptive field in the directions shown by the filled arrows. The orientation of the bar was always perpendicular to the direction of motion. For convenience, the receptive field is illustrated with the preferred axis of movement arranged vertically on the page. The size of the response field was $3\frac{1}{2} \times 1\frac{1}{2}$ deg. P = preferred direction, N = null direction. Reponses are displayed (positive upwards) next to the stimulus which evoked them. Luminance of the dark part of the screen was approximately 5 cd m^{-2} and of the light part of the screen approximately 20 cd m^{-2}. (Reproduced with permission from Blakemore and Van Sluyters, 1975.)

were comparable with those found in visually inexperienced kittens. The term "immaturity" has been introduced to describe this situation in the visual cortex of the binocularly-deprived cat (Sherman, 1972; Blakemore, 1974). Studies on orientation in visual space by neonatal vertebrates and its modification by monocular and binocular pattern deprivation have been reviewed by Ganz (1975); see also Ganz *et al.* (1972) and Ganz and Haffner (1974).

Other studies have revealed that the *in vivo* ^{3}H-dopamine (DA) release from the retinae of cats differed as a function of their visual adaptation. In datk-adapted cats, the rate of DA release and the amount released per light flash of this possible retinal inhibitory

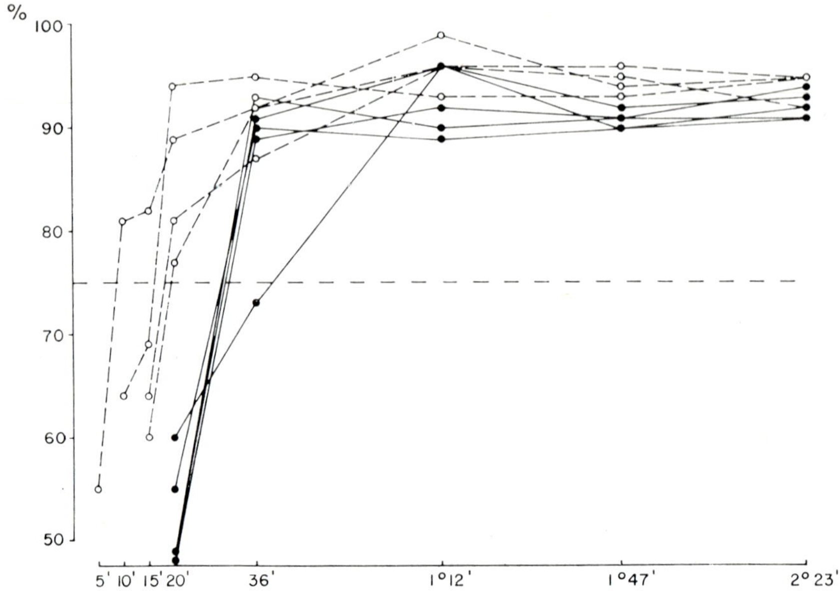

FIG. 18. Visual acuity in light-deprived and normal rabbits. Vertical *vs* horizontal striations were placed 30 cm in front of the gates. By reducing the width of the black and white stripes visual acuity was determined. Each stripe width was offered 100 times on 10 subsequent days. The percentage of correct choices during the last 500 presentations was plotted against the stripe width. (○) The results obtained by Van Hof (1967) in four normal animals; (●) those from five rabbits deprived of light for 7 months after birth. (Reproduced with permission from Van Hof and Kobayashi, 1972.)

transmitter increased (non-linearly) with increasing flash frequency, whereas in light-adapted animals the rate of DA release was high regardless of the flash frequency used (Kramer, 1971). GABA might also play a role in mediating the effects of monocular deprivation (Duffy *et al.*, 1976; see above). At the macromolecular level, Manocha and Perachio (1973) showed that cerebral monoamine oxidase and lactate dehydrogenase activities in squirrel monkeys were altered by varying environmental lighting conditions.

3. *A Model for "Short-term" Memory*

Van Harreveld and Fifkova (1974, 1975a) have postulated a mechanism for "short-term" memory and for synaptic potentiation based on the "swelling" of hippocampal dendritic spines that is produced by electrical stimulation of an afferent pathway to this region (see also Chapter 7). They have suggested that increased neuronal activity (such as that which occurs during spreading depression (SD), causes a non-specific release of glutamate which, in turn, could increase the uptake of water and ions by the spines, an effect that would decrease their longitudinal electrical resistance (Diamond *et al.* 1970; Rall, 1970; Bliss and Lømo, 1973) and thereby enhance the efficacy of synapses on dendritic spines, or the effectiveness of their synaptic excitation (Van Harreveld and Fifkova, 1974, 1975a; Fig. 19). These workers provided further support for this model by showing that L-proline (1·5–2·0 mM), which suppresses SD in chicken retina, also

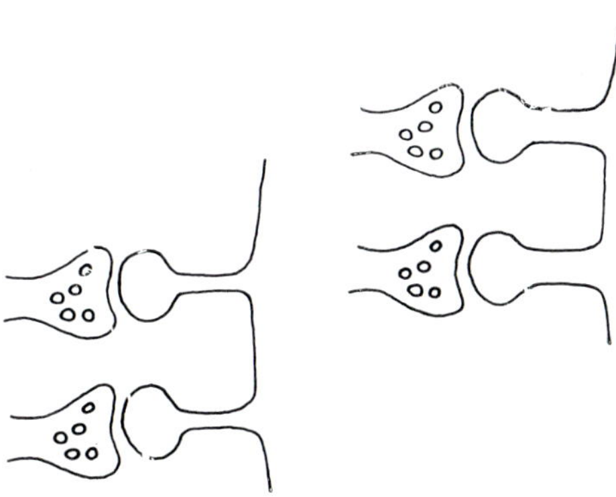

FIG. 19. Possible involvement of glutamate in memory formation with proposed change in dendritic spines caused by glutamate release. On the right two spines connected with a dendrite before, and at the left after, the glutamate action. (Reproduced with permission from Van Harreveld and Fifkova, 1974.)

prevented the ionic and water movements elicited by glutamate. More recently, Van Harreveld and Fifkova (1975b) have shown that the mean area of distal dendritic spines of the fascia dentata of mice was significantly increased after stimulation of perforant fibers. They proposed that this change in dendritic spines, which decreased their length and increased their length constant, might be a mechanism for the long-lasting post-tetanic potentiation observed in this monosynaptic pathway.

4. *Hormonal Effects*

Hormones also appear to affect dendritic morphology. The dorsal preoptic area of female rats appears to contain more spine-type stria terminalis connections than the same area of male or testosterone-treated female rats (Raisman and Field, 1973). Also, Schapiro and Vukovich (1970) have shown that early post-partum stressful stimulation increased synaptic spine number in rat cerebral cortex, whereas injections of cortisol impeded spine formation (see also Schapiro *et al.*, 1973). Administration of thyroxine during the same period enhanced dendritic spine formation (Schapiro, 1971; Schapiro *et al.*, 1973), whereas thyroid deficiency, produced by daily treatments of mother rats with propylthiouracil (commencing at the eighteenth day of gestation), severely reduced the dendritic arborization of Purkinje cells (Legrand, 1967). It has been suggested that this change in Purkinje cells causes the "death" of cerebellar granule cells during thyroid deficiency (Lewis *et al.*, 1976). From these findings it becomes easy to understand that hormones mediate some of the effects that experiential stresses produce on postnatal neural development and adaptation. However, some other effects caused by environmental "enrichment" may not be mediated by hormones of the pituitary or by those regulated by the pituitary (Riege and Morimoto, 1970; Rosenzweig *et al.*, 1972b). The involvement of hormones with behavior is, of course, well known (see, e.g. Levine, 1972; Eleftheriou and Sprott, 1975a, b; Gispen *et al.*, 1975a; Martini and Ganong, 1976; Sachar, 1976, see Chapter 4, Part E).

5. *Studies at the Primate Level*

Functional changes in humans, characterized by impairment of learning and memory, have also been correlated with swelling of synaptic boutons and with losses of synaptic vesicles and mitochondria from presynaptic boutons of the neocortex (Gonatas, 1967; Fig. 20). Swelling of apical dendrites of the cerebral cortex (with or without swelling of axon terminals), astrocytes, or neuronal cell bodies, has been observed in electron micrographs under various experimental conditions (including respiratory acidosis) and in several human diseases (see Schlote *et al.*, 1975). Also, Tatetsu (1974) has claimed that apical dendrites and axons were unusually thick and that nuclear and cytoplasmic dimensions were changed in human schizophrenic cortex. Swelling of cell perikarya and dendrites of the human neocortex also occur during ageing (Scheibel *et al.*, 1975).

An example of *external* environmental agents which act upon the *internal* environment to produce dendritic changes and behavioral modifications are the viruses that produce subacute spongiform encephalopathies (e.g. Scrapie, kuru and Creutzfeldt-Jakob disease; see, e.g. Lampert *et al.*, 1972). Swelling and vacuolation of dendrites and neuronal perikarya and swelling of astrocytic processes characterize the cerebral cortices of chimpanzees afflicted with these diseases (Lampert *et al.*, 1975; Fig. 21). Morphological changes in the CNS of monkeys have also been produced by intracerebral inoculations with a virulent strain of Mumps virus (Yuzepchuk *et al.*, 1975). The notion

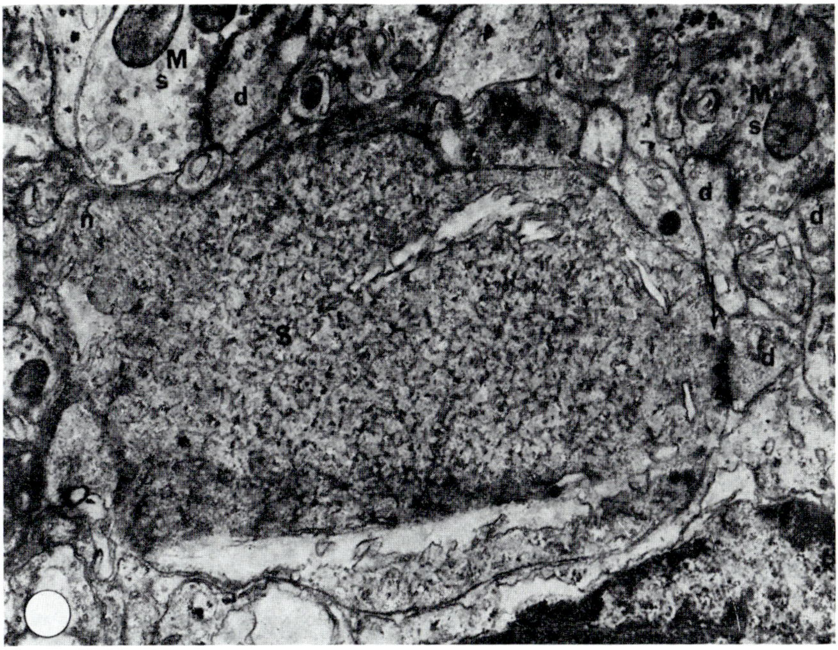

FIG. 20. Subcellular morphological lesions in the neocortex of a 30-month-old boy with psychomotor retardation of unknown origin. Note the enlarged synaptic bouton (S) and the absence of mitochondria from the abnormal terminal. The arrow points to one or two synaptic vesicles near the intersynaptic cleft. *d*, dendritic spine; *n*, neurotubules; *s*, normal presynaptic terminals; *M*, mitochondria (×24 000). (Reproduced with permission from Gonatas, 1967.)

that such morphological changes could be associated with changes in possible neurotransmitter systems gains added impetus when one considers that viral-induced granule cell loss in the hamster cerebellum, has been correlated with a marked reduction of glutamate concentration (Young *et al.*, 1974) and with a selective and marked loss of GABA "binding" (Simantov *et al.*, 1976b).

At the genetic level, Marin-Padilla (1972, 1974) showed gross abnormalities in the cerebral cortical dendritic spines of infant humans afflicted with trisomic chromosomal aberrations that are associated with mental retardation. However, similar types of dendritic spine abnormalities have been shown in the cerebral cortices of children afflicted with mental retardation of unknown origin who possessed normal karyotypes (Purpura, 1974). A relationship between these findings and maturation was brought out by Purpura (1975) who, recognizing that the dendritic spines of normal pre-term infants were morphologically similar to those found in cases of mental retardation, suggested

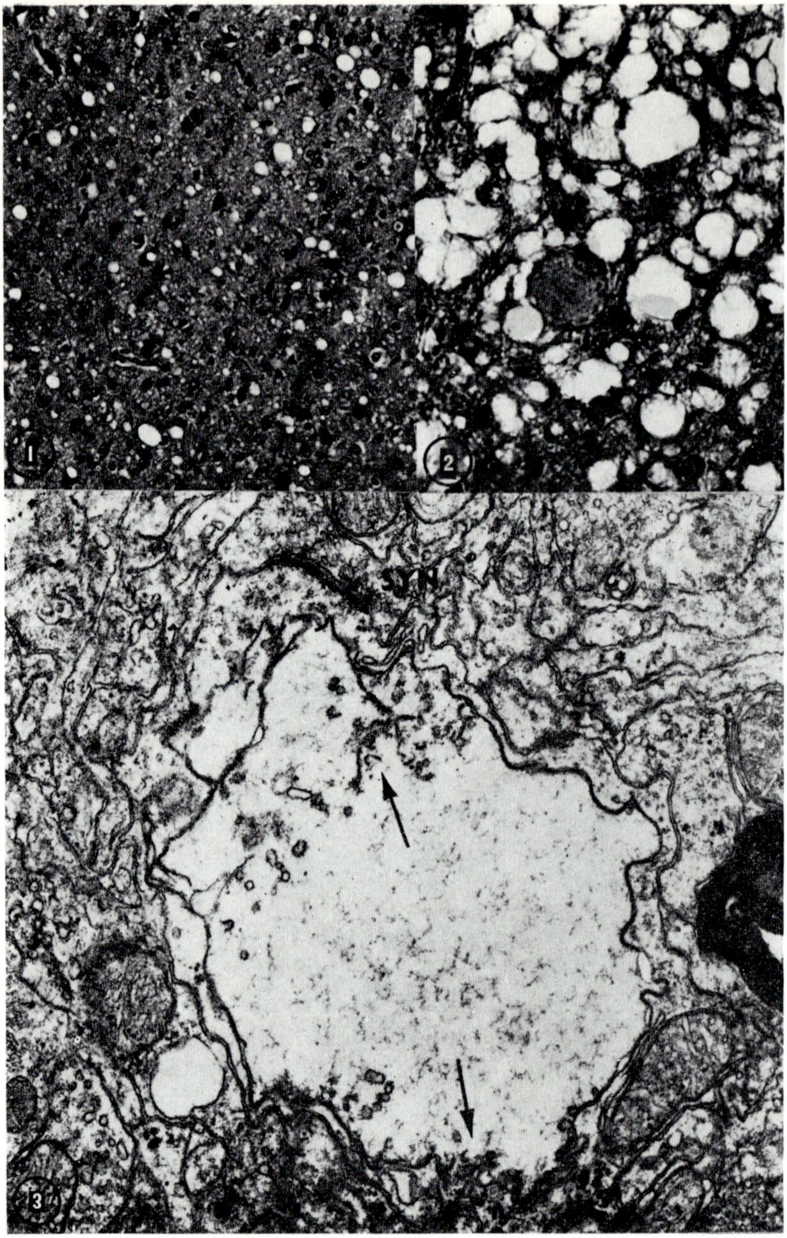

FIG. 21. 1, Status spongiosus and glial proliferation in the cerebral cortex of a chimpanzee afflicted with experimental Creutzfeldt-Jakob disease (×165). 2, Swollen nerve cell showing chromatolysis in the vacuolated cerebral cortex of a chimpanzee afflicted with experimental kuru (×400). 3, Vacuolated dendrite in the cerebral cortex of a chimpanzee with experimental Creutzfeldt-Jakob disease. Note fragments of membranes (*arrows*) and finely granular material within the vacuole. SYN, synapse. (×30 000). (Reproduced with permission from Lampert *et al.*, 1975.)

that "maturational arrest" might explain such spine abnormalities in mentally retarded children. A similar hypothesis has been derived in Chapter 8 (Part G) to explain the effects of environmental impoverishment on neuromorphology; i.e. that *biological maturation* does not keep pace with *chronological maturation* under this condition.

C. Glial Cells

Virchow (1846) first defined glia as a special type of brain cell, and Weigert (1895) and Ramón y Cajal (1910) then characterized various types of glia on a morphological basis. Holmgren (1902) described the intimate relationship that exists between glial cells and neurones which has led to the present involvement of these cells in neural function and in behavior. The role of glia as components of the "blood-brain barrier", a complex system involved in the regulation of the environment of brain cells, is well known (see, e.g. Davson, 1972).

Significant roles for glial cells in information processing have been proposed (e.g. Galambos, 1961, 1965; Pribram, 1966, 1972; Pevsner, 1972; Hydén, 1973b) which were based on Hydén's conception that neurones and glia are united as single functional units (Hydén, 1959a, b, 1967; Hydén and Pigon, 1960).

Krech *et al.* (1962a, b, 1966) found an increased number of neocortical glial cells in rats subjected to "enriched" environments, whereas the number of neurones was not affected (but see Geller, 1971, and Chapter 2, Part B). Increases in oligodendroglia and glial intermediates and a decrease in astroglia of the hippocampus have been produced by subjecting rats to environmental complexity (Walsh *et al.*, 1969; Table 7), and a decrease in cerebellar glial number occurred in mice subjected to "long-term" individual housing (Essman, 1971a, d). Also, the greater proportional water content of the brains of "aggregated" mice as compared with their "isolated" littermates (DeFeudis, 1972a) could reflect an alteration in the size of intra- and/or extra-cellular cerebral compartments which might explain the suggested increase in glial cell number in the brains of animals subjected to "enriched" environments. The fine structure of cerebral glia (or "glia-like" elements) also appears to be modified by environmental changes (Diamond *et al.*, 1964, 1966; Altman, 1967; Essman, 1971d, 1972).

Astroglial swelling produced by chemically-induced convulsions (e.g. Harris, 1964; DeRobertis *et al.*, 1969) and the hypertrophy and proliferation of glial cells that occurred in chronic "mirror" epileptic foci of rat cerebral cortex (Bogolepov and Pushkin, 1975) further support a behavioral role for glia. Glial cells appear to serve as an homeostatic "buffering system" against the increases in external K^+

TABLE 7. Differential cell counts in the medial area of the hippocampus.

Cell type	Mean number of cells per field: Environmental complexity	Mean number of cells per field: Isolated control	Difference (%)	p-value
Neurone	0·79	0·79	0	N.S.
Astroglia	9·72	10·78	−9·8	<0·02
Oligodendroglia	2·46	1·77	39	<0·001
Intermediates	0·645	0·178	262	<0·001
Total macroglia	12·83	12·73	0·8	N.S.

(Reproduced with permission from Walsh *et al.*, 1969.)

that are produced during neuronal activity, thereby ensuring functional integrity of mechanisms of synaptic transmission and nerve conduction (see, e.g. Kuffler and Nicholls, 1966; Dennis and Gerschenfeld, 1969; Trachtenberg and Pollen, 1970). This "isolating" role for glia, which is now associated with their K^+-buffering action, was first proposed by Ramón y Cajal (1909). Such an action also reflects glia-neuronal metabolic coupling (see, e.g. Hydén, 1959a, b. 1967; Benjamin and Quastel, 1972, 1975). Also, rapidly synthesized S-100 protein(s) and the "training-sensitive" glycoprotein 10B, which may be involved in learning and memory processes, appear to be constituents of glial cells (e.g. Benda *et al.*, 1968; Bogoch, 1968; Perez *et al.*, 1970; Hydén, 1973b), and therefore, could be involved in glial-neural intercommunication and/or in other synaptic events.

The above-mentioned findings, along with others which have indicated that glia serve a prominent role in ischaemic processes, lead the writers to believe that changes in neural function and mild forms of neural injury may share some common mechanisms. This notion, together with the findings that many central synapses are closely invested by glia (e.g. Colonnier and Guillary, 1964; Peters and Palay, 1965; Pappas *et al.*, 1966; Pappas and Waxman, 1972), and that neurones and glia may be metabolically coupled by the glutamate-ammonia-glutamine system (see, e.g. Benjamin and Quastel, 1975), indicate that glia play a significant role in synaptic function, memory, and behavior (see also Chapter 4, Part B). Recent microscopic studies which have shown that norepinephrine (10^{-7}–10^{-5} M) induced morphological changes in cultured C_6 glioma cells (Oey, 1975) have further supported this notion. It seems especially pertinent to note that glia appear to transport K^+ away from the extra-cellular space after its release during neuronal activity, thus preventing undesirable increases in neuronal excitability. Thus, a pathophysiological explanation of post-traumatic focal epilepsy, which is characterized by glial scar formation and gliosis (Penfield, 1927; Penfield and Humphreys, 1940), might involve impared glial control of extra-cellular K^+ (Pollen and Trachtenberg, 1970).

D. Studies with Subcellular Particles

Subcellular fractionation studies have revealed that "long-term" changes in environment that alter behavior (presumably by an influence on learning and memory) can modify the "binding" and uptake of neurotransmitter suspects and the "binding" of certain psycho-active agents in cerebral synaptic fractions (DeFeudis, 1972d, e, 1975a; DeFeudis and Marks, 1973a; Hendley *et al.*, 1973; Welch *et al.*, 1974b; DeFeudis *et al.*, 1976a, b, c). Although neurotransmitter theory, itself, provides information about only one major element of the *Behavioral Code* (see Fig. 80), studies of the changes in macromolecular binding sites for possible transmitters that accompany environmental modification should provide further insight into its more relevant structural elements. Such an approach should be quantitative and should include consideration of the "active" (free) concentrations of both small ligands and macromolecular binding sites within the framework of metabolic and subcellular compartmentation (see Sols and Marco, 1970).

The "binding" of GABA and glycine to cerebral synaptosomal fractions of mice occurred to a lesser extent when the animals were kept "isolated" than when they were kept "aggregated" for several weeks (DeFeudis, 1972d; DeFeudis *et al.*, 1976a, b; Fig. 22). The numbers of cerebral uptake-plus-receptor sites for GABA and glycine were about 40% lower in the "isolated" mice (see Table 8). Since no difference between the two groups of mice was evident when results were expressed as "mole amino acid/mg protein" (Fig. 23a, b), a reduction in the number of inhibitory synapses may have occurred. These synaptic changes were correlated with dramatic behavioral changes (e.g. hyperexcitability; increased aggressiveness) in these animals. Since inhibitory pathways are known to be essential for avoiding incompatible actions (Sherrington, 1906), such a decrease in inhibitory "binding" sites could be explained by an effect of environment on the "arousal" mechanism; i.e. "isolated" animals would be less capable of controlling afferent input to the reticular-activating system.

Other experiments by Hendley *et al.*, (1973) showed that the affinity of norepinephrine (NE) uptake was decreased (K_m was increased) and

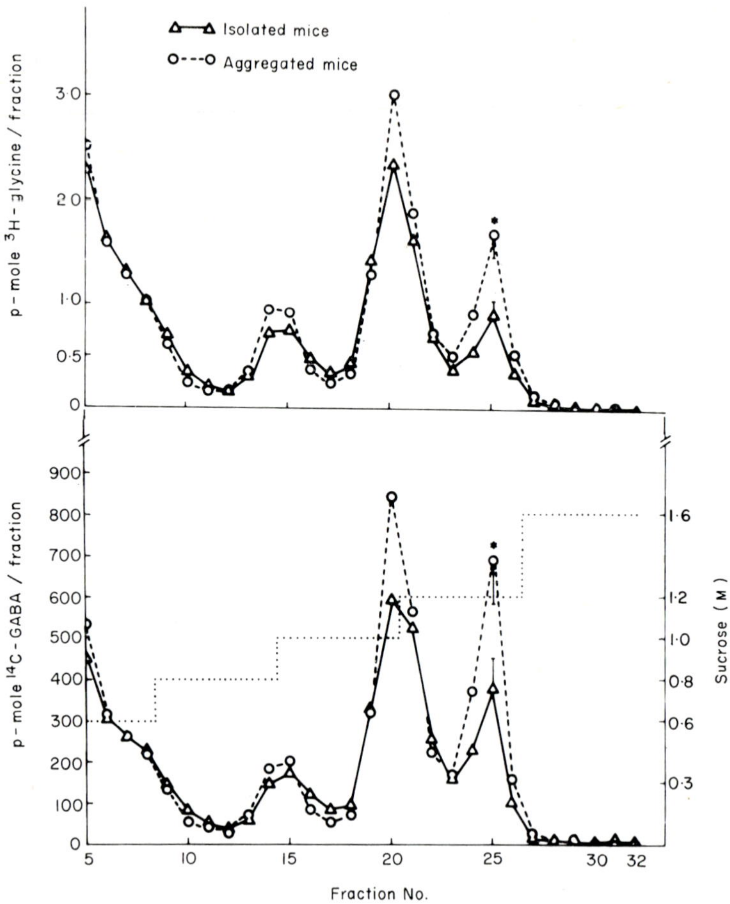

FIG. 22. Profiles of the "binding" of ^{14}C-GABA and ^{3}H-glycine to gradient-fractionated "synaptosomal-mitochondrial" fractions of the brains of differentially housed mice, expressed as mole-amino acid/0·5 ml fraction. Significant differences (*) and relevant standard errors are indicated; 5 "isolated" or 5 "aggregated" mice per point (Student's *t*-test). Sucrose molarities of the gradient are also indicated. Fractions are numbered from top to bottom of the tubes. (Reproduced with permission from DeFeudis *et al.*, 1976a.)

that the number of NE "binding" sites (V_{max})* was increased in cerebral cortical homogenates of pre-isolated mice that had been allowed

---

* K_m and V_{max} refer to constants of Michaelis-Menten kinetic theory. K_m is the value for a transport process which permits it to proceed at half maximal velocity, and its reciprocal provides a measure of the affinity of the ligand for its transport sites. V_{max} is the maximal velocity that can be obtained by elevating ligand (substrate) concentration, and provides a measure of the number of uptake sites.

TABLE 8. Protein contents and the "binding" of γ-aminobutyric acid (GABA) and glycine in the peak region situated at the 1·2:1·6 M interface of discontinuous sucrose gradients run on "synaptosomal-mitochondrial" fractions of the brains of differentially housed male Swiss albino mice.

Housing condition	Total protein (μg)	p-mol amino acid in peak		p-mol amino acid/mg protein in peak	
		^{14}C-GABA	^{3}H-glycine	^{14}C-GABA	^{3}H-glycine
"Aggregated"	1091 ± 157	1223 ± 207	3·12 ± 0·39	3494 ± 325	10·0 ± 1·1
"Isolated"	629 ± 84†	720 ± 159*	1·98 ± 0·28†	3975 ± 829	11·6 ± 1·4

Sums of the values for fractions Nos 24–26 in peak regions of gradients were calculated for each mouse; means ± S.E.M. (N = 5 in all cases); * and † indicate, respectively, $p < 0{\cdot}05$ and $p < 0{\cdot}02$ for comparisons between these values and corresponding values for "aggregated" mice (Student's *t*-test). Note that differences in the "binding" of GABA and glycine existed between "isolated" and "aggregated" mice when the data were expressed in total mol amino acid, but not when expressed in mol amino acid/mg protein. (Reproduced with permission from DeFeudis *et al.*, 1976a.)

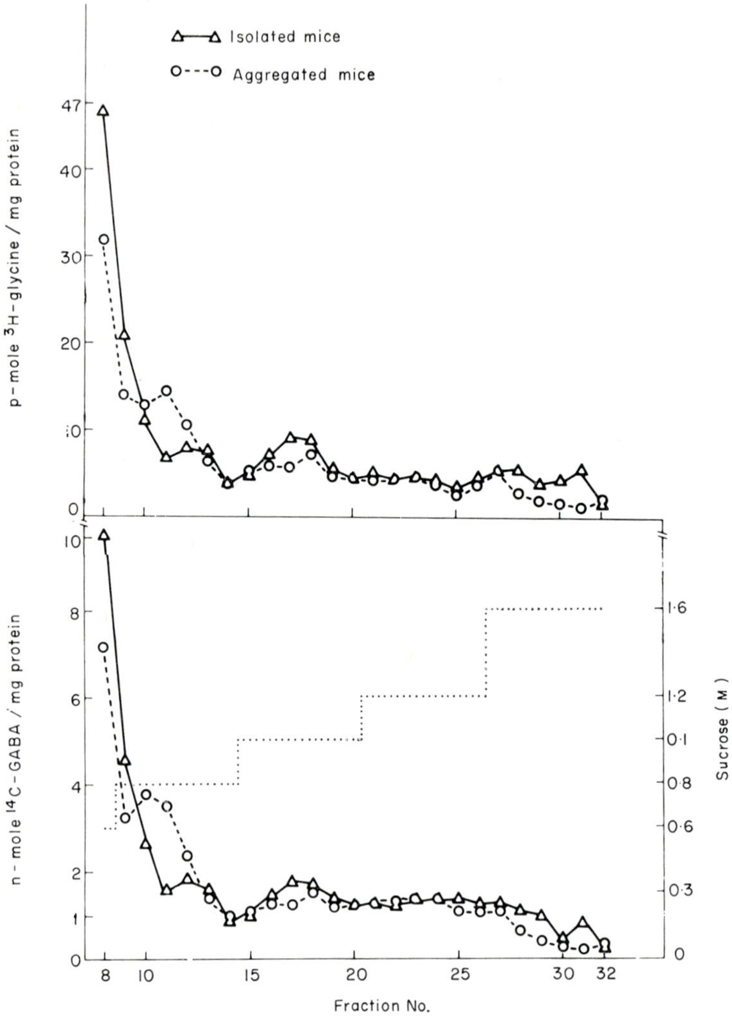

FIG. 23a. "Binding" of ^{14}C-γ-aminobutyric acid and ^{3}H-glycine to gradient-fractionated "synaptosomal-mitochondrial" fractions of the brains (rostral to inferior colliculi; excluding cerebellum) of differentially housed male Swiss albino mice, expressed on a protein basis. Weanling mice (21–22 days of age) were kept either "isolated" or "aggregated" for 6–7 weeks prior to experiments. Each point represents the mean value from 5 mice. Gradient profiles were obtained by layering 1·2 ml aliquots of re-suspended "synaptosomal-mitochondrial" fractions (representing about 0·15 g of fresh brain) onto the gradients and centrifuging these at 53 000 × g, 60 min; 0·5-ml fractions of each tube were collected. Molarities of the sucrose gradient are shown on the right-hand ordinate, and fractions are numbered from top to bottom of the tubes. (Reproduced with permission from DeFeudis *et al.*, 1976a.)

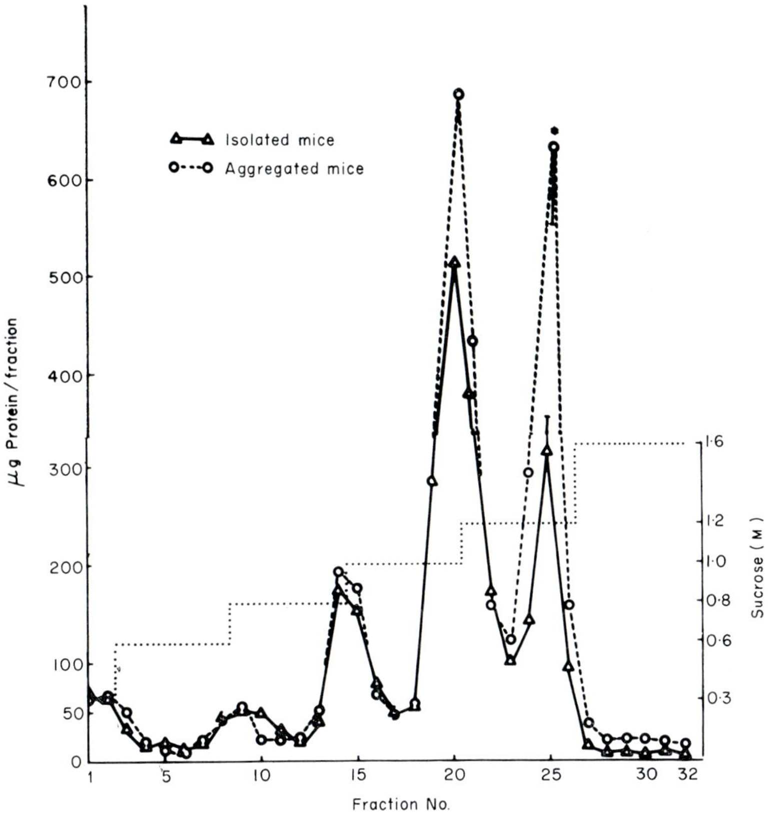

FIG. 23b. Protein profiles of "synaptosomal-mitochondrial" fractions of the brains of differentially housed mice. Each point represents the mean value of 5 mice; the significant difference in fraction No. 25 (*) and relevant S.E.M. values are shown. For more experimental detail, see legend to Fig. 23a (from DeFeudis *et al.*, 1975; reproduced with permission from *Naturwissenschaften*).

to fight (see Table 9). Further studies revealed that the fighting-induced decrease in NE affinity was similar to that caused by exposure of mice to five minutes of ECS, that both effects were completely reversed in 18–20 h, but that no significant increase occurred in uptake velocity (Welch *et al.*, 1974b). Most recently, Hadfield and Weber (1975) examined the effects of fighting among spontaneously aggressive male mice on whole brain synaptosomal uptake kinetics of NE. Both K_m and V_{max} were increased by fighting, in accord with the results of Hendley *et al.* (1973). This fighting-induced increase in NE uptake by nerve terminals was decreased by addition of 10^{-4} M diphenylhydantoin (DPH) to the medium. Interestingly, DPH had been shown previously to prevent fighting in electroshocked mice (e.g. Chen *et al.*, 1963).

TABLE 9. Uptake of *dl*-^{3}H-norepinephrine by homogenates of mouse cerebral cortex.

Animals	Apparent K_m (μM)	V_{max} (n-mol/g/5 min)	K_i for *d*-amphetamine (μM)
		"Group-housed"	
Controls[b]	0·279 ± 0·040 (6)	1·67 ± 0·24 (6)	0·059 ± 0·008 (5)
Fight-stressed	0·457 ± 0·055 (6)	2·20 ± 0·54 (6)	0·089 ± 0·007 (5)
p, paired	$p < 0{\cdot}07$[a]	$p > 0{\cdot}05$	$p < 0{\cdot}005$
		"Isolated"	
Controls[b]	0·397 ± 0·049 (15)	1·79 ± 0·27 (15)	
Fighters	0·646 ± 0·093 (15)	3·21 ± 0·68 (15)	
p, paired	$p < 0{\cdot}005$	$p < 0{\cdot}025$	

Group-housed male mice were used at 3–4 months of age; controls were animals among which no fighting occurred; "fight-stressed" animals were those among which spontaneous fighting repeatedly occurred. Individually housed male mice were either left undisturbed until killed ("controls"), or were placed together with 3 to 10 other "isolated" mice for 10–15 min of intense fighting per day for 14 consecutive days, then killed at 18–20 h after the last fight ("fighters"). Regarding the individually housed mice, data from two separate experiments with similar results were combined; one with mice at 5–6 months of age, and the other with mice 8–9 months old. [a] $p < 0{\cdot}05$ by Student's unpaired two-tailed *t*-test; [b] "group-housed" and "isolated" controls are not directly comparable since they were not assayed concomitantly. (Reproduced with permission from Hendley *et al.*, 1973.)

These studies, which have provided examples of environmentally induced changes in cerebral morphology and protein content that are reflected by changes in the number of transmitter "binding" sites and which correlate with dramatic changes in behavior, are in accord with those of Hydén (1961) which indicated that the visual and auditory cortices of rats kept in dark quiet boxes contained less protein and RNA than the same brain areas of rats kept under "enriched" conditions. Differences in NE uptake kinetics in cerebral synaptosomal fractions have also been correlated with the differences in behavior displayed by various mouse strains (Moisset *et al.*, 1975; Fig. 2). The above studies have revealed that adaptable, plastic responses of the CNS can be shown at the level of transmitter–membrane interaction, without using inhibitors of RNA or protein synthesis.

Regarding psycho-active agents, *d*-amphetamine (+ metabolites), dihydromorphine, and Li^+ were retained to a greater extent by nerve-ending fractions of the brains of "isolated" mice than by those of their "aggregated" littermates (DeFeudis, 1972b, 1975b; DeFeudis and Marks, 1973a; DeFeudis *et al.*, 1976b; Fig. 24; Table 10; see also Chapter 8, Part F). These studies have indicated that "binding" sites for exogenous ligands on macromolecular constituents of synaptic structures are sensitive to environmental change and have provided evidence that there exists an environmental basis for drug addiction (e.g. DeFeudis and Marks, 1973a; DeFeudis, 1975a; see Chapter 8, Part F).

These results, which have revealed changes in macromolecular binding or uptake sites for possible transmitters and psycho-active agents, remind one of the results of recent studies concerning trans-synaptic enzyme induction that have been conducted on elements of the peripheral sympathetic nervous system. Much evidence supports an ACh-mediated increase in tyrosine hydroxylase (TH) activity (see refs in Thoenen, 1975). More recent studies have indicated that an actual increase in number of active enzyme sites occurred (Mueller *et al.*, 1969a) which was probably due to an increased synthesis of new enzyme protein since the increase in TH activity was abolished by inhibition of protein synthesis (Mueller *et al.*, 1969b; Otten *et al.*, 1973). Since both cycloheximide and actinomycin-D were effective inhibitors, it was concluded that the regulation occurred at the level of transcription.

E. General Conclusions

The studies discussed in this chapter reveal that the cellular and subcellular morphologies of the CNS are continuously changing in response to the amount of environmental stimulation (experience) received by

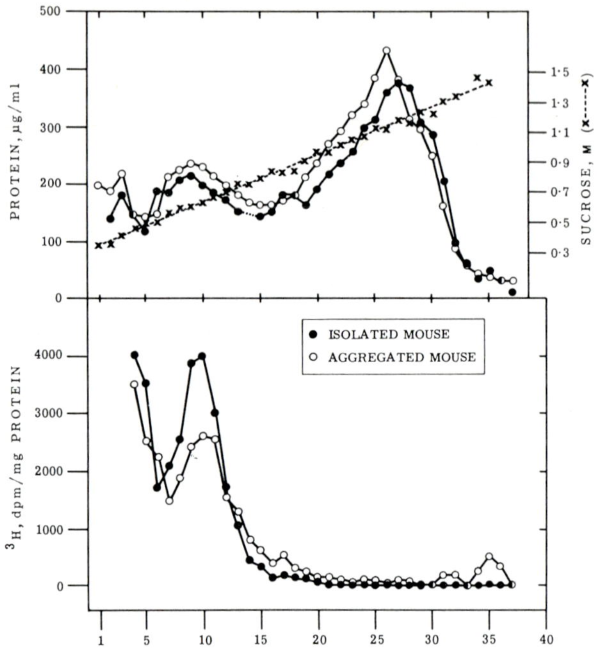

FIG. 24. Subcellular profiles of ^3H and protein contents of synaptosome-enriched "P_2" fractions prepared from the brains of "isolated" and "aggregated" C-57 Black mice. Both mice were housed differentially for 5 weeks. About 17 min after injection of ^3H-*d*-amphetamine-SO_4 (50 mg/kg; 0·1 *µCi*/g), the mice were decapitated and their brains (rostral to the inferior colliculi; excluding cerebellum) were homogenized in 10 vols 0·32 M sucrose solution. "P_2" fractions, prepared by the method of Gray and Whittaker (1962), were re-suspended in 3·5 ml 0·32 M sucrose solution, and 2·0 ml of each sample were layered onto linear gradients (0·4–1·8 M sucrose) and centrifuged at 137 000 × g, 1·5 h (Whittaker *et al.*, 1964). Small fractions of the gradients were collected using a gradient fractionator and analyzed for their contents of ^3H and protein. *Upper graphs* indicate the subcellular distributions of total protein and the linearity of the gradient; *Lower graphs* indicate the difference in the incorporation (or "binding") of ^3H which existed between the brains of the "isolated" and the "aggregated" mouse, as well as the localization of ^3H to regions of the gradients which contained fragments of synaptic membranes (from DeFeudis and Marks, 1973a; reproduced with permission from *Experientia*).

the animal. Hence, the interaction of an organism with its environment plays an important role in governing neural development.

Studies discussed in Part B1 have provided evidence that the vesicular release of CNS neurotransmitter(s) is related to functional activity, and are in accord with other reports which have revealed that changes in synaptic structure accompany changes in functional activity at peripheral ganglionic and neuromuscular synapses (e.g. Parducz and

TABLE 10. Binding of ^{3}H-dihydromorphine to brain regions of differentially housed mice.

Condition of mice	Brain region	mole ($\times 10^{14}$) dihydromorphine/mg protein	mg Protein/g pellet	$\left[\frac{\text{dpm } ^3\text{H/g } P_2}{\text{dpm } ^3\text{H/g } S_2}\right]$
Isolated	cerebral cortex	4·49 ± 0·41 (18)*	90·5 ± 5·4 (18)*	3·27 ± 0·22 (18)
Aggregated	cerebral cortex	4·12 ± 0·29 (17)	97·5 ± 9·0 (17)	3·21 ± 0·32 (17)
Isolated	whole cerebellum	10·12 ± 1·58 (16)	70·2 ± 5·6 (17)	5·60 ± 0·73 (17)
Aggregated	whole cerebellum	9·73 ± 2·37 (18)	77·9 ± 14·4 (18)	5·90 ± 1·07 (18)
Isolated	brain stem	4·70 ± 0·27 (18)	78·0 ± 4·4 (18)†	2·87 ± 0·25 (17)†
Aggregated	brain stem	4·76 ± 0·30 (17)	83·4 ± 3·7 (17)	3·15 ± 0·20 (17)

Male Swiss albino mice (34–36 days of age) were housed either singly or in a group of 24 for 5–6 weeks. Synaptosome-enriched (P_2) fractions were prepared and allowed to stand in contact with ^{3}H-dihydromorphine ($1{\cdot}4 \times 10^{-9}$ M) for 10–15 min before final centrifugation. Brain regions from 4 mice were pooled for each sample, and 8 or 9 separate determinations were made per sample; means ± standard deviations; numbers of determinations in parentheses; * and † indicate, respectively, $p < 0{\cdot}01$ and $p < 0{\cdot}001$ when these values are compared with corresponding values for "aggregated" mice (Student's *t*-test; two-tailed). (Reproduced with permission from DeFeudis *et al.*, 1976b.)

Fehér, 1970; Ceccarelli *et al.*, 1972, 1973; Korneliussen, 1972; Heuser and Reese, 1973; Pysh and Wiley, 1974). Studies reviewed in Part B2 certainly indicated that cerebral connectivity can be affected by visual stimuli. This could mean that the interneuronal connections in the visual system may only be partly pre-coded (pre-wired) so that their organization can be influenced by changes in retinal input (see, e.g. Van Hof, 1974). The morphological basis of learning and memory could be related to such environmental modification of cortical cells. Other studies (Part C) indicated that glial cells exhibit changes in response to environmental stimulation. Further studies about the roles of glial elements in neural function (e.g. DeRobertis and Carrea, 1965; Lasansky, 1971; Hydén, 1973b), and about the changes in glial structure and function caused by various environmental insults (e.g. Pevsner, 1972; Vaughn and Skoff, 1972; Hydén, 1973b), have been reviewed. The findings discussed in Part D provide rationale for further studies with subcellular fractionation and macromolecular methods.

Since synaptic alterations have been shown to occur as early as within two days after initial exposure of rodents to light (Cragg, 1967; Valverde, 1971), such changes appear to be within the time scale associated with processes of "long-term" memory consolidation. These morphological changes may be associated with changes in the inductive role of presynaptic nerve endings on the development of dendritic spines (Hamori, 1973) that could be produced by genetic and/or by environmental modification. Similar changes may characterize the neuronal plasticity that underlies the normal learning and memory processes which govern behavior. Excellent reviews of the effects of sensory deprivation or other environmental insults on synaptic morphology have been recently published (Cragg, 1972a; Chow, 1973; Fairén and Valverde, 1973; Globus, 1975; Barlow, 1975; Hirsch and Jacobson, 1975; see also refs in Vrensen and DeGroot, 1975).

4. Macromolecular Changes

A. Introduction

Since neural structure is composed of macromolecules, it is obvious that these substances should play roles in the processing, storage, and retrieval of information by the nervous system. These considerations have given rise to a research area of intense activity and much enthusiasm (see, e.g. Glassman, 1969, 1974; Rose, 1969a; Bloom *et al.*, 1970; Jakoubec and Semiginovsky, 1970; Gaito, 1972; Nakajima and Essman, 1973; Hydén, 1973a, b, c; Ansell and Bradley, 1973; Izquierdo, 1975a, b; Hudgens *et al.*, 1975; Entingh *et al.*, 1975; Delgado and DeFeudis, 1976; Rosenzweig and Bennett, 1976b) which seems to be at the basis of the *Behavioral Code.* Changes in the cerebral contents and syntheses of RNA and proteins have been correlated with the retention of acquired behaviors and may be required for learning and for the consolidation of "long-term" memory (see, e.g. Agranoff, 1967, 1974; Flexner *et al.*, 1967; Barondes and Cohen, 1967; Hydén and Lange, 1968, 1970; Rose *et al.*, 1970; Glassman and Wilson, 1970; Hydén, 1973a, b, c; Izquierdo, 1975a, b; Dunn, 1975).

It seems likely that all changes in cerebral morphology will involve RNA and proteins. The recent finding that the total protein contents (on a wet weight basis) of many structures of the feline CNS are rather constant (Battistin *et al.*, 1969; Gervas-Camacho *et al.*, 1976; Table 11) could provide a chemical basis for the "general constancy law" of Breuer and Freud (1892) which states that "the nervous system endeavors to keep constant something in its functional condition that may be described as the sum of excitation". In this regard, it seems noteworthy that changes in total steady-state protein content and protein patterns of whole brain or brain regions have not generally been observed following nontraumatically-induced changes in behavior, though some such changes have been reported (see, e.g. Barraco and Irwin, 1976, and refs therein). Hence, the changes in protein of CNS regions that accompany behavioral modification perhaps represent

TABLE 11. Protein contents of 46 regions of the feline CNS.

Tissue	Protein (mg/g, wet wt.)
Cerebral cortex (combined values)	113·9 ± 1·4 (58)
Gyrus proreus	107·4 ± 2·8 (5)
Parasplenic gyrus	107·3 ± 6·4 (3)
Cingulate gyrus	113·8 ± 3·7 (4)
Gyrus rectus	94·3 ± 2·7 (5)
Frontal gyrus	112·7 ± 2·7 (5)
Suprasplenic gyrus	120·9 ± 7·2 (3)
Sigmoid anterior gyrus	110·5 ± 3·0 (5)
Sigmoid posterior gyrus	117·2 ± 3·2 (5)
Lateral gyrus	116·9 ± 3·9 (4)
Coronal gyrus	108·4 ± 3·1 (4)
Posterior ectosylvian gyrus	117·5 ± 6·7 (4)
Medial ectosylvian gyrus	116·4 ± 8·0 (4)
Anterior ectosylvian gyrus	124·0 ± 6·6 (4)
Parahippocampal gyrus	124·5 ± 7·4 (4)
Hippocampus proper	114·4 ± 4·3 (4)
Olfactory bulb	109·8 ± 6·5 (4)
Cerebellar cortex	107·0 ± 3·5 (4)
Cerebellar white matter	110·6 ± 4·4 (5)
Hypothalamus	113·0 ± 5·0 (4)
Inferior colliculus	113·3 ± 4·5 (4)
Superior colliculus	112·4 ± 5·9 (4)
Inferior cerebellar peduncle	105·1 ± 3·9 (4)
Middle cerebellar peduncle	111·9 ± 2·1 (5)
Superior cerebellar peduncle	109·1 ± 5·9 (5)
Medial geniculate body	111·4 ± 4·9 (4)
Lateral geniculate body	118·2 ± 3·4 (4)
Thalamus	113·0 ± 7·0 (4)
Pons	114·0 ± 4·5 (3)
Trigeminal nerve	114·3 ± 1·8 (4)
Inferior olivary nucleus	118·8 ± 4·5 (4)
Pyramidal tract	115·6 ± 4·9 (4)
Optic chiasma	95·8 ± 7·2 (6)
Optic tract	103·6 ± 7·1 (5)
Olfactory tract	114·7 ± 7·6 (4)
Fornix	115·6 ± 2·9 (4)
Corpus callosum	117·2 ± 5·8 (6)
Head of caudate nucleus	114·0 ± 6·1 (4)
Corona radiata	118·0 ± 6·2 (4)
Internal capsule	113·5 ± 8·1 (4)
Globus pallidus	112·5 ± 6·0 (4)
Whole medulla oblongata	106·2 ± 8·3 (4)
Cerebral peduncle	118·7 ± 1·0 (4)
Hypophysis	179·9 ± 10·5 (4)
Cervical cord	101·5 ± 2·5 (15)
Thoracic cord	98·2 ± 2·8 (15)
Lumbar cord	102·0 ± 2·3 (15)

Protein was determined by the method of Lowry *et al.* (1951). Transverse sections of whole pons and whole regions of the spinal cord were used.
Means ± S.E.M.; numbers of determinations in parentheses; only the value for hypophysis differed significantly from that for gyrus proreus ($p < 0{\cdot}001$), a finding which is likely related to its high blood content; Student's *t*-test; two-tailed. (Reproduced with permission from Gervas-Camacho *et al.*, 1976.)

shifts of certain protein species among various tissue compartments (e.g. nerve endings) rather than changes in the total protein contents of these regions. Cerebral protein synthesis might also be involved in the sequential processes of sleep (e.g. Oswald, 1969; Pegram *et al.*, 1973), which in turn, may regulate "long-term" memory storage via the paradoxical (REM) sleep state (see Feldman and Dement, 1968; Pegram *et al.*, 1973).

Although macromolecular changes and other metabolic events that occur during various environmental insults appear to be "non-specific" since they seem to involve all areas of the nervous system and some peripheral tissues, certain elements of specificity have emerged. Quantitative differences have been shown in relation to various anatomical, behavioral and temporal aspects of environmental change (see, e.g. Dunn, 1975; Bennett, 1976). Perhaps specificity will be shown best at the level of alterations in cellular structure which govern cell connectivity (e.g. Rose *et al.*, 1976). Although such changes may not be regionally specific on a qualitative basis, such specificity may exist on a quantitative basis.

B. Nucleic Acids

1. *Deoxyribonucleic Acid (DNA)*

a. Introduction

The nuclear DNA of differentiated "macroneurones" (i.e. long-axoned neurones) does not appear to undergo metabolic turnover. However, a slow rate of DNA turnover may exist in some non-dividing cells (Pelc, 1970), and cerebral mitochondrial DNA turns over with a half-life of about 31 days (Gross *et al.*, 1969). Also, certain brain cells (e.g. glia, some interneurones) continue to divide into infancy or even adulthood and thus contribute to DNA turnover (see Altman, 1969; Mahler, 1972). Some evidence also supports the existence of "informational-DNA" that could function like messenger-RNA (see Yuwiler, 1971).

Although mammalian brain cells vary greatly in size, they all contain about 6–7 picograms of DNA, i.e. about the same amount of

DNA as is present in other diploid cells of the body (e.g. Mandel *et al.*, 1948; Heller and Elliott, 1954; Kissane and Robins, 1958). Cerebral DNA is indistinguishable from the DNA of other body cells, and like RNA, it undergoes species-dependent changes in its cerebral concentration during development. As examples, rabbit brain contains about 4, 6 and 10 mg DNA/brain at 10 days, 1 month and adult ages, respectively, whereas there exist about 1·2, 1·8 and 1·8 mg DNA/brain in rats of 10 days, 1 month and adult ages (see Mandel *et al.*, 1964; Rappoport *et al.*, 1969). Cerebral DNA and RNA both increase during mitosis, but when mitosis ceases, RNA continues to increase as the cells enlarge, whereas no further increase occurs in DNA. After all cellular growth has ceased, DNA and RNA contents remain constant throughout the remaining life of the animal. Although RNA turnover occurs after cessation of cell growth (Leslie and Davidson, 1951; Winick and Noble, 1965), the turnover of DNA is still subject to controversy.

Although differentiated neurones never divide, neuroblast division may occur in some species (Altman, 1969; see below). Since DNA in non-dividing cells is not renewed, neuronal DNA appears to be markedly stable, but DNA repair mechanisms perhaps operate in neurones. Although all types of cerebral glial cells divide, their rates of division are generally not high. However, astroglia and microglia may proliferate extensively during certain pathologic conditions and during mechanical injury to the brain. In many brain regions gliogenesis is mainly a postnatal phenomenon (e.g. Bryans, 1959; Altman, 1969; Patel *et al.*, 1976). In this regard, it is noteworthy that glia could be agents of "long-term" memory (see also Chapter 3, Part C).

DNA synthesis and turnover usually reflect processes of cellular repair, and appear to be correlated with the ability of cells to divide. Two different DNA polymerases from rat brain have been shown, and during development the decrease in their activity is disproportionate (Chiu and Sung, 1973), indicating that one enzyme might be used for DNA replication while the other might be concerned with DNA repair (which implies "turnover"). Certain conditions (e.g. mechanical injuries, viral infections, vascular diseases, thyroid deficiency, ionizing radiation, malnutrition, "enriched" environments) can alter cerebral cell proliferation and DNA content, thus providing further evidence for DNA turnover and synthesis.

The use of fine-resolution ^{3}H-thymidine autoradiography has revolutionized this area of investigation. Systemically-injected ^{3}H-thymidine becomes readily incorporated into all cell nuclei that are synthesizing DNA in their preparation for cell division; i.e. cells that are in the interphase stage of mitosis (e.g. Amano *et al.*, 1959). Since DNA is metabolically stable (Hughes, 1959) such labelled cells, unless undergoing further division, will retain their label indefinitely. The dilution of the amount of radioactivity present over nuclei and time course of increase in number of labelled cells can thus be used to study changes in cellular proliferation. Using this method, Altman and colleagues (see Altman, 1969, 1972), showed that considerable DNA synthesis and cell proliferation, migration and differentiation occurred in the brains of infant rats. These events seemed to be initiated by cells of the outer granular layers of cerebellum, hippocampus, olfactory bulb and ventral cochlear nucleus. Most hippocampal cells were postnatally formed. These workers concluded that though long-axoned neurones are formed prenatally, many short-axoned neurones and glia are postnatally formed. With respect to human brain, two principal periods of cell proliferation have been detected using DNA measurements. One commences at about 15–20 weeks of gestation and corresponds to neuroblast proliferation, and the other begins at about 25 weeks and continues into the second year of postnatal life and corresponds to glial multiplication and perhaps also to a second wave of neurogenesis (see Dobbing and Sands, 1973, and refs therein).

b. Relationship to Environment

Gaito (1967) hypothesized that during learning: (1) DNA is activated to form RNA; (2) upon activation, DNA causes the synthesis of several RNAs in the presence of RNA polymerases; (a) ribosomal-RNA; (b) mRNA for synthesis of ribosomal protein; (c) mRNAs for synthesis of synaptic proteins of ChAc and AChE; (d) mRNAs for synthesis of structural protein of nerve endings; (e) mRNA for membrane protein; (f) mRNA for synthesis of the enzymes involved in the synthesis of membrane lipids. Another theory by Altman (1967) emphasized that learning and memory, if related to environmentally induced changes in cerebral morphology, might be explained by postnatal neurogenesis (see above). In contrast, Verzar (1968) proposed that a non-renewal, or lack of DNA turnover in differentiated cerebral cells might underly the permanency of "long-term" memory.

Altman and Das (1964) showed that glial DNA synthesis was higher and that there were more glia in the cerebral cortices of "enriched" (EC) than of "impoverished" (IC) rats after 110 days of differential rearing (see also Diamond *et al.*, 1964). Other experiments have revealed that rats kept in "enriched" (or "grouped") environments possessed a lower DNA concentration in whole brain after 30 days (Geller, 1971) and in occipital and somesthetic cortex after 155 days (Rosenzweig *et al.*, 1972a). After a year, "grouped" rats possessed a lower amount of DNA only in telencephalon, among the several brain regions examined by Welch *et al.*, (1974). Decreases in the DNA contents of cerebellum, diencephalon and cerebral cortex were also produced by individual housing of mice at various post-weaning ages (Essman, 1971d), implying that such "isolation" might cause a decrease in the formation of brain cells during this period of development. Also, whole brain DNA was decreased by increasing group size, regardless of group density in mice (Bell *et al.*, 1971), whereas in rats, brain DNA content appeared to vary with living space (Welch *et al.*, 1974). These findings indicate that certain environmental conditions might alter DNA metabolism and/or the possible age-associated loss of cerebral neurones.

With more specific regard to learning and memory, Hubscher *et al.*, (1975) have shown that neuronal nuclei of the cerebral cortices of rats subjected to 10 days (postnatal days 21–30) of training at a complex behavioral task possessed significantly more DNA-polymerase activity than control animals. In other studies, mice injected intraperitoneally or intracranially with ^{3}H-thymidine before training at a "step-through" task, incorporated more ^{3}H into cerebral DNA than did shocked or handled controls for approximately a week after training (Reinis and Lamble, 1972). Also, 5-iodo-2′-deoxyuridine (a substance that becomes incorporated into DNA) interfered with the performance of mice in a passive avoidance task, but this effect appeared to be related more to maintenance and retrieval than to learning itself (Reinis *et al.*, 1972). Although these findings indicated that DNA might be involved in the alterations of cerebral metabolism that occur during learning, inhibition of DNA synthesis by cytosine arabinoside (an anti-leukaemic cytotoxin which appears to act by inhibiting DNA polymerase activity) had no effect on learning in goldfish (Casola *et al.*, 1968). As suggested by Reinis and Lamble (1972), the existence of mechanisms for postnatal neurogenesis and for changing DNA structure in the absence of mitosis indicates that further study in this area should be encouraged.

2. *Ribonucleic Acid (RNA)*

a. *Introduction*

Neurones possess prominent nucleoli and active mechanisms for synthesizing ribosomal-, messenger-, and transfer-RNA. Neuronal RNA content varies with respect to cell type and among cells of the same type. Glia contain less RNA than most neurones, but their RNA synthesis occurs about twice as rapidly as that of neurones (see Daneholt and Brattgard, 1966). After Nissl (1891, 1894) had noticed that ribosomes ("Nissl substance") disintegrated following nerve transection-induced chromatolysis, many detailed microscopic studies of this reaction were conducted (see, e.g. Cammermeyer, 1969, 1972, 1975; Chapter 1, Part G). Cerebral RNA turnover appears to be necessary throughout life for the maintenance of structural and enzymatic proteins and for the restoration of polypeptides that are continuously released (e.g. "releasing factors"). Ribonucleoprotein particles, localized to dendritic positions subjacent to synaptic boutons (Bodian, 1965), could regulate the synthesis of proteinaceous material associated with synaptic contacts (see Talwar and Iqbal, 1972). For a review of the roles of RNA in nervous tissue, see Koenig (1972).

b. Relationship to Environment

Evidence for a role of RNA in information processing stems from Hydén's work (see, e.g. Hydén, 1943, 1959a, b, 1964a, b, 1973a, b, c). Hydén (1943) showed that mild electrical stimulation or physical exercise increased the RNA content of spinal motoneurones, whereas more intense stimulation caused it to decrease. Later, Hydén (1959a, b) postulated that the uniqueness of each remembered event might be carried by a specific encoding of RNA molecules in nerve cells. However, Eccles (1964) suggested that this hypothesis was unacceptable, since it disregarded organizational specificity of the nervous system and instead assumed that the specificity attached to the frequency pattern of the nerve impulses, rather than to the lines of communication of the nerve fibers.

Further experiments revealed that rats which were trained to obtain food by balancing on a wire perched at an angle possessed significantly greater contents of nuclear RNA in neurones and glia of Deiters' nucleus, but not in neuronal cytoplasmic RNA (Hydén and Egyhazi, 1962, 1963). Adenine-to-uracil (A/U) ratios of RNA were greater in trained than in control animals. Also, in cortical neurones of rats that had been trained to use the non-preferred paw, Hydén and Egyhazi (1964) found the guanosine + cytosine/adenine + uracil ratio of RNA to be decreased, indicating that the RNA associated with learning might be of the messenger type (see also Hydén, 1973c). Also, in this context, Dingman and Sporn (1961) found that pre-trial administration of 8-azaguanine (an analog of guanine which becomes incorporated into RNA, thus interfering with its synthesis) interfered with maze-learning of rats while not impairing previously learned experiences, and Zemp *et al.* (1966, 1967) demonstrated a short-lasting increase in incorporation of uridine into whole brain RNA of mice during a "short-term" training experience (but see Entingh *et al.*, 1974, below). Recently, the possible informational specificity of RNA molecules has been de-emphasized, and now it is not strongly believed that molecules can record specific memories within their structures (see, e.g. Hydén, 1973a).

Increases in RNA concentration (Nasello and Izquierdo, 1969; Gattoni and Izquierdo, 1974) and synthesis (Zemp *et al.*, 1967; Bowman and Strobel, 1969; Bowman and Kottler, 1970) and in RNA polymerase activity (Catravas *et al.*, 1973) of the hippocampus have been correlated with various learning tasks. However, no specific type of RNA showed increased labelling (Zemp *et al.*, 1966). More recently, Entingh *et al.* (1974) have observed that avoidance training of mice produced no change in the incorporation of radioactivity of injected 5-^{3}H-uridine into brain, but reduced the amount of radioactivity present in uridine monophosphate (UMP). Such changes in the labelling of UMP could have accounted for the changes in RNA labelling observed by Zemp *et al.* (1966, 1967) since the radioactivity in UMP was used as a correction factor in these experiments.

Using a two-way active avoidance response, Gattoni and Izquierdo (1974) showed that both conditioning and "pseudoconditioning" (achieved by presenting buzzers and shocks in a random sequence with no temporal association to rats in a shuttle-box so that they performed pseudoconditioned barrier-crossing responses to the buzzer) caused changes in hippocampal RNA concentration and synthesis, but that these changes differed in both their nature and timing. Conditioning appeared to stimulate precursor (uridine) incorporation primarily into hippocampal RNA, whereas pseudoconditioning appeared to stimulate acid ribonuclease activity first and RNA labelling only secondarily in both hippocampus and neocortex. These findings and others which have revealed that RNA concentration and synthesis in various nervous structures can be altered by afferent electrical stimulation, electroshock, high K^+ stimulation, or by systemic administration of stimulant drugs (e.g. Geiger *et al.*, 1956; Talwar *et al.*, 1961; Sacktor *et al.*, 1966; Berry, 1969; Izquierdo *et al.*, 1969; Nasello and Izquierdo, 1969; Kernell and Peterson, 1970; Jakoubec and Semiginovsky, 1970; Gisiger, 1971; Peterson and Erulkar, 1973; Daroqui and Orsingher, 1973), all seem to have a common basis since oxygen consumption and cerebral blood flow would be altered in all cases to meet the modified cellular demands associated with the changes in function. In other studies, intensive electroshock treatments of rabbits elicited seizures which were accompanied by a 20% disaggregation of cerebral polysomes to monomeric ribosomes (Vesco and Giuditta, 1968). More recently, administration of the convulsant, methionine sulphoximine (MSO) together with methionine (which prevented the convulsant action of MSO) led to a transient disaggregation of polysomes of rat cerebral cortex to mono- and di-somes (Sellinger and Azcurra, 1970). Since this effect was accompanied by an increased rate of protein synthesis, these investigators postulated that such changes might be related to cerebral mechanisms of defense against seizures. Such insults (e.g. electrical stimulation) of neurones also cause changes in the synthesis of proteins (see Part C, below) and phospholipids (see, e.g. Burt and Larrabee, 1973).

Other recent studies have revealed that though uridine incorporation into RNA was inhibited by both 5-bromotubercidin and 5-fluorouridine, the latter agent did not affect memory (see Agranoff, 1974). Since bromotubercidin prevents the synthesis of both messenger- and ribosomal-RNA, whereas fluorouridine prevents only the latter process (Brdar *et al.*, 1973), memory formation might depend mainly on messenger-RNA. Other results which showed that intracranial injection of camptothecin within 1·5 h of training greatly diminished memory (conditioned avoidance in goldfish) that was tested one week later, whereas injection of the drug at 5 or 24 h after training caused no detectable effect on this response have also indicated that the cell nucleus might be involved in "long-term" memory (Neale *et al.*, 1973b). Camptothecin is a less toxic and a more reversible inhibitor of RNA synthesis than actinomycin, which also produces amnestic effects when injected intracerebrally into rodents (Squire and Barondes, 1970) and chicks (Codish, 1971), and appears to exert a more selective action than actinomycin on nuclear RNA (see refs in Neale *et al.*, 1973b).

Ferchmin *et al.* (1970) showed that the RNA content of the cerebrum was higher in rats subjected to "enriched" conditions (EC) than in "impoverished" animals (IC) during the period of fast acquisition of the difference in brain weight, but not in the ensuing 60 days. These effects were obtained under conditions that were considerably more complex than those used by the original Berkeley group. Rosenzweig *et al.* (1972a) found, with rats reared in both their usual EC-IC paradigm and in light *vs* darkness, that the RNA/DNA ratio of the cerebral cortex was higher in the more stimulated animals. Such increases in total RNA and in RNA/DNA ratios of rat cerebral cortex have been shown under EC conditions in experiments enduring from 4 to 165 days (see Bennett, 1976). The findings that one year of "grouping" of post-weanling rats increased their cerebral RNA concentration (Welch *et al.*, 1974a), and that the cerebral RNA content of mice decreased with increasing group density (Bell *et al.*, 1971) further support these results. Bennett (1976) has suggested that such changes in RNA could represent the integration of many individual learning and memory events which cause increased incorporation of RNA precursors, since RNA is environmentally sensitive and since the persistence of RNA/DNA effects is less than that of brain weight effects (see also Table 12).

TABLE 12. Effects of successive environments on RNA and DNA measures (% differences).[a]

	Duration (days)		Occipital cortex			Dorsal (including somesthetic) cortex		
	EC	IC	RNA/mg	DNA/mg	RNA/DNA	RNA/mg	DNA/mg	RNA/DNA
EC	45	0	−2·8*	−8·9‡	6·7‡	−1·4	−4·8†	3·6*
EC · IC	30	15	−4·5†	0·8	−5·3†	−2·8*	0·3	−3·1
EC	58	0	−2·2*	−7·5‡	5·8‡	−0·4	−5·0‡	4·9†
EC · IC	30	28	−3·3†	−1·8	−1·6	−2·2†	−2·5	0·3
EC	109	0	−0·2	−6·2‡	6·3‡	0·2	−4·4†	5·1†
EC · IC	95	14	−3·1*	−2·2	−0·9	−1·9	−1·2	−0·8
EC	119	0	0·9	−4·7†	5·8‡	0·8	−2·1*	2·9‡
EC · IC	91	28	−3·8†	−0·5	−3·2*	−1·4*	2·0	−3·3†

[a] All experiments were begun at weanling (25–28 days) age. Percentages are (EC − IC)/IC or (EC · IC − IC)/(IC · IC). Each value represents the average of 12 littermate comparisons. * $p < 0{\cdot}05$; † $p < 0{\cdot}01$; ‡ $p < 0{\cdot}001$. (Reproduced with permission from Bennett, 1976.)

Using a RNA:DNA hybridization technique, Mushynski *et al.* (1973) found no significant qualitative differences between RNA from the brains of "enriched" and "impoverished" rats (see Fig. 25),

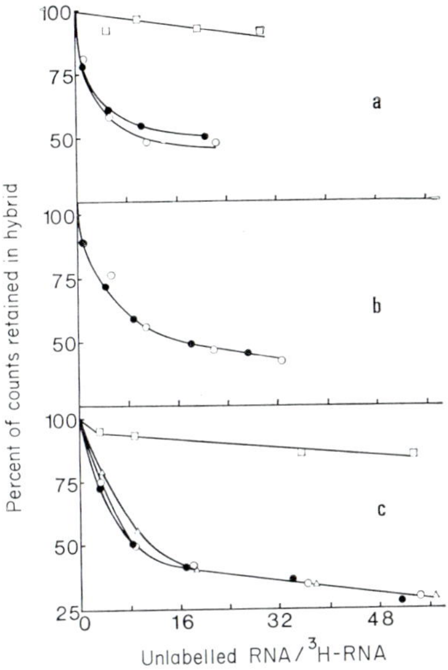

FIG. 25. Competition hybridization between labelled brain RNA from "enriched" rats and unlabelled brain RNA from "enriched" or "deprived" rats at various times after assignment to the environments. The experiments were performed by using the following RNA preparations: (a) ^{3}H-RNA (2300 dpm/μg) from "enriched" rats at day 7 following assignment to the environments *vs* unlabelled competitor RNA from *E. coli* (□) or from "enriched" (●) or "deprived" (○) rats of the corresponding period. The ratio of ^{3}H-RNA to DNA was 1·0; 100% hybridization was equal to 6200 dpm/100 μg of rat brain DNA. (b) ^{3}H-RNA (5700 dpm/μg) from 20-day "enriched" rats *vs* unlabelled competitor RNA from "enriched" (●) or "deprived" (○) rats of the corresponding period. The ratio of ^{3}H-RNA to DNA was 0·5; 100% hybridization equalled 9800 dpm/100 μg of rat brain DNA. (c) ^{3}H-RNA (6150 dpm/μg) from 35-day "enriched" rats, *vs* unlabelled competitor RNA from *E. coli* (□) or from "enriched" (●), "deprived" (○) or "deprived-stressed" (△) rats of the corresponding period. The ratio of ^{3}H-RNA to DNA was 0·5; 100% hybridization equalled 9550 dpm/μg of rat brain DNA. (Reproduced with permission from Mushynski *et al.*, 1973.)

which indicated that no change had occurred in the number of genes expressed in these RNAs. However, in an earlier hybridization study, K. A. Bonnet (unpublished results) observed an apparent increase in gene expression in "split-brain" cats on the side of the brain associated with the learning of a visual discrimination task. This technique exploits the complementarity of sequences present in the double-stranded structure of DNA, and provides sensitive measures for the similarity of base sequences in DNA or RNA.

In differentially housed mice, the RNA contents of subcellular fractions of cerebral cortex were found to be lower in "isolated" animals than in their "aggregated" counterparts (Essman, 1972). Also, although both neuronal and glial cell numbers of the cerebellum were found to be higher in "aggregated" mice, the neuronal fraction of "aggregated" mice contained more RNA, whereas the glial fraction contained less RNA, than in "isolated" mice (Essman, 1971a; Table 13). These differences were most significant in synaptosomal fractions.

Other studies have revealed that the anterior forebrain roof of the chick increases its incorporation of precursors into RNA (Rose *et al.*, 1970a, b; Bateson *et al.*, 1972) and protein (see Part C2 below) during "imprinting".* Three groups of chicks were used in these studies: (1) experimental group exposed to the imprinting stimulus; (2) light-exposed controls; (3) dark-exposed controls. Experimental birds were placed in pens from which they could see a rotating flashing light (the imprinting stimulus). Following ^{3}H-uracil injections, the imprinting stimulus caused an increased incorporation of label into RNA of the forebrain roof (which contains the optic tectum), but not into forebrain base or midbrain regions. *In vitro* determinations revealed that the activity of RNA polymerase in the forebrain roof was selectively increased in imprinted chicks (Haywood *et al.*, 1970). Hence, the increased uracil incorporation appeared to represent an increase in RNA synthesis, rather than a "precursor pool artifact". However, it should be noted that such a change in RNA could have been mediated by the arousal or attentional changes which appear to increase blood flow in this visual system (see Bondy *et al.*, 1974). In any case, these results indicated that an environmental change can alter cerebral metabolism in a region of the brain that is associated with the modality stimulated (see also Horn *et al.*, 1973; Rose *et al.*, 1973a). Using a "split-brain" preparation and an eye patch to eliminate the effects of generalized arousal and attentional phenomena, it was found that the forebrain roof of the imprinted side (contralateral to the imprinted eye), but not that of the covered eye, showed a significantly increased incorporation of ^{3}H-uracil, which further substantiates the view that the increase in RNA synthesis was associated with increased visual function (Horn *et al.*, 1971). These changes in RNA and those reported

* "Imprinting" may be regarded as a special case of more general learning phenomena (see, e.g. Bateson, 1971). For a limited period, beginning shortly after hatching, a chick will follow the first moving object it perceives, thereafter forming an attachment to that object.

TABLE 13. Percentage difference in cell number and estimated RNA concentration for neuronal and glial fractions from mouse cerebellum following differential housing at different ages of onset.

Age of onset of differential housing (days)	Percent difference in cell number[a]		Estimated mean (±) cellular RNA difference	
	Neuronal fraction	Glial fraction	Neuronal fraction	Glial fraction
21	15	27[b]	15·4 (1·9)	−33·1 (2·5)[b]
23	11	20[b]	10·5 (1·3)	−25·1 (1·7)[b]
32	4	7	9·4 (1·1)	−7·6 (1·0)

[a] After differential housing (aggregation-minus-isolation); [b] $p < 0{\cdot}02$. (Reproduced with permission from Essman, 1971a.)

for protein (see Part C2, below) in the visual system could be associated with some of the morphological and electrophysiological changes that accompany various modifications of visual input (see also Chapter 3, Part B2).

3. *General Comments on Nucleic Acids*

These studies, especially those on RNA, have strengthened the notion that nucleic acids and protein synthesis are important elements of the *Behavioral Code.* However, a direct relationship between learning or memory and the synthesis of specific species of RNA does not appear to exist. Also, no evidence exists to support the contentions that these macromolecules, in themselves, can encode or store experiential information or that RNA can function other than as a mediator between DNA and protein synthesis. Furthermore, macromolecular changes that occur during learning and/or memory formation could be secondary to behavioral events, or both could be parallel, non-interacting processes (Pevsner, 1966). It should also be remembered that changes in RNA and protein synthesis represent only one step between environmental input and functional synaptic activity (see Fig. 81). Nevertheless, the above mentioned results, along with those which have indicated that the number and structure of ribosomes and polysomes and the RNA contents of cerebral subcellular fractions can be altered by experiences that affect learning and memory (e.g. Appel *et al.*, 1967; Bennett and Rosenzweig, 1971; Essman, 1972; Welch *et al.*, 1974a; Table 14; Wenzel *et al.*, 1975; Murthy *et al.*, 1976) and that changes in protein content and synthesis in nervous structures can be produced by various environmental conditions (see Part C, below) certainly support the hypothesis that experience produces changes in cerebral structure.

C. Proteins

1. *Introduction*

In 1950, an hypothesis assigning a significant role to nucleoproteins in learning was developed (Katz and Halstead, 1950; Halstead, 1951), and a few years later Gerard (1953) proposed that changes in nerve proteins produced synaptic events which facilitated learning and maintained memory. In support of these contentions, it has been shown that changes in cerebral proteins occur during excitation and inhibition (see, e.g. Lajtha, 1964; Jakoubec and Semiginovsky, 1970), that phosphoproteins appear to be associated with cerebral Na^+, K^+-ATP-ase systems (e.g. Heald, 1962; Rodnight, 1966), and that

TABLE 14. Differences in RNA metabolism between trained and control mice.

	Experiment	N pairs	Nuclei			Polysomes	
			RNA/UMP	RNA	UMP	RNA/UMP	RNA
1	Zemp *et al.* (1966)	10	28 ± 5	—	—	40 ± 7	—
2	Entingh *et al.* (1974)	17	26 ± 5	2 ± 7	−20 ± 6	—	—
3	Adair *et al.* (1968)	40	—	—	−12 ± 6	38 ± 4	22 ± 6
4	Uphouse *et al.* (1972)	10	—	—	—	—	28 ± 5

Results are expressed as the percentage difference between 1·00 and the geometric mean for paired Trained/Control radioactivity ratios, with associated S.E.M. Geometric means were used in these comparisons to avoid errors arising from calculating the arithmetic means of series of ratios. (Reproduced with permission from Entingh *et al.*, 1974.)

cerebral protein synthesis may be involved in learning and memory (e.g. Agranoff *et al.*, 1965, 1966; Flexner *et al.*, 1965, 1967; Cohen and Barondes, 1968; Bogoch, 1968, 1974; Hydén and Lange, 1968, 1972; Hydén, 1973a; Agranoff, 1973).

Although it has not been possible to demonstrate conclusively that the total protein contents of whole brain regions are altered by conditions that affect learning and memory (but, see Hydén, 1961; refs in Barraco and Irwin, 1976), several studies have indicated that the total protein content of cerebral synaptosomal fractions tends to be increased in animals subjected to environmental "enrichment" or "grouping" (e.g. Irwin, 1969a, b; DeFeudis, 1972c; DeFeudis *et al.*, 1975). However, changes in the rate of protein synthesis appear to be easier to demonstrate (see below).

It has been claimed that the proteins most enhanced in synthesis by learning and sensory stimulation appear to be the acidic species, S-100 and 14–3–2 (e.g. Singh and Talwar, 1969; Hydén and Lange, 1972; Hydén, 1973a), which are membrane constituents (e.g. Rusca *et al.*, 1972; Haglid and Stavrou, 1973; Donato *et al.*, 1975; Hydén and Rönnbäck, 1975) that could be involved in the regulation of ionic fluxes and other synaptic phenomena. The S-100 protein(s) undergoes axoplasmic flow (Miani *et al.*, 1972), its interaction with membranes is strictly Ca^{2+}-dependent (Calissano and Bangham, 1971; Calissano, 1973; Moore, 1973; Donato and Michetti, 1974), and it binds to synaptosomal membranes of guinea pig brain by a specific, saturable and reversible process (Donato *et al.*, 1975). Furthermore, protein synthesis appeared to be increased to its greatest extent in the limbic system during information processing, and one of the responsive elements was S-100 protein (see Beach *et al.*, 1969; Izquierdo, 1975a). With respect to plasticity, Perez and Moore (1968) showed that Wallerian degeneration of rabbit tibial nerve caused a progressive decrease in S-100 protein(s), indicating that it was of neural origin. However, further studies on rabbit optic nerve during degeneration indicated that the level of S-100 was slightly increased, whereas that of 14–3–2 was decreased as a function of degeneration (Perez *et al.*, 1970). These findings, together with those of Hydén and McEwen (1966), Mihailović and Hydén (1969) and Cicero *et al.* (1970) indicated that S-100 was mainly glial and that 14–3–2 was neuronal (see Moore, 1973). Increases in cerebral nuclear acidic phosphoproteins have also been produced by "short-term" training experiences (e.g. Machlus *et al.*, 1974a, b). Such changes in phosphorylation, if produced in postsynaptic membranes, could alter membrane conformation and thus affect ion conductance and postsynaptic potentials (Dunn, 1975).

Although examination of the changes in cerebral proteins that occur during environmental "enrichment" and during learning and memory has led to the contention that these appear to be general rather than specific phenomena (see, e.g. Levitan *et al.*, 1972b; Dunn, 1975; Conde and DeFeudis, 1977), the important result is that such environmentally induced changes in cerebral protein do indeed occur.

2. *Studies with Precursor Amino Acids*

Using rats, Altman (1963) tested the effects of forced exercise, using two different speeds (7 and 12 rpm) of a motor driven wheel, on the incorporation of intraperitoneally injected 4–5–^{3}H-DL-leucine into brain. With an autoradiographic-microdensitometric method, quantitative indices of the utilization of the ^{3}H-labelled precursor were obtained. Appreciable increases in the utilization of this amino acid occurred in motor cells during mild activity, whereas a slight decrease in its uptake occurred immediately after activity (Fig. 26; Altman, 1963). Even more interesting was the significant increase in the labelling of all cell types (not only in motor, but also in sensory regions) that occurred (see Table 15). Although such changes can be explained by an effect of exercise on the sizes of local amino acid pools (and hence, their specific activities), or by the increases in local blood circulation which are caused by functional activation (Serota and Gerard, 1938; Sokoloff, 1961), these findings indicated that protein synthesis had been changed in the *entire* brain.

In a further study, Altman and Das (1966) showed that exposure of rats to a similar sensorimotor task did not alter the utilization of ^{3}H-leucine in several brain regions when the stress produced by exercise had been reduced or eliminated by pre-adaptation to the task. Thus, the stress associated with the novelty of the situation appeared to be the essential prerequisite to changes in cerebral protein synthesis. In other experiments, a significant increase in incorporation of ^{3}H-leucine into proteins of several brain regions which occurred in a "restricted" group of rats also appeared to be caused by the stresses associated with handling and injection procedures (Das and Altman, 1966; Fig. 27). However, no significant differences in cerebral ^{14}C-leucine incorporation existed between rats trained in a maze on a series of visual pattern discrimination tasks and visually deprived rats, and the slight increase in ^{3}H-leucine utilization found in visually deprived rats was attributed to the greater stress produced by the injection procedure in these previously unhandled animals (Altman *et al.*, 1966). In support of these findings, Rees *et al.* (1974) have shown that brief exposures of mice to a variety of sensory stimuli increased the incorporation of ^{3}H-lysine into total protein. These responses were also greatly diminished in mice that had been previously exposed to the treatments, and this effect of sensory stimulation also occurred in

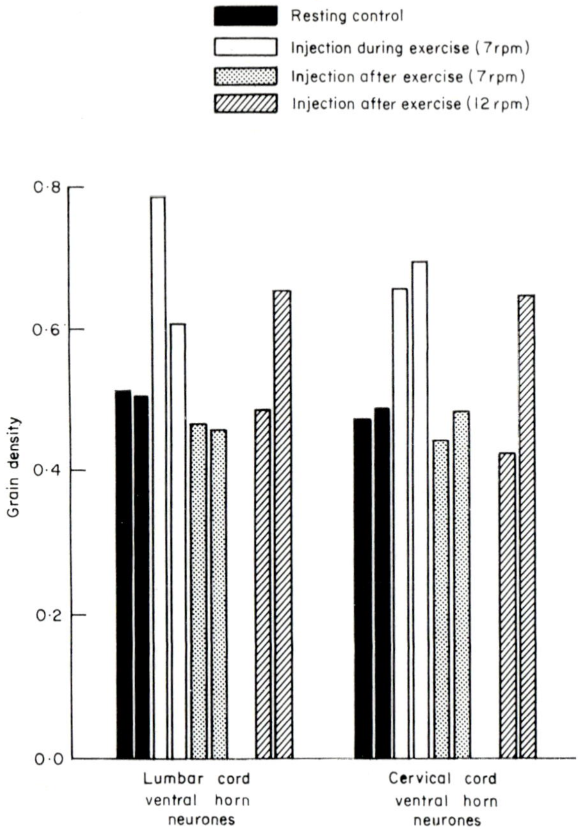

FIG. 26. Mean density of single neurones (N = 20) in the lumbar and cervical spinal cords of rats. All animals were injected intraperitoneally with 2 m*Ci* of DL-leucine-4,5-^{3}H. Each bar represents one animal. (Reproduced with permission from Altman, 1963.)

several brain regions as well as in liver. Thus, no apparent regional specificity existed with respect to the effects of these various environmentally triggered responses. The metabolic alterations produced by the various stimuli used (e.g. shocks, buzzers, lights) appeared to depend, to a great extent, upon their novelty in relation to stress. Note that in the study of Rees *et al.* (1974) the ^{3}H-lysine was subcutaneously injected 20 min after behavioral treatments and its incorporation into total protein was measured during a 10-min pulse (see Fig. 28). These studies have indicated that a stress reaction which could be associated with the release of hormones and/or with changes in rate of cerebral blood flow is an important element of the *Behavioral Code* (see step 2 of Fig. 80).

Table 15. Mean autoradiographic grain density of single neurones in several brain structures in a control group and in animals exercised at 7 rpm and injected during exercise.

Structure	Density				% Increase	p
	Control		Exercised			
	N	Mean ± S.D.	N	Mean ± S.D.		
Motor cortex, neurones in second layer	40	0·185 ± 0·036	40	0·307 ± 0·043	66	<0·001
Cervical cord, ventral horn, neuropil	80	0·063 ± 0·024	80	0·099 ± 0·034	57	<0·001
Motor cortex, pyramidal neurones in fifth layer	40	0·282 ± 0·042	40	0·421 ± 0·055	49	<0·001
Posterior cortex, neurones in second layer	40	0·215 ± 0·053	40	0·315 ± 0·037	47	<0·001
Cervical cord, ventral horn cells	40	0·476 ± 0·042	40	0·671 ± 0·062	41	<0·001
Posterior cortex, neurones in fifth layer	40	0·274 ± 0·044	40	0·374 ± 0·044	36	<0·001
Lumbar cord, ventral horn cells	40	0·512 ± 0·063	40	0·693 ± 0·113	35	<0·001
Mammillary body, neurones	40	0·331 ± 0·046	40	0·442 ± 0·084	34	<0·001
Cerebellum, Purkinje cells	40	0·458 ± 0·052	40	0·612 ± 0·061	34	<0·001
Medulla, large reticular neurones	40	0·512 ± 0·058	40	0·674 ± 0·080	32	<0·001
Choroid plexus (third ventricle)	40	0·333 ± 0·052	40	0·426 ± 0·062	28	<0·001
Hippocampus, pyramidal neurones	40	0·262 ± 0·040	40	0·330 ± 0·039	26	<0·001
Deiters' nucleus, neurones	40	0·573 ± 0·051	40	0·718 ± 0·044	25	<0·001
Hippocampus, granule cells	40	0·176 ± 0·027	40	0·218 ± 0·035	24	<0·001
Cochlear nucleus, neurones	40	0·639 ± 0·064	40	0·760 ± 0·095	19	<0·001

Rank ordering of structures in terms of percentage increase of uptake in the exercised animals. (Reproduced with permission from Altman, 1963.)

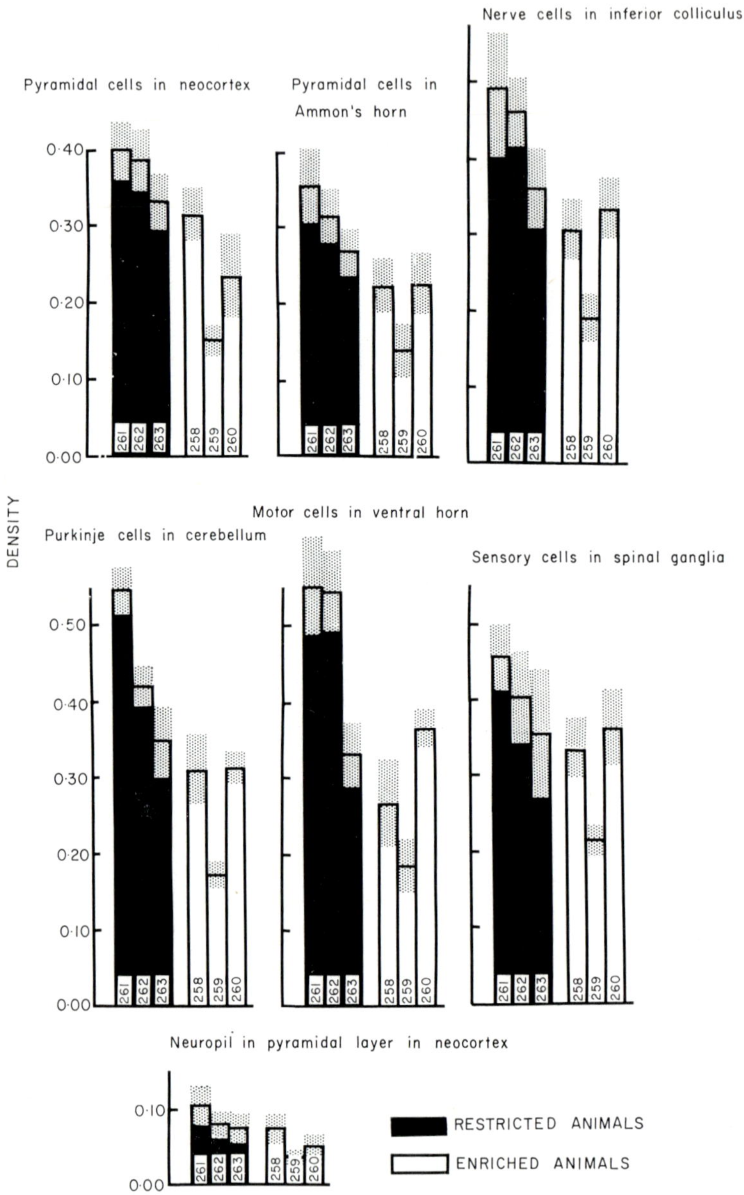

FIG. 27. Mean microdensitometric measures of grain density over a variety of brain structures in three "restricted" and three "enriched" rats. Stippled areas represent standard deviations for the individual animals. (Reproduced with permission from Das and Altman, 1966.)

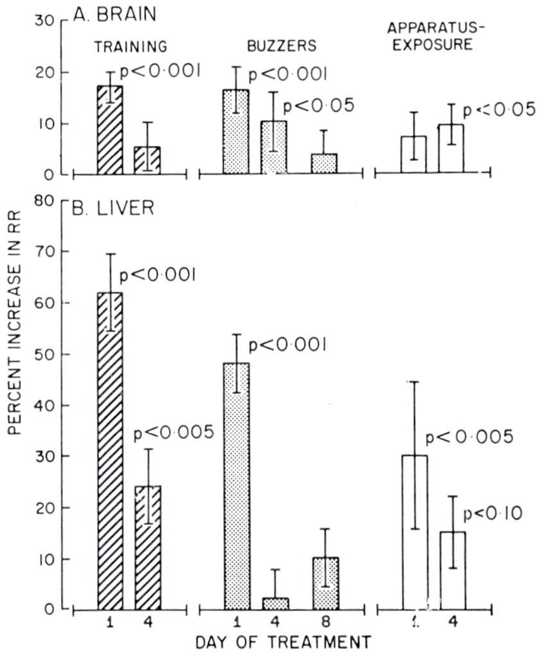

FIG. 28. Effects of prior experience upon lysine incorporation response of brain (A) and liver (B) to various behavioral treatments. Mice were given avoidance training, or exposed to 30 buzzers, or left undisturbed in the apparatus during a 15-min session. For some mice these treatments were administered daily for 4 or 8 consecutive days. Twenty minutes after the session on either day 1, 4, or 8, mice received an injection of ^{3}H-lysine and were sacrificed 10 min later. The relative radioactivity (RR) of protein of the two organs was measured and compared to similar measures for quiet mice. Mean per cent increases $\pm$ S.E.M. are presented; significance levels are for two-tailed comparisons; there was a minimum of 6 subjects per group. (Reproduced with permission from Rees *et al.*, 1974.)

A subcellular study revealed that incorporation of intraventricularly injected ^{3}H-leucine into protein occurred to a greater extent into nuclear fractions of "enriched" rats than into other fractions examined (Levitan *et al.*, 1972a). Also, synaptosomal proteins of the hippocampus incorporated this precursor at a much higher rate, and those of cortex at a lower rate, in "enriched" rats than in environmentally impoverished controls (Table 16). Regarding subcellular localization, proteins of cerebral nuclear and synaptosomal fractions appeared to be selectively affected by environmental modification. However, all protein species of hippocampal nuclear and synaptosomal fractions seemed to be affected equally by the environmental change (see Fig. 29; Levitan *et al.*, 1972b). In a related study, rats trained for 4 days to use the non-preferred paw to retrieve food incorporated more ^{3}H-leucine than

TABLE 16. Effects of environmental complexity on the *in vivo* incorporation of ^{3}H-leucine into the proteins of sub-synaptosomal fractions of rat cerebral cortex and hippocampus.

Sub-synaptosomal fraction	Corrected specific radioactivity (counts/min/mg)		Enriched/deprived[a]
	Enriched	Deprived	
Cortex			
synaptic soluble	1720 ± 90	2900 ± 220	0·59
synaptic mitochondria	810 ± 90	1380 ± 110	0·59
synaptic membranes	1270 ± 100	2700 ± 290	0·47
Hippocampus			
synaptic soluble	11300 ± 1000	7500 ± 810	1·51
synaptic mitochondria	4940 ± 620	2600 ± 190	1·90
synaptic membranes	8790 ± 680	5200 ± 530	1·69

[a] All the ratios differed significantly from 1·0 ($p < 0{\cdot}001$); Means ± S.E.M. (9 brains). (Reproduced with permission from Levitan *et al.*, 1972a.)

control animals into protein of cytosol, microsomes, synaptosol, synaptosomal mitochondria and synaptosomal membranes, and less into perikaryal mitochondria of the hippocampus; training did not affect protein synthesis in the cerebral cortex, and, as in the above mentioned study, hippocampal changes were not restricted to the synthesis of any given protein species (Levitan *et al.*, 1972c; but see Hydén and Lange, 1968, 1970).

In rats which had been dark-reared from birth to seven weeks of age and then exposed to light, transient increases in incorporation of ^{3}H-lysine occurred in TCA-insoluble fractions of visual cortex, lateral geniculate body, and retina, whereas no change was noted in liver and a slight depression of incorporation occurred in motor cortex (Rose, 1967; Richardson and Rose, 1972). Further studies revealed that this increase in protein synthesis in visual cortex occurred primarily in the neuronal (primarily perikaryal) fraction, as opposed to the neuropil (primarily glia, but also containing dendritic and axonal processes and nerve terminals) fraction (Sinha and Rose, 1972; Rose *et al.*, 1973b) and that it could be reversed by continued light exposure (see Fig. 30). Such changes were associated with several of the protein fractions of visual cortex and retina that could be separated by polyacrylamide gel electrophoresis (Rose, 1967; Richardson and Rose,

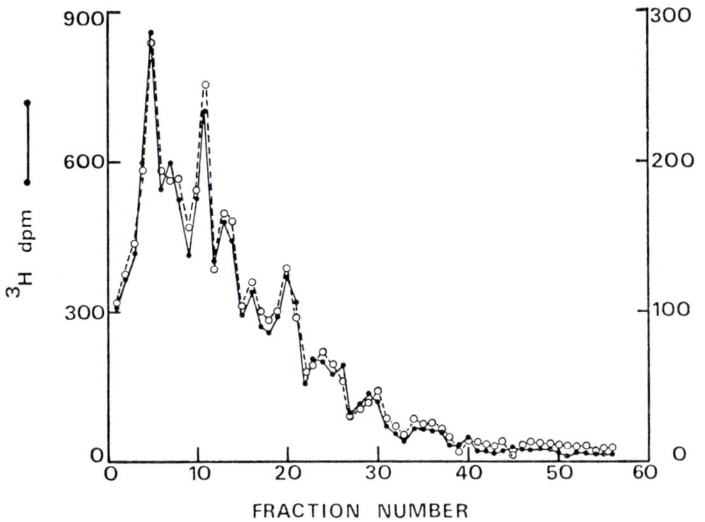

FIG. 29. Distribution of radioactivity following electrophoresis of the nuclear acidic (phenol-soluble) proteins of "enriched" and "deprived" rats. "Enriched" rats received L-^{3}H-leucine, and "deprived" rats received L-^{14}C-leucine, and the nuclear acidic proteins were extracted from the pooled non-cortical portions of the brains. The proteins were electrophoresed in polyacrylamide gels (16% acrylamide) containing 0·1% sodium dodecyl sulphate, and the gels were sliced and counted. The ordinate scales for ^{3}H and ^{14}C dpm are in the same ratio as the mean ^{3}H dpm/^{14}C dpm ratio along the electropherogram, in order to allow the easy visualization of any regional differences. ●———●, ^{3}H dpm ("enriched" rats); ○– – – –○, ^{14}C dpm ("deprived" rats); origin is at the left. (Reproduced with permission from Levitan *et al.*, 1972b.)

1973a, b; Figs 31 and 32). Examination of the time course of incorporation of ^{3}H-lysine into protein revealed that the neuronal proteins of visual cortex whose syntheses were inhibited by dark-rearing form a rapidly labelling, water-insoluble pool (Rose and Sinha, 1974). The increased incorporation of ^{3}H-lysine into protein caused by first exposure to light was confined to the ribosomal pellet of the visual cortex and it appeared to involve several specific protein species, some of which were ribosomally bound (Jones-Lecointe *et al.*, 1976).

Studies on the incorporation of radioactive lysine into retinal protein fractions in dark-reared rats by Richardson and Rose (1975) revealed that one hour of exposure to light produced significant changes in 17 of the 41 fractions separated. This finding indicated that, at least in the retina, initial light exposure might increase the synthesis of a number of particular protein species (which included AChE and

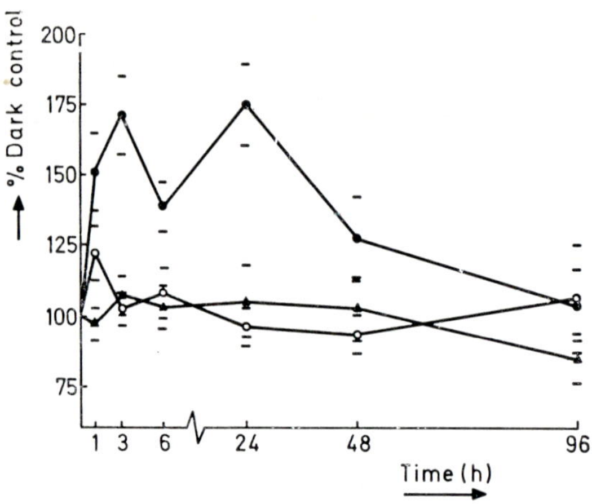

FIG. 30. Effects of length of light exposure on protein incorporation into cell fractions. Dark-reared rats were exposed for varying lengths of time to light, injected with ^{3}H-lysine, and killed after 1 h. Incorporation as a percentage of dark control values is shown for suspension (○), neuropil (▲) and neuronal (●) fractions of visual cortex. Horizontal lines show S.E.M.; 6–17 experiments per point; the result for suspension at 1 h was significant at $p < 0{\cdot}02$; results for neuronal fraction at 1, 3, 6 and 24 h were significant at $p < 0{\cdot}01$ (paired *t*-test). (Reproduced with permission from Rose *et al.*, 1973b.)

glycoprotein material) rather than producing a general elevation in all proteins (see also Richardson and Rose, 1973a; Shashoua, 1975; Barraco and Irwin, 1976).

An increased incorporation of 4,5-^{3}H-lysine into protein of the forebrain roof occurred in "imprinted" chicks, as compared to dark controls, but this increase did not differ from that which occurred in light-exposed controls; changes in free lysine were in a similar direction for both groups of chicks (Bateson *et al.*, 1969, 1972). Such effects could have been mediated by changes in arousal or attention which could increase regional blood flow and thereby alter precursor uptake by the brain (see Bondy *et al.*, 1974). Other studies have indicated that changes in AChE activity and possibly in ChAc activity, might be caused in certain brain regions by the imprinting procedure (Haywood *et al.*, 1973; Rose *et al.*, 1976).

The above studies have dealt with effects of the *external* environment on cerebral protein synthesis. It should also be noted that various insults to the *internal* environment also affect cerebral protein synthesis. As examples, estradiol or progesterone administration to adult ovariectomized guinea pigs increased the incorporation of radioactive leucine into proteins of the cerebral cortex and other brain areas (Garcia

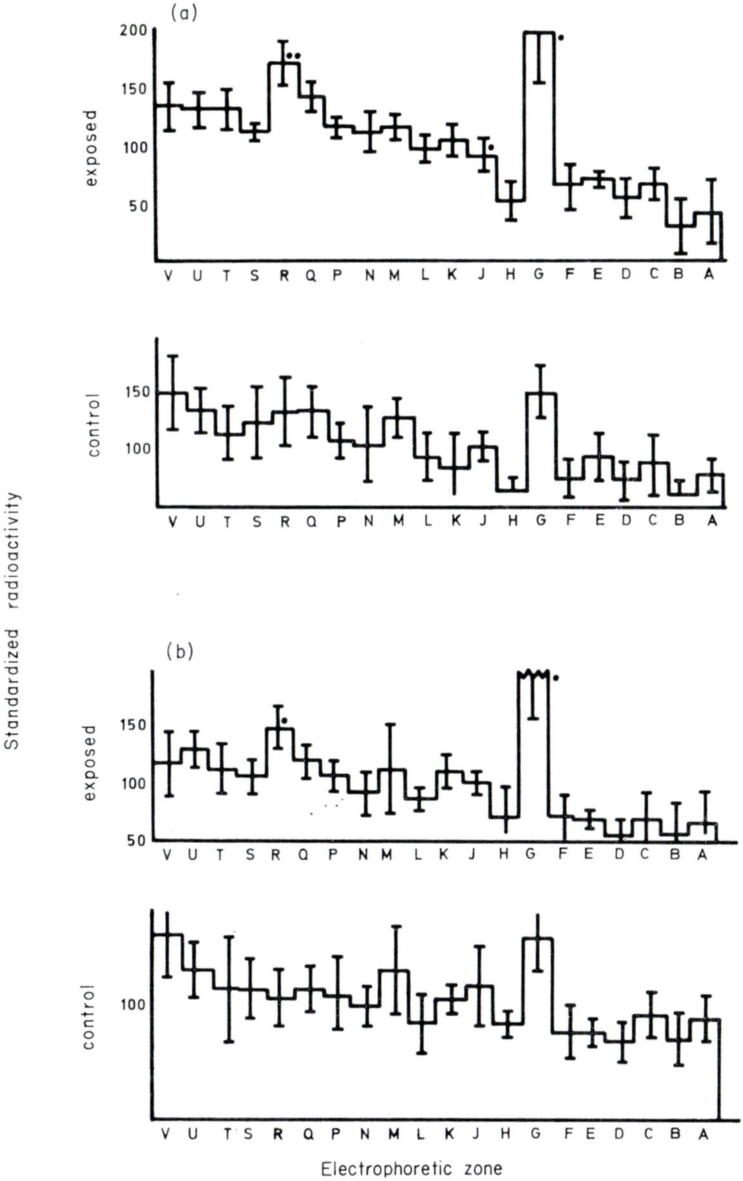

FIG. 31. Histograms showing standardized radioactivities of gel zones for soluble proteins of visual cortex from two series of experiments: (a) ^{14}C-lysine in the light-exposed rats; (b) ^{3}H-lysine in the light-exposed rats. Means $\pm$ standard deviations; 6 gel runs; • and • • indicate, respectively, $p < 0{\cdot}05$ and $p < 0{\cdot}01$ by comparison with controls. (Reproduced with permission from Richardson and Rose, 1973b.)

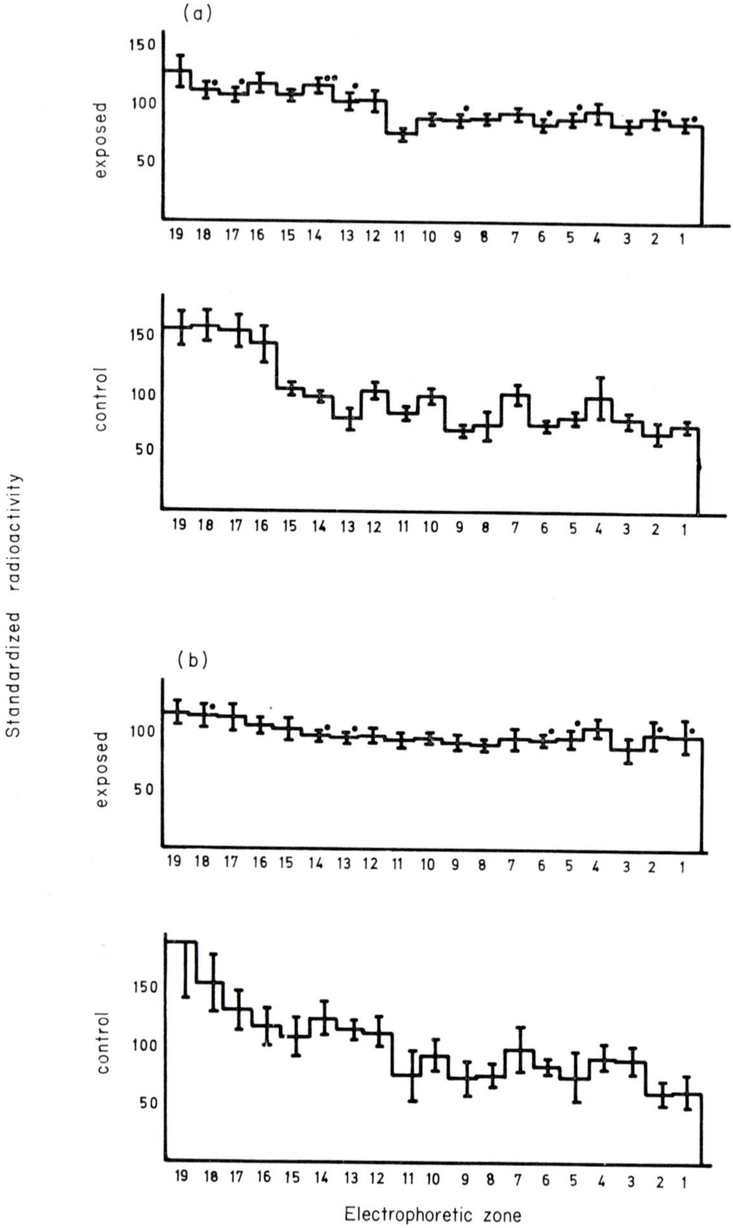

FIG. 32. Histograms showing standardized radioactivities for the insoluble proteins of visual cortex: (a) ^{14}C-lysine in the light-exposed rats; (b) ^{3}H-lysine in the light-exposed rats. Means $\pm$ standard deviations; 5 gel runs in a and 4 gel runs in b; • and •• indicate, respectively, $p < 0{\cdot}05$ and $p < 0{\cdot}01$ by comparison with controls. (Reproduced with permission from Richardson and Rose, 1973b.)

et al., 1973; Wade and Feder, 1974) and increased the total protein content of cerebral cortex (Garcia *et al.*, 1973), whereas chronically administered estradiol to adult ovariectomized rats decreased ^{14}C-lysine incorporation into proteins of specific hypothalamic nuclei (Litteria and Timiras, 1970). More recently, Litteria and Thorner (1976) have shown that injections of 17-β-estradiol to neonatal male and female rats during the "critical period" of brain development decreased the incorporation of ^{3}H-lysine into neuronal proteins of various regions of the cerebral cortex when these animals attained adulthood. Such effects might be explained by the presence of a cortical estrogen receptor in neonatal rats (Barley *et al.*, 1974).

3. *Protein Phosphorylation*

Dunn *et al.* (1974a) studied the phosphorylation of nuclear proteins in rats using intracranial injections of either ^{32}P or ^{33}P, and showed that non-histone acid-extractable nuclear protein (NAEP) from the brains of "trained" rats contained about twice as much radioactivity as that from quiet ("untrained") rats (see Table 17). This increase in NAEP labelling appeared to be a rapid and transient response to training, and was accounted for by covalently bound phosphoserine. Further

TABLE 17. Incorporation of ^{32}P or ^{33}P into rat brain nonhistone acid-extractable nuclear protein (NAEP) during a one-way step-up avoidance task.

Treatment	N	NAEP/AMP Increase (%)
Trained/quiet	30	+104 ± 11*
Yoked/quiet	20	−5 ± 59
Shocked/quiet	4	−11 ± 14
6-day prior-trained performing/quiet	8	+95 ± 4*
6-day prior-trained reminded/quiet	8	+78 ± 14*
6-day prior-trained handled/quiet	4	+84 ± 6*

A dual-isotope method was used in which rats were injected intracranially with 100 μCi [^{32}P] H_3PO_4 or [^{33}P] H_3PO_4. Thirty minutes later they were trained for 5 min and sacrificed immediately. Each brain was homogenized with the brain of a quiet rat that had received the other radioisotope. NAEP and AMP were isolated from the homogenate and their radioactivities determined. Means ± standard deviations; * indicates $p < 0{\cdot}01$ (two-tailed *t*-test). Note that "yoked" animals received the same stimuli but were unable to avoid the footshock, and "quiet" animals were left undisturbed after injection in individual cages in a separate room but in a normal laboratory environment. (Data of Machlus, 1971; reproduced with permission from Dunn *et al.*, 1974a.)

experiments indicated that 15 min of "jump-box" training produced an increase (about 30%) in the incorporation of phosphate into mouse brain synaptosomal protein that was associated with membranes (see Dunn *et al.*, 1974a). Previously, Machlus (1971) had shown that histones of trained animals contained only about 60% as much radioactive phosphorous as those of quiet animals.

Such a protein phosphorylation mechanism could be at the basis of memory processes since it could be involved in alterations of membrane conformation. Several recent observations have supported this contention and have indicated further that cyclic-AMP, cyclic-GMP and dopamine (DA) may play roles in the phosphorylation of protein constituents of synaptic membranes. Since this type of mechanism could explain the transduction of neurotransmitter action into changes in ionic permeability and synaptic membrane structure, it will be discussed in some detail (see also Dunn and Bondy, 1974).

First, it should be realized that such studies were stimulated by the discovery of cyclic-AMP-dependent protein kinases in mammalian brain (Miyamoto *et al.*, 1969a, b), particularly in synaptic membrane fractions (Maeno *et al.*, 1971). Subsequently, synaptic membrane fractions were found to be enriched in phosphoprotein phosphatase activity (Maeno and Greengard, 1972), and the phosphorylation of a specific protein of synaptic membrane fractions was shown to be markedly stimulated by cyclic-AMP (Johnson *et al.*, 1972).

Ueda *et al.* (1973) then showed, using polyacrylamide gel electrophoresis, that cyclic-AMP could regulate the phosphorylation of two specific membrane proteins of rat brain; these were designated Protein I (mol. wt. $\simeq$ 86 000 daltons) and Protein II (mol. wt. $\simeq$ 49 000 daltons). Phosphorylation of these proteins was rapid, reaching maximal levels within 5 seconds. Protein I underwent cyclic-AMP-stimulated phosphorylation only in membrane fractions of neural tissues known to contain synapses, whereas the phosphorylation of Protein II was observed in several neural and non-neural tissues (see Fig. 33). Protein D of toad bladder membranes (DeLorenzo *et al.*, 1973) appears to be identical to Protein II. In the presence of Zn^{2+}, cyclic-AMP inhibits the phosphorylation of this protein by an action on endogenous protein kinase (Ueda *et al.*, 1973; DeLorenzo and Greengard, 1973). Cyclic-AMP also stimulates the dephosphorylation of this protein (DeLorenzo and Greengard, 1973; Maeno *et al.*, 1975) by an endogenous protein phosphatase (see Fig. 34). Further study indicated that Protein II (also termed "Protein 49") occurs in a wide variety of vertebrate tissues along with the regulation by cyclic-AMP of its endogenous phosphorylation and dephosphorylation (Malkinson *et al.*, 1975; Fig. 35). Since this protein could be a regulatory subunit of cyclic-AMP-dependent protein kinases in all vertebrate tissues, its further study could help to explain the general somatic effect of environmental changes (see Chapter 8, Part C).

Using data obtained in the above mentioned studies, Greengard 1975, 1976) proposed a model by which cyclic nucleotides may mediate

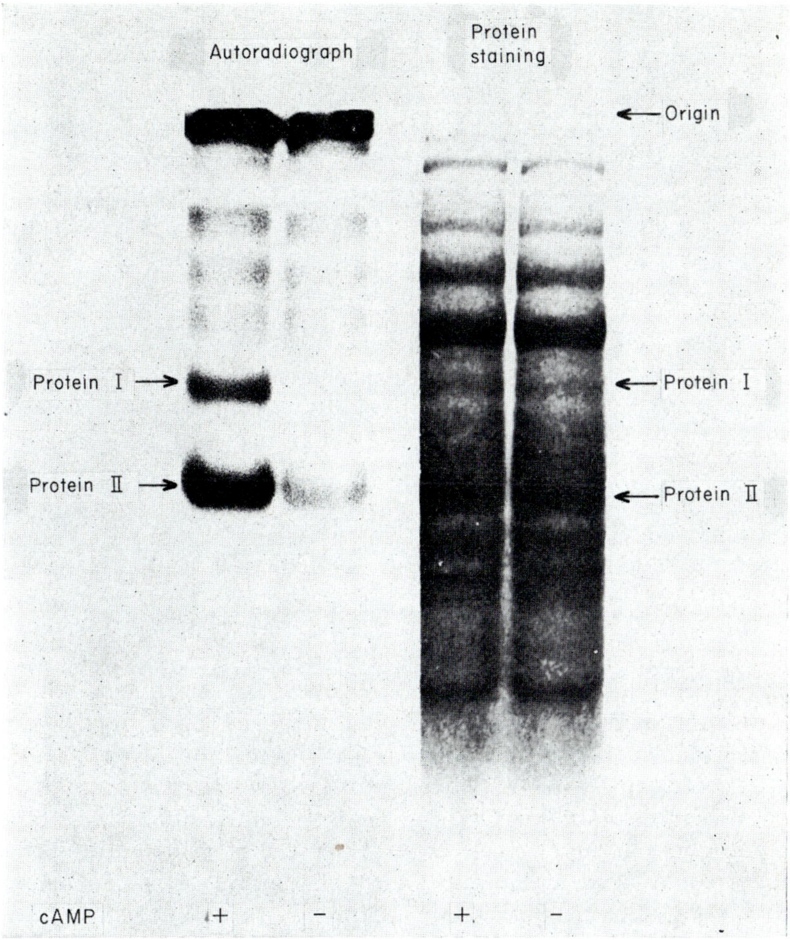

FIG. 33. Effect of cyclic-AMP on the endogenous phosphorylation of a synaptic membrane preparation from rat cerebrum. The synaptic membrane preparation was incubated with [$\gamma-{}^{32}P$]ATP in the presence or absence of cyclic-AMP. The reaction was stopped by the addition of sodium dodecyl sulfate, and the solubilized protein was subjected to slab gel electrophoresis, protein staining, and autoradiography. The resulting autoradiograph is shown on the *left side*, and a photograph of the protein staining pattern of the dried gel is shown on the *right side*. The + and − signs at the *bottom* of the gel indicate the presence and absence, respectively, of 5×10^{-6} M cyclic-AMP in the incubation mixture. The *arrows* identify the positions on the gel of Protein I and Protein II, the phosphorylation of which was markedly increased in the presence of cyclic-AMP. (Reproduced with permission from Ueda *et al.*, 1973.)

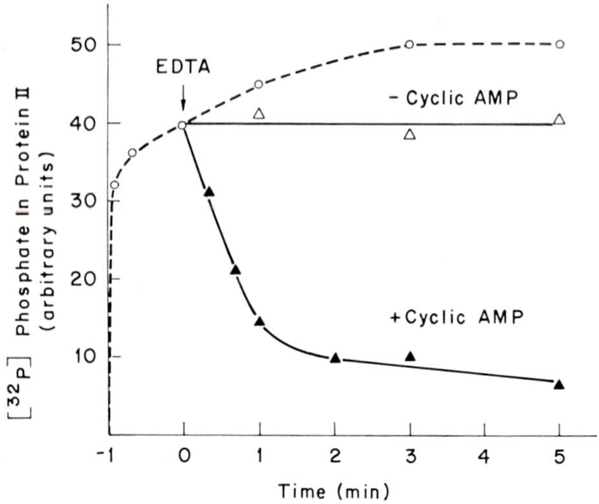

FIG. 34. Effect of cyclic-AMP on the endogenous dephosphorylation of Protein II as a function of incubation time. The synaptic membrane preparation was phosphorylated by preincubation for 1 min. EDTA (5 mM) was added at zero time, without (△———△) or with (▲———▲) 10^{-5} M cyclic-AMP. Open circles (○ – – – ○) indicate the time course of the phosphorylation of Protein II in the absence of added EDTA or cyclic-AMP. The amount of ^{32}P-phosphate in Protein II was determined by a densitometric method. In this experiment, 50 arbitrary units correspond to about 800 cpm of radioactivity. (Reproduced with permission from Maeno *et al.*, 1975.)

synaptic transmission at some synapses (see Fig. 36). Dopamine released from neurones might activate adenyl cyclase in the postsynaptic membrane, thereby inducing the formation of cyclic-AMP, which could then hyperpolarize postsynaptic neurones and thus modify their excitability. Cyclic-AMP could then act by regulating the phosphorylation of postsynaptic membrane proteins, and such changes could alter membrane ionic permeability. This model may also be useful for explaining the widespread use and beneficial effects of L-DOPA treatment in cases of Parkinsonism (see, e.g. Sandler, 1972; Malitz, 1972; Yahr, 1974; refs in Lloyd *et al.*, 1975a, and in Calne *et al.*, 1975). L-DOPA, the direct precursor of DA formation in the brain, could act by altering the DA-induced phosphorylation of postsynaptic (or presynaptic) membrane proteins by the above described mechanism. Such an effect on the brain regions that are unaffected by Parkinsonism could provide an increased "take-over" function of the brain for its lesioned pathways, thus restoring extrapyramidal function and relieving Parkinsonian symptoms. This type of mechanism could be generally involved in CNS function; i.e.

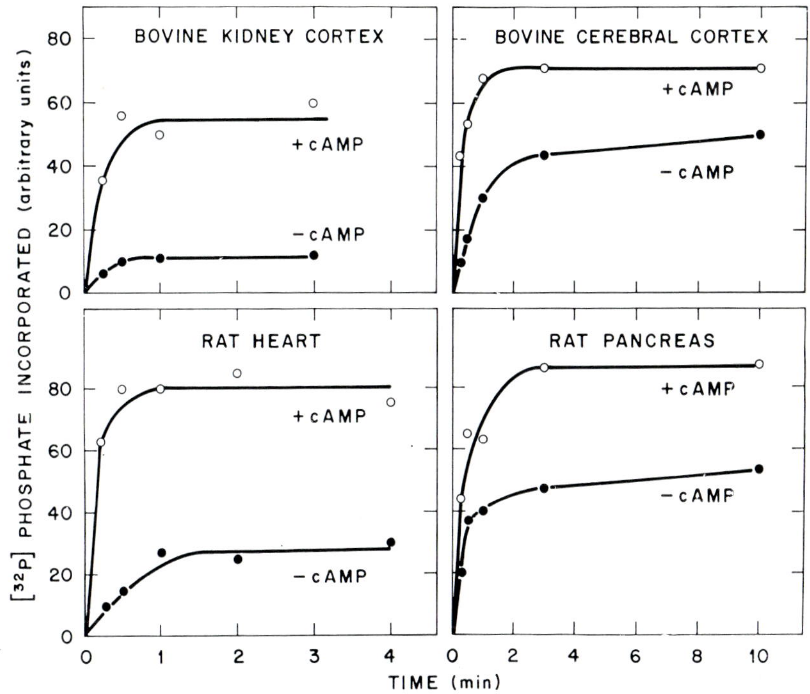

FIG. 35. Effect of cyclic-AMP (10 μM) on the endogenous phosphorylation of protein 49 in cytosol fractions from various mammalian tissues, assayed in the presence of 10 mM $MgCl_2$. (Reproduced with permission from Malkinson *et al.*, 1975.)

other transmitters could also act by altering the syntheses and structures of synaptic macromolecules which are necessarily involved in continued normal function and behavior.

Even though cyclic-AMP-mediated phosphorylation of membrane proteins occurs rapidly (in 5 sec; see above), this mechanism could be involved in the enduring conformational changes in membranes that appear to be related to memory and behavior. In support of this notion, Routtenberg *et al.* (1975) have shown that *in vitro* phosphorylation patterns of a phosphoprotein of synaptic plasma membranes of rat caudate nucleus and temporal cortex were altered by a 10-min passive avoidance training procedure,* even in animals that were

* *Passive avoidance:* a conditioning technique whereby an animal learns *not* to make a response that has been punished (e.g. by shock).
Active avoidance: a conditioning technique whereby an animal learns to make a response in order to avoid punishment (e.g. shock).

sacrificed 24 h after learning. Of the three proteins, designated as D, E and F by Ehrlich and Routtenberg (1974), protein F showed increased phosphorylation in osmotically-shocked P_2 membrane preparations from trained animals, as compared to controls. The effect of training on protein F, though shown clearly in caudate nucleus and temporal cortex, was not shown readily in cerebral cortex of rats. No consistent effect of training was shown on proteins D or E (which

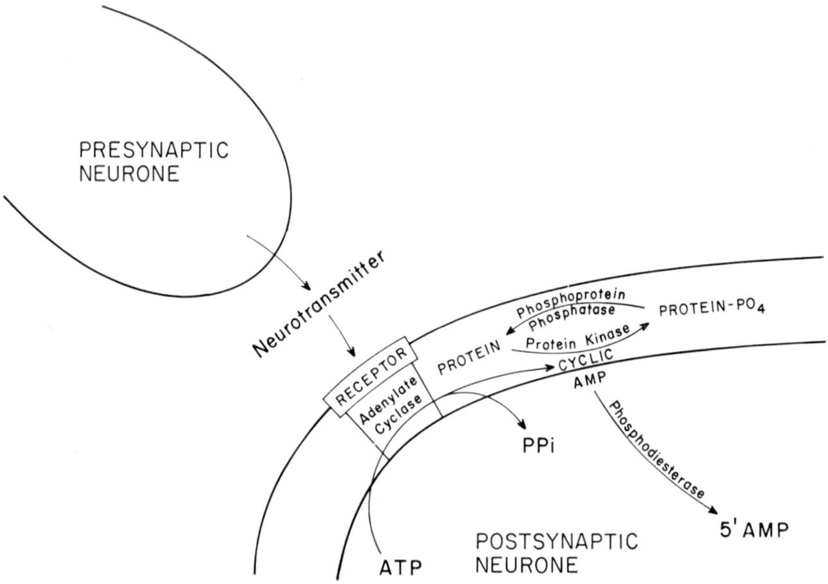

FIG. 36. Model of the proposed mechanism by which cyclic-AMP may mediate synaptic transmission at certain types of synapses. (Reproduced with permission from Greengard, 1975.)

may be, respectively, Proteins I and II of Ueda *et al.*, 1973). It was concluded that such effects of training, which can be measured *in vitro*, are long-lasting and specific to particular membrane phosphoproteins in particular brain regions (Routtenberg *et al.*, 1975).

Other studies along this line have revealed that four membrane-bound proteins of rat cerebral cortex (two of which may be identical to Proteins I and II of Ueda *et al.*, 1973), can undergo endogenous cyclic-AMP-stimulated phosphorylation (Routtenberg and Ehrlich, 1975). Since the time courses of phosphorylation of each of these proteins (designated as D, E, F, G) were affected differently by (a) the presence of cyclic nucleotides or Zn^{2+}, and (b) the concentration of ATP and Mg^{2+}, it was suggested that many different conformational changes could occur in membranes by mechanisms related to these proteins. With relation to the mechanism of axoplasmic transport, cyclic-AMP has been shown to stimulate the phosphorylation of both tubulin, the major microtubule protein (Goodman *et al.*, 1970; Lagnado *et al.*, 1972) and a microtubule-associated protein which could be involved in microtubule assembly or function (Sloboda *et al.*, 1975; Fig. 37).

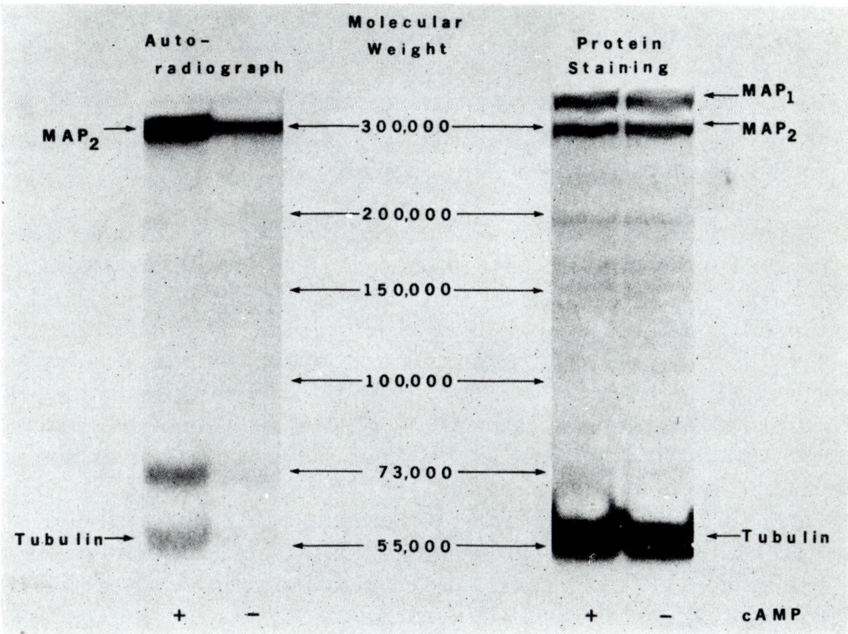

FIG. 37. Effect of cyclic-AMP on the endogenous phosphorylation of microtubule proteins. Fifty micrograms of 3×-polymerized microtubule protein were phosphorylated for 30 sec, in the standard reaction mixture, in the presence or absence of 10^{-5} M cyclic-AMP. An aliquot containing 25 μg of the sodium dodecyl sulfate-solubilized protein was then subjected to polyacrylamide gel electrophoresis. The autoradiograph showing ^{32}P incorporation is on the left, and the protein staining pattern is on the right. MAP, microtubule-associated protein. (Reproduced with permission from Sloboda *et al.*, 1975.)

Electroconvulsive shock (ECS) can also cause a decrease in the phosphorylation of synaptosomal membranes *in vitro*, which could be a reflection of conformational changes (Appel and Locher, 1973). More recently, the anti-convulsant drug, diphenylhydantoin (DPH), has been shown to affect the phosphorylation of a specific protein of rat cerebrum (DeLorenzo and Glaser, 1976). DPH decreased the incorporation of ^{32}P into a radioactive protein band (mol. wt. $\simeq$

58 000 daltons), this effect being shown both by autoradiography and by scintillation counting of polyacrylamide gel slices. Thus, the anti-convulsant action of DPH could be caused by its effect on the enzymes that control the phosphorylation of this cerebral protein. This effect of DPH on protein phosphorylation may underly its action as a "membrane stabilizer" (Woodbury and Kemp, 1971). Membrane stabilization of DPH has been correlated with its effects of increasing presynaptic inhibition in the frog spinal cord (Davidoff, 1972a) as well as postsynaptic inhibition in the crayfish stretch receptor (Ayala *et al.*, 1976) and feline cerebral cortex (Raabe and Ayala, 1976). Other evidence has indicated that DPH may act by affecting "GABA-ergic" mechanisms (e.g. Vernadakis and Woodbury, 1960; Davidoff, 1972b; Raabe and Ayala, 1976). Effects on protein phosphorylation could be involved in the actions of many other neuro-active agents (see DeLorenzo, 1975).

Recent results have indicated that cyclic-AMP also plays a role in seizures. Dibutyryl-cyclic-AMP produced seizure activity when given intraventricularly (Gessa *et al.*, 1970) or when applied topically to the exposed cerebral cortex (Walker *et al.*, 1975), and epileptogenic cortical foci induced by freezing lesions contained increased levels of cyclic-AMP (Walker *et al.*, 1973). Also, electrical stimulation of cortical slices *in vitro* increased their cyclic-AMP contents (Kakiuchi *et al.*, 1969). More recently, it has been shown that the convulsant agent, pentylenetetrazol, increased the content of cyclic-AMP of rat cortical slices incubated *in vitro*, by an action that was partially reversed by theophylline (Lewin *et al.*, 1976). These relationships between cyclic-AMP and seizures may be related to cyclic-AMP-mediated changes in protein phosphorylation.

These studies have suggested that both phosphoproteins and cyclic-AMP are elements of the *Behavioral Code*. Changes in protein phosphorylation which can be induced by released transmitters and which appear to be mediated by cyclic-AMP, could regulate both the changes in membrane ionic fluxes produced during synaptic transmission and the changes in membrane conformation which constitute the basis for enduring changes in neural function and behavior.

4. *Studies with Inhibitors of Protein Synthesis*

ECS and antibiotics, though sharing some properties with respect to memory processes, could act by different mechanisms (see Agranoff, 1974). Recent studies by Dunn and Bergert (1976) have indicated that although both ECS and subcutaneously injected cycloheximide caused marked decreases in amino acid incorporation into total mouse brain protein as well as a selective increase of amino acid incorporation into the microsomal fraction, only cycloheximide produced an increase in labelling of the synaptosomal fraction and a decrease in the cell-sap fraction. It was concluded that ECS and cycloheximide inhibit cerebral protein synthesis by different mechanisms, and that ECS-induced amnesia might not be mediated by an inhibition of protein synthesis. ECS may act by depleting cerebral ATP or by causing an imbalance of cations (Dunn, 1973) or by disrupting polysomes (see also Part B2, above), whereas the action of cycloheximide could be mediated by its direct effect on peptide bond synthesis (Obrig *et al.*, 1971). Puromycin, like ECS, produces retrograde amnesia. This supports the contention that both treatments cause a disruption of memory formation, presumably by an effect on synaptic structure. However, an anterograde amnesia (memory impairment resulting from an amnestic agent administered prior to training) has also been produced by both ECS (Kopp *et al.*, 1968; Zerbolio, 1969; Gardner *et al.*, 1972) and puromycin injection (Springer *et al.*, 1975; see Fig. 38). It is difficult to reconcile such effects on anterograde amnesia as being caused by changes in synaptic structure.

Some negative evidence exists concerning the roles of proteins in information processing that have been proposed on the basis of studies

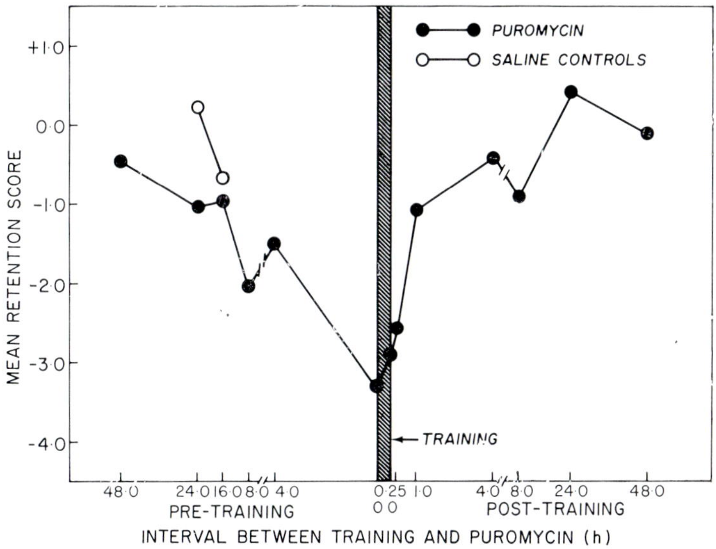

FIG. 38. Anterograde and retrograde effects of puromycin on memory formation in the goldfish. Dependent, one-tailed *t*-tests were used to determine the significance of the difference between the number of avoidances achieved in ten retraining trials and the number of avoidances predicted to occur during the retraining for each treatment ($A - P$); mean $A - P$ retention scores are shown as a function of puromycin-training interval. (Reproduced with permission from Springer *et al.*, 1975.)

with antibiotics. Schwartz *et al.* (1971) showed that "short-term" habituation (a simple form of behavioral plasticity; see, e.g. Peeke and Herz, 1973) occurred in Aplysia after > 95% of protein synthesis had been inhibited, and extensive inhibition of RNA or protein synthesis in goldfish brain (Agranoff *et al.*, 1968) or in mouse brain (Barondes and Cohen, 1967) did not hinder the acquisition of certain conditioned avoidance tasks. However, it should be noted that in these, experiments changes could have occurred mainly in the conformations, rather than in the syntheses, of macromolecules. Since amnestic responses can be produced by administration of antibiotics just before, during, or just after training (e.g. Flexner *et al.*, 1963, 1967; Brink *et al.*, 1966; Cohen and Barondes, 1968; Nakajima, 1969; Geller *et al.*, 1969; Agranoff, 1970, 1973, 1974; Mark and Watts, 1971; Quinton, 1971; Squire and Barondes, 1972, 1974), the formation of "long-term" memory does appear to be associated with macromolecular synthesis (see Fig. 39). The study of Flood *et al.* (1973), which indicated that longer durations of protein synthesis inhibition by anisomycin in mice produced greater percentages of amnestic subjects, also supports a role for protein synthesis in the formation of "long-term" memory; furthermore, with greater training strengths, less amnesia was induced. Recently, Squire and Becker (1975) have shown that "long-term" habituation in mice can be impaired by injections of anisomycin.

Other criticisms of studies involving inhibitors of protein synthesis have been based on the notion that the effects of these agents on memory

---

FIG. 39. Mice were given saline, cycloheximide (CXM; 120 mg/kg), or anisomycin (150 mg/kg) 30 min before training for 21 trials (A) or 27 trials (B). Separate subgroups of mice from each drug group were retested 1, 7, or 14 days later by giving them an additional 21 or 27 trials. Retention scores were submitted to a two-way analysis of variance, and individual comparisons were made between specific groups at each retention interval (Winer, 1962). The number of mice in each group is indicated in parentheses. A, for mice given 21 trials, the overall effect of drug treatment was significant ($F = 9{\cdot}1$, $p < 0{\cdot}01$). The effect of retention interval ($F = 2{\cdot}4$) and the interaction of drug × retention interval ($F = 0{\cdot}2$) were not significant. Individual comparisons indicated that the retention scores for mice given inhibitors were significantly different from the scores of saline control mice in all but two cases. The retention score of CXM-treated mice tested at 1 day after training was not significant, presumably because of the small number of mice tested. The mice given anisomycin did not differ significantly from control mice at 14 days. Two additional groups were given anisomycin 30 min after training (ANISOMYCIN AFTER) and tested for retention 1 day or 7 days later. Retention scores of these groups were significantly better than the scores of mice that were given anisomycin *before* training and tested for retention after the same interval ($t = 2{\cdot}1$, $p < 0{\cdot}03$, at 1 day; $t = 2{\cdot}3$, $p < 0{\cdot}03$, at 7 days). The scores of the ANISOMYCIN AFTER group were not measurably different from the scores of saline control mice ($p > 0{\cdot}1$). B, for mice given 27 trials, the main drug effect ($F = 32{\cdot}8$), the effect of retention interval ($F = 20{\cdot}7$), and the interaction of drug × retention interval ($F = 5{\cdot}3$) were signifi-

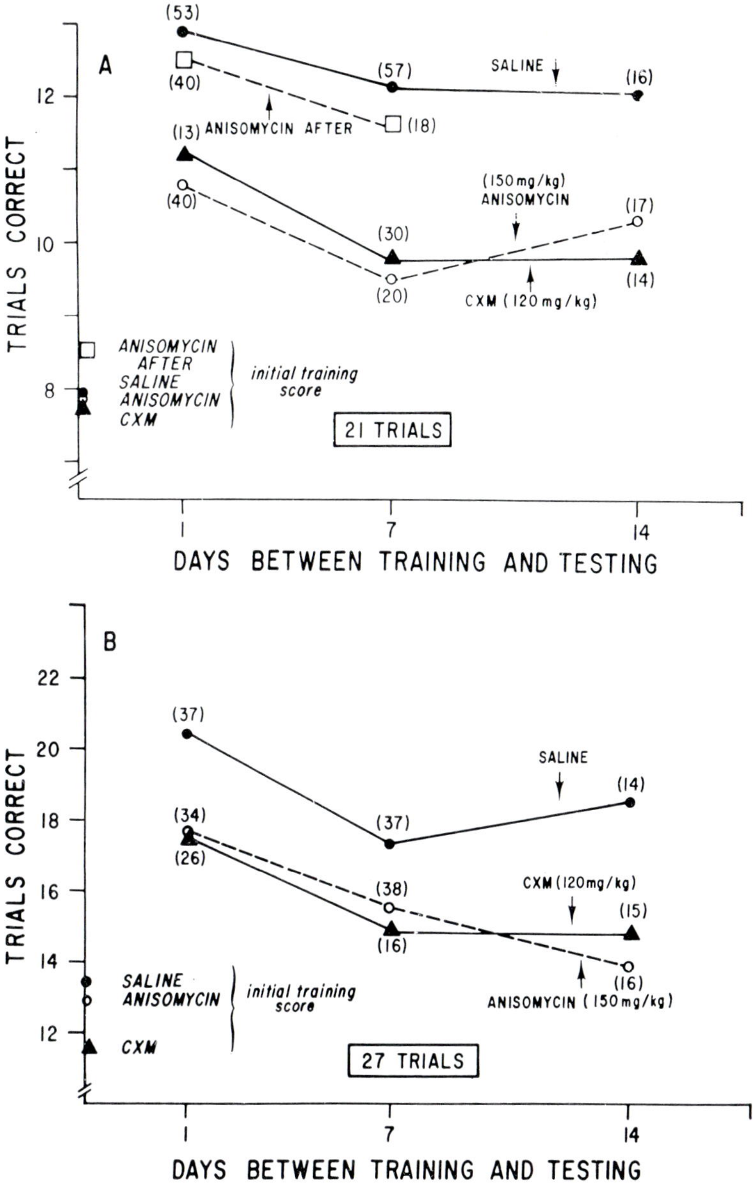

cant ($p < 0{\cdot}01$). Individual comparisons indicated that mice given the inhibitors performed significantly worse than control mice at each retention interval ($p < 0{\cdot}01$). Saline mice performed as well 14 days after training as they did 1 day after training ($p > 0{\cdot}1$). The performance of mice given the inhibitors was significantly worse 14 days after training than it was 1 day after training ($p < 0{\cdot}05$), indicating a decline in performance during this interval. (Reproduced with permission from Squire and Barondes, 1974.)

might be mediated by their actions on mechanisms other than that of protein synthesis (see, e.g. Dunn and Bondy, 1974). For example, Booth and Simson (1973) showed that cycloheximide can produce aversion to an odor if it is given after ingestion of odorized food in rats. They suggested that such conditioned aversion could account for at least part of the cycloheximide-induced amnesia for discrimination tasks. However, more recently Squire *et al.* (1975) have shown, in mice, that though amnestic doses of cycloheximide and anisomycin caused gustatory aversion, lithium chloride did not affect memory while producing at least as much aversion as the protein synthesis inhibitors (Fig. 40; Table 18). Therefore, the aversive effects of these

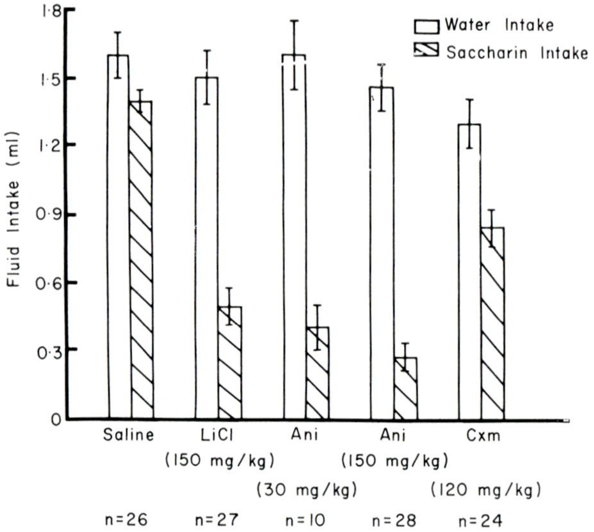

FIG. 40. Mice were given 10 min access to saccharin (S) or water (W) on alternate days according to the schedule SWSWS and were given one of the indicated drugs after each experience with saccharin. Results presented are averages for the two days with water and for the last two days with saccharin. Cxm = cycloheximide; Ani = anisomycin; LiCl = lithium chloride. Vertical bars indicate ± S.E.M. (Reproduced with permission from Squire *et al.*, 1975.)

protein synthesis inhibitors, as measured, did not explain their effects on memory. Also, the data shown in Fig. 39 regarding injection of anisomycin 30 min after training indicated that the sickness which occurs in animals several hours after the drug was not responsible for the amnestic effect (Squire and Barondes, 1974).

Puromycin can induce seizure activity in the hippocampus (e.g. Cohen *et al.*, 1966; Cohen and Barondes, 1967) which could impair learning (see Izquierdo, 1975a). In a sense, this intended criticism serves to strengthen the notion that changes in the content and synthesis of cerebral protein and RNA are involved in memory, since seizures produced by ECS (which also cause retrograde amnesia; e.g. McGaugh, 1966), also produce decreases in cerebral protein synthesis (e.g. Cotman *et al.*, 1971; Dunn *et al.*, 1971, 1974b; Dunn, 1971; Essman, 1973, 1974; Dunn and Bergert, 1976) and cerebral RNA content (e.g. Chitre *et al.*, 1964; Tannhauser and Izquierdo,

TABLE 18. Effects of drugs on performance seven days after training.[a]

Drug	N	Trials correct training ± SEM	Trials correct retention ± SEM
Saline	80	7·2 ± 0·3	12·8 ± 0·3
LiCl (150 mg/kg)	20	6·5 ± 0·5	12·6 ± 0·7
LiCl (150 mg/kg)[b]	17	6·2 ± 0·4	13·3 ± 0·8
Anisomycin (30 mg/kg)	38	6·9 ± 0·4	11·0* ± 0·6
Anisomycin (150 mg/kg)	20	7·4 ± 0·5	10·7* ± 0·7
Cycloheximide (120 mg/kg)	20	7·9 ± 0·5	10·5* ± 0·7

[a] Mice were given the indicated drugs 30 min before training for 20 trials. Retention was tested by giving an additional 20 trials 7 days later.
[b] This group was injected 10 min after training.
* Significantly different from saline group, $p < 0{\cdot}05$.
(Reproduced with permission from Squire *et al.*, 1975.)

1974) and can cause a disaggregation of cerebral polysomes (Vesco and Giuditta, 1968; MacInnes *et al.*, 1970; Dunn *et al.*, 1974b). However, it should be noted that other environmental insults (or lesions) such as increased intracranial pressure (Wasterlain, 1974), cerebral ischaemia (Kleihues and Hossmann, 1971), administration of certain antibiotics (McInnes and Luttges, 1973) and hypophysectomy (Gispen *et al.*, 1971) can also alter cerebral polysome structure and protein synthesis. Such changes could be mediated by a release of lytic enzymes from lysosomal membranes (Koenig and Jibril, 1962; Sellinger and Nordrum, 1969) and/or by increases in the activities of cerebral ribonucleases (see Roberts and Morelos, 1976).

Further criticism of the actions of antibiotics on memory has been provided (see Nakajima, 1972, 1973; Dunn and Bondy, 1974). However, in a recent study in which cycloheximide was applied intracranially to the pia mater overlying the fore-limb sensorimotor cortex, rats previously trained to use the non-preferred paw reversed their paw preference but failed to retain it for more than 3 days, whereas control animals receiving saline injections continued to show the preference for weeks (Pilcher and Booth, 1975). This effect of cycloheximide on "long-term" retention was accompanied by an inhibition of cerebral protein synthesis (by about 20–36%) that occurred maximally in the region containing the crucial fore-limb sensorimotor cortex. Other recent studies by Eichenbaum *et al.* (1976) have revealed that a regional difference in the effects of cycloheximide on retention of learned passive avoidance (see footnote p. 99) occurred; maximal amnesia followed injections of the drug into the striatum, hippocampus, amygdala and posterior lateral thalamus of mice (Fig. 41). Thus, in

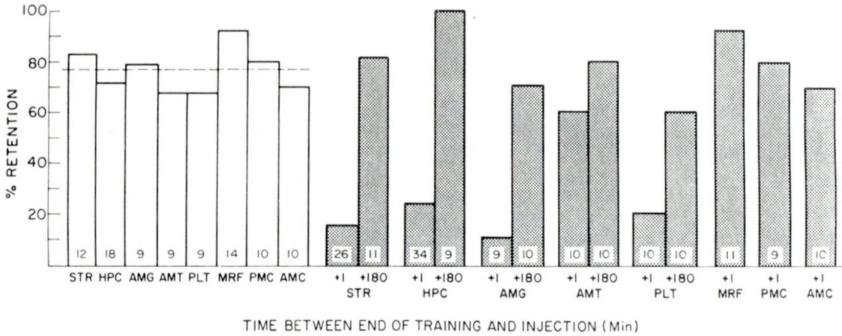

FIG. 41. Retention of groups injected bilaterally into various brain regions with saline immediately after training or CXM (3 μl, 10 μg/μl) immediately or 180 min after training. Open bars indicate saline-treated groups. Dashed line indicates retention score of pooled saline-treated group. Injection sites included striatum (STR), hippocampus (HPC), amygdala (AMG), anterior medial thalamus (AMT), posterior lateral thalamus (PLT), midbrain reticular formation (MRF), posterior medial cortex (PMC), and anterior medial cortex (AMC). Numbers of subjects in each group are stated inside the bars. (Reproduced with permission from Eichenbaum *et al.*, 1976.)

a one-trial passive avoidance task in the mouse the hippocampus was implicated as a probable site of action. Since these studies were performed with small, stereotaxically injected amounts of cycloheximide (see Fig. 42), the non-specific effects of the drug were minimized.

Drugs such as cycloheximide and anisomycin do not render mice *completely amnestic* (e.g. Quartermain *et al.*, 1970; Squire and Barondes, 1974; Squire and Davis, 1975), and after the drug effect decays in goldfish these animals can learn and can store memories normally (see Agranoff, 1974). Since these latter findings indicate that the effects of these drugs are reversible, they may not act by inducing changes in cerebral synaptic morphology which may indeed be the natural mechanism for memory storage. Instead, these agents may block

FIG. 42. Cannulated rat, illustrating the relative size and appearance of the apparatus used for studying the effect of intracranial injection of cycloheximide on the long-term retention of reversed paw preference. (Unpublished photograph; reproduced with permission of Pilcher and Booth, 1975.)

temporarily the processes involved in "long-term" memory formation (e.g. by slowing the turnover of protein constituents of synaptic membranes) while not affecting synaptic growth. In this regard, it should be noted that the extent to which protein synthesis is inhibited at the time of training of the animals and the dose of inhibitor given are likely to be the major factors that determine the severity of the induced amnesia (see Flood *et al.*, 1973; Squire and Davis, 1975). Also, these drugs appear to be more effective at blocking those memories associated with more complex tasks (see Agranoff, 1974).

Studies by Neale *et al.* (1973a) on temperature-dependent consolidation of puromycin-sensitive memory in goldfish revealed that protein synthesis, *per se*, could be dissociated from memory formation. Reducing the temperature to 4·5°C immediately after acquisition consistently produced no greater retention deficit than a similar treatment 24 h after learning a shock-avoidance task in these animals. Such a reduction in temperature produces very marked decreases in the rate of cerebral protein metabolism in goldfish (Lajtha and Sershen, 1975) and in rats (Lajtha and Dunlop, 1974). Hence, the important difference between temperature- and chemical (antibiotic)-mediated inhibition of protein synthesis appears to be that cerebral metabolism is depressed in the former, whereas protein synthesis is apparently more selectively inhibited in the latter. A potential conclusion is that production of an amnestic effect by protein synthesis inhibitors may require ongoing turnover (catabolism) of cerebral protein in the absence of its new synthesis.

Even though many of the above mentioned criticisms are justified, there exists overwhelming evidence that drugs such as puromycin and cycloheximide can disrupt memory processes for several different tasks in several species. However, it should be remembered that many other drug-induced conditions that cause a decrease in protein synthesis do not appear to affect memory; e.g. pentobarbitone-induced narcosis (Gaitonde and Richter, 1956), insulin-induced hypoglycaemia (Gaitonde and Richter, 1956) and chlorpromazine administration (Piha *et al.*, 1963).

5. *Neurotransmitter-associated Enzymes*

Small environmentally induced changes in the enzymes involved in ACh inactivation, AChE and ChE, have been well documented, especially in "long-term" studies (see, e.g. Bennett and Rosenzweig, 1971; DeFeudis, 1975a; Bennett, 1976). The change in cerebral AChE activity caused by environmental "enrichment" (EC) persisted for a longer time (after transfer of the rats to an impoverished environment) when it was caused by 80 days exposure to EC than when it was caused by 30 days of EC (Fig. 43; Bennett *et al.*, 1974).

The activities of cholinergic enzymes also appear to be sensitive to nutritional modification (e.g. Sereni *et al.*, 1966; Adlard *et al.*, 1970; Im *et al.*, 1971). Im *et al.* (1971) showed that subjecting rats to severe protein-calorie deficiency in early life produced a long-lasting increase in the specific activity of their cerebral ChE which paralleled long-term increases in emotional reactivity that had been observed in adult rats subjected to the same nutritional treatment (Levitsky and Barnes, 1970). Sustained decreases in emotional reactivity (Denenberg and Morton, 1964) and in subcortical ChE activity (Tapp and Markowitz, 1963) have also been produced by early handling of rats.

Changes in enzymes associated with the synthesis (rather than the inactivation) of neurotransmitter suspects are perhaps more closely related to changes in behavioral adaptation (see also Chapter 6). Such changes have been produced in cerebral tyrosine hydroxylase (TH) activity by various psycho-active agents and by modification of the *external* environment (see Mandell, 1970, 1975b; Fig. 4). Chronic administration of reserpine increased TH activity in rat midbrain, whereas antidepressant and stimulant drugs caused the opposite

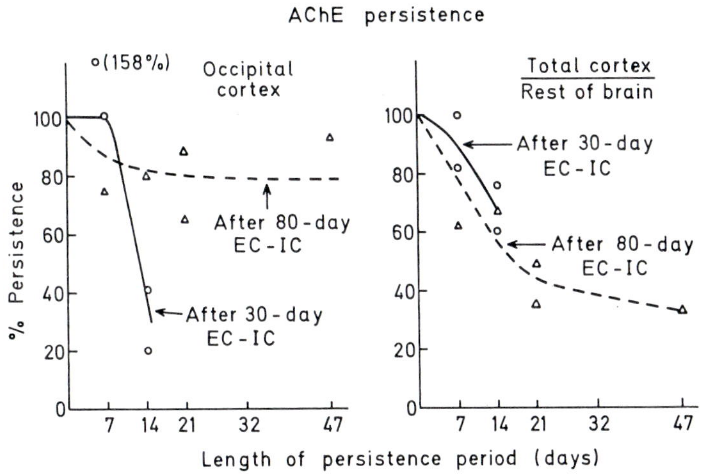

FIG. 43. Persistence of AChE activity for occipital cortex and for the ratio Total Cortex/Rest of Brain of rats transferred to IC after 30 days in EC (○———○) and after 80 days in EC (△----△). (Reproduced with permission from Bennett *et al.*, 1974.)

effect (see Mandell, 1975b). These studies have revealed that the activity of this enzyme (possibly the rate-limiting enzyme of catecholamine synthesis; see Spector *et al.*, 1967) can adapt to induced changes in synaptic function and that this adaptation (regulation of enzyme induction) can be mediated by the available store of transmitter (see Fig. 4).

6. *Antibodies*

Plasma proteins, designated as antibodies, have also been implicated in behavioral control. Antibodies directed against certain brain components (e.g. soluble protein extracts from caudate nucleus; synaptic membrane fractions) can produce electrographic and behavioral changes in mammals (e.g. Mihailović and Janković, 1961; Heath and Krupp, 1967; Janković *et al.*, 1968; Wald *et al.*, 1968;

Hydén and Lange, 1970; MacPherson and Chinerman, 1970; MacPherson and Shek, 1970; Mihailović and Curpić, 1971; Janković, 1972; Karpiak and Rapport, 1975). Recently, Karpiak and Rapport (1975) showed that the 2-month-old male offspring of rats that had been treated on the nineteenth day of gestation with antiserum to a synaptic membrane fraction displayed behavioral deficits such as perseveration and poor retention which could be measured using a differential reinforcement at low rates (DRL)* training paradigm (see Fig. 44). Such perseverative behavior indicated that some organic brain damage might have occurred (see Karpiak and Rapport, 1975), providing a tentative morphological explanation for these behavioral changes. Membrane antigens to the injected antibody appeared to be specific and behaviorally-sensitive, since antiserum to galactocerebroside did not produce such behavioral effects. More recently, Karpiak *et al.* (1976) showed that rats which had received subdural injections of antisera to a synaptic membrane fraction or to S-100 protein were inhibited in their performances on a Lashley III maze, whereas injections of antiserum to myelin had no effect (see Fig. 45).

D. Glycoproteins and Glycolipids

1. *Introduction*

Glycoproteins and glycolipids (especially gangliosides) are found in high concentration in synaptosomal membrane fractions and are concentrated at areas of high synaptic density (e.g. Wolfe, 1961; Wherrett and McIlwain, 1962; Seminario *et al.*, 1964; Quamina and Bogoch, 1966; Lapetina *et al.*, 1967; Wiegandt, 1967; Derry and Wolfe, 1967; Gombos *et al.*, 1972). The receptors for some possible neurotransmitters of the brain (e.g. the glutamate receptor, see Michaelis *et al.*, 1974; Michaelis, 1975; the nicotinic acetylcholine receptor, see Mattsson and Heilbronn, 1975) and some neurotransmitter-synthesizing enzymes (e.g. dopamine-β-hydroxylase; see Wallace *et al.*, 1973; Lovenberg *et al.*, 1975) appear to be glycoproteins, and the serotonin receptor may be a ganglioside (Woolley and Gommi, 1964, 1965).

Gangliosides are water-soluble lipids possessing physicochemical properties which permit their non-polar portions to be fixed to membranes and their polar carbohydrate groups to remain exposed for interaction with ions and polyelectrolytes (see, e.g. Svennerholm, 1970).

---

* *Differential reinforcement at low rates* (*DRL*) refers to a schedule of reinforcement in which low rates of responding are differentially reinforced (i.e. reinforcement is contingent upon the length of the interval between responses) so that the rat learns to emit relatively few responses.

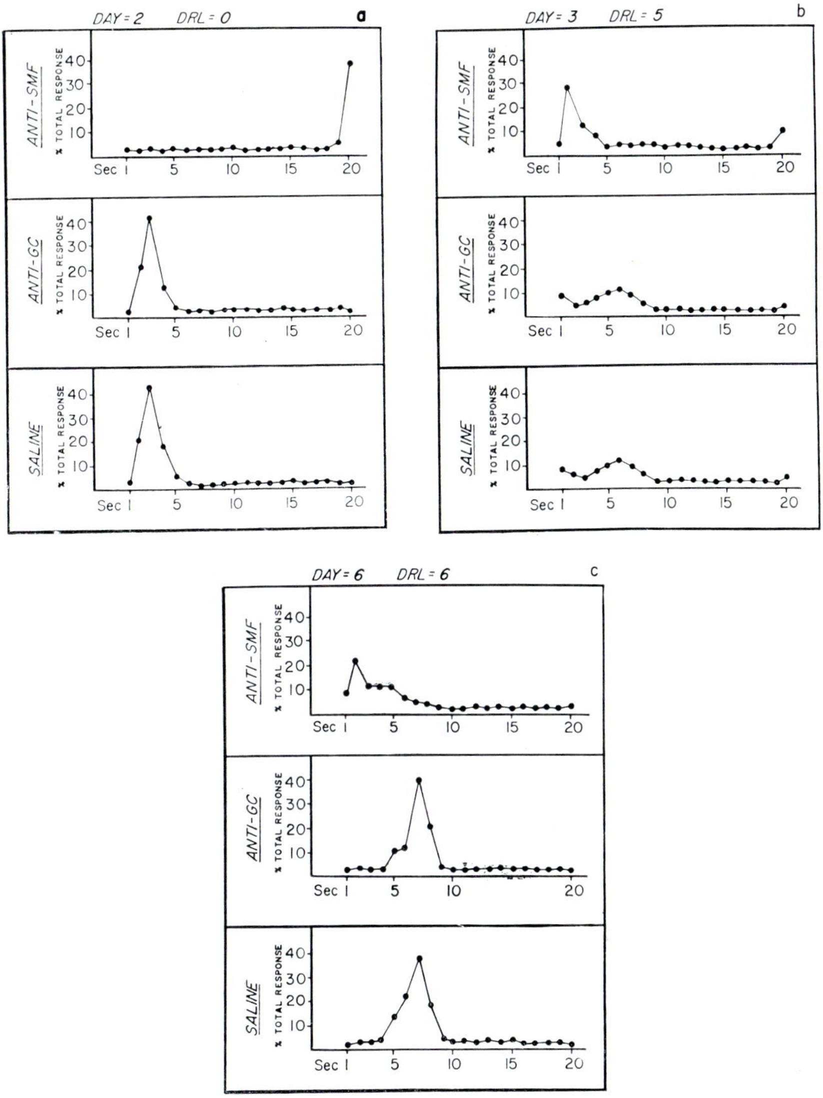

FIG. 44. Response latencies on continuous reinforcement training (a) and initial DRL training schedules (b and c) of 2-month-old male offspring of pregnant rats injected intravenously with antiserum either to synaptic membrane fraction or galactocerebroside, or with isotonic saline. (Reproduced with permission from Karpiak and Rapport, 1975.)

Glycoproteins and glycolipids, because of their negatively charged, displaceable *N*-acetyl-neuraminic acid (NANA) groups, may be involved in receptor, excitational, and informational properties of neural cells (e.g. McIlwain, 1963; Suzuki, 1965; Brunngraber, 1969;

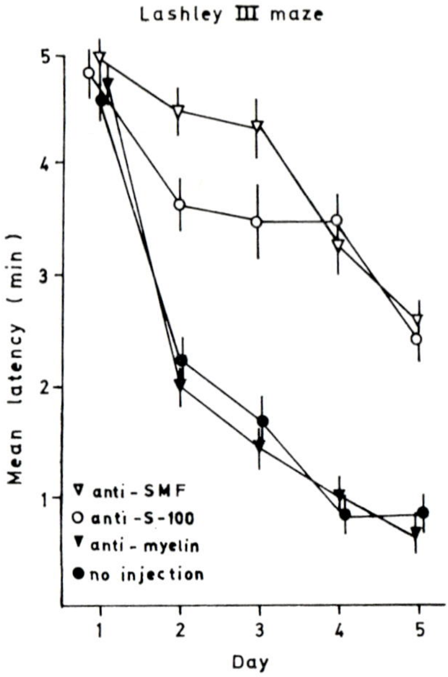

FIG. 45. Graph of mean latency responses (time spent running the maze) for control rats and for three groups of rats that received injections of antisera. Vertical lines indicate S.E.M. Significant inhibition of behavioral performance occurred only in rats injected with antiserum to the synaptic membrane fraction and to S-100 protein. (Reproduced with permission from Karpiak *et al.*, 1976.)

see below). These substances have been implicated in many cerebral pathologic conditions, especially in inherited storage diseases, and their metabolism and regional and subcellular distributions have been studied in various mammals (see, e.g. Suzuki, 1972; Irwin, 1974). A relationship between cerebral glycoproteins and certain psychotic states has also been proposed (e.g. Bogoch *et al.*, 1964; Campbell *et al.*, 1967; Bogoch, 1974). The glycolipids termed cerebrosides and simple sialoglycolipids and some glycoproteins are mainly myelin constituents and are therefore found in high concentrations in white matter and in glia-enriched tissues.

A comprehensive review of the roles of glycoproteins and glycolipids in brain function (Irwin, 1974) and a comprehensive study of the localization and origin of neuronal gangliosides (Ledeen *et al.*, 1976) have recently appeared. Figure 46 illustrates some proposed roles for glycolipids and glycoproteins in neural function.

2. *Behavioral Studies; Effects of Environment*

Changes in the concentration and synthesis of cerebral glycoproteins occurred during training in pigeons, indicating that these substances

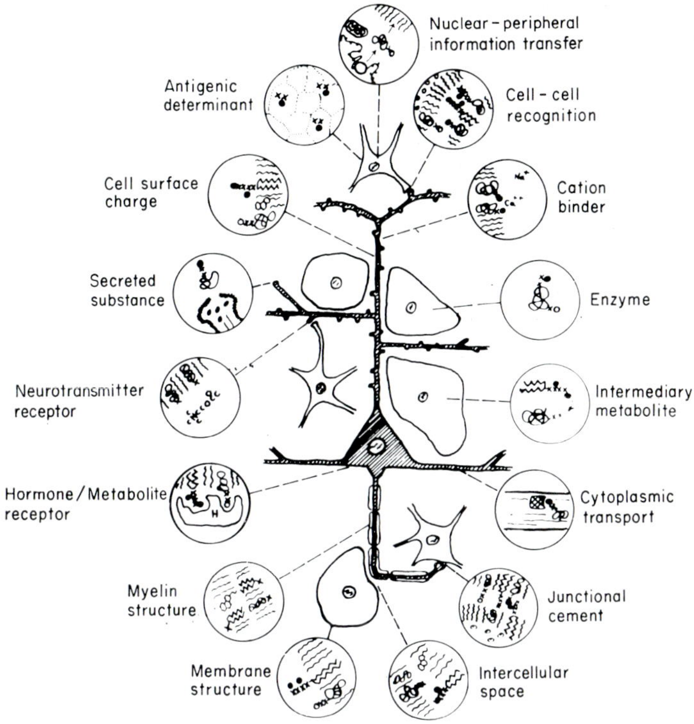

FIG. 46. Schematic representation of postulated roles for glycolipids and glycoproteins in brain function. These possible roles have been further grouped into five general categories: structural; metabolic; excitational; transductional; informational (Reproduced with permission from Irwin, 1974.)

may be involved in surface recognition phenomena that mediate information, communication, and contact functions in the nervous system (see, e.g. Bogoch, 1968, 1973, 1974; Barondes, 1969a). "Training-sensitive" glycoproteins 10B, 11A and Group 2 may act as switching mechanisms ("sign-posts") that route transmission, and therefore could be associated with the establishment of specific brain circuitry (e.g. Bogoch, 1965, 1968, 1972, 1974). Two of these behaviorally induced changes (i.e. in fractions 10BII and 11A of Bogoch, 1968) have been confirmed (Barraco and Irwin, 1976). This proposed role for glycoproteins is supported by their marked heterogeneity. Various glycoproteins containing D-glucosamine, D-galactosamine, D-galactose, D-mannose, D-glucose or L-fucose, as well as different types of sialic acid, can form chains of different lengths which can be organized in different ways on the protein backbone. Also, many readily soluble glycoproteins could be "brain specific", and therefore could be precursors for specific membrane constituents (e.g. Moore and McGregor, 1965; Warecka *et al.*, 1972).

According to Bogoch (1974), changes in the structure and function of glycoproteins should accompany both expression of, and regression of, higher nervous functions. "Short-term" experiments have revealed a more active incorporation of radioactive glucose into pigeon brain glycoproteins during training than under resting conditions (Bogoch, 1972, 1973; Irwin and Barraco, 1975; Fig. 47), and "long-term" experiments revealed a correlation between the absolute amount of cerebral glycoprotein 11A and the "amount learned" by pigeons (Bogoch, 1968, 1973). These substances, which appear to be major constituents of the outer surfaces of cerebral membranes (e.g. Pease,

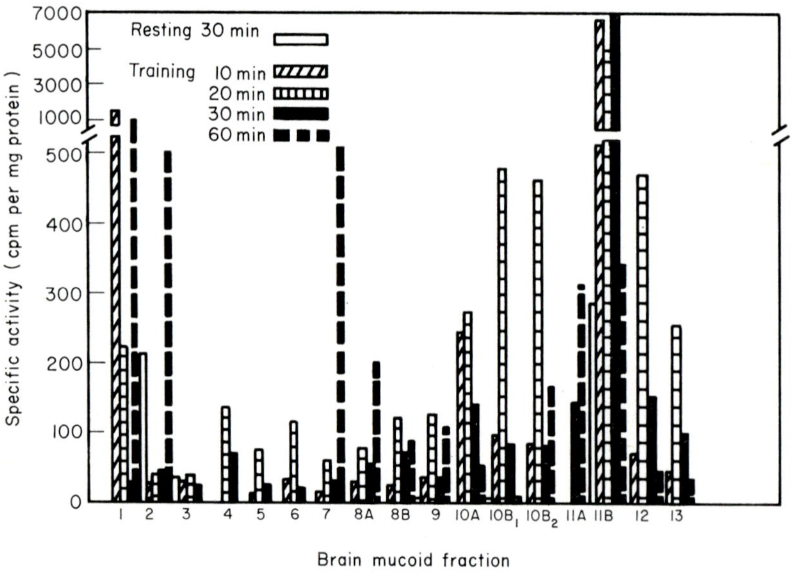

FIG. 47. Incorporation of ^{14}C-glucose into brain mucoids of pigeons. Glucose labelled in the first carbon was injected intravenously into pigeons at rest. They were then allowed to rest, or to engage in a training procedure for varying periods of time before sacrifice in a dry ice-acetone bath. Brain glycoproteins were extracted and separated as described by Bogoch (1965, 1968). It is noteworthy that at rest for 30 min, only slight incorporation was observed in groups 2, 3 and 11B and that much more incorporation was observed at only 10 min of training. Furthermore, the nature of the most actively labelled groups was a function of time. Thus, at 10 min training, groups 1, 10A, and 10B were most prominently labelled. At 20 min, groups 10A, 10B, and 10B2, 11B, 12, and 13 were most active. At 60 min, groups 1 and 2 were again active, together with groups 7, 11A, and 11B. It is clear that there is an extremely active turnover of carbohydrates in the brain glycoproteins of training pigeons, and that much of this activity occurs within minutes. (Reproduced with permission from Bogoch, 1972.)

1966; Jones, 1970), including postsynaptic membranes (e.g. Cotman, 1975), are thus likely to be elements of the *Behavioral Code*.

In other studies aimed at elucidating the interrelationships of glycoproteins, glycolipids and behavior, no changes in NANA content were found in cerebral synaptosomal fractions of rats subjected to simple or complex maze training (Irwin, 1969a, b). However, subjecting rats to a "swim-escape" task ("experimental rats") did cause a slight decrease in the NANA content of microsomes which could have been due to a decrease in membrane-bound NANA outside the synaptic region (Irwin, 1969a, b). Since the same change was found in rats that were forced to swim, but not taught to escape ("functional controls"), this change did not appear to be specific to learning itself. Other experiments (Irwin and Samson, 1969) revealed that behavioral stimulation altered the uptake and loss of radioactive metabolites (e.g. U-^{14} C-glucose) by cerebral glycolipid and glycoprotein fractions. Since purification of ganglioside fractions tended to obscure these differences, precursor pools rather than gangliosides were perhaps altered by the behavioral modification. However, further experiments by Irwin and Samson (1971) revealed that the brains of both experimental rats and functional controls incorporated less radioactivity of intracranially-injected D-1-^{14}C-glucosamine into ganglioside M_1, non-ganglioside glycolipids, and a lipid-insoluble (glycoprotein) residue than rats which were simply left in their home cages. Hence, a general deceleration of glycolipids and glycoprotein turnover may be involved in behavioral stimulation. Also, since the turnover of some gangliosides (D_{1a} and D_{1b}) differed between experimental rats and functional controls, it was suggested that such changes could be related to the learning situation, itself. Further studies by Maccioni *et al.* (1971) revealed that glucosamine incorporation into gangliosides was decreased in synaptosomes and microsomes, but increased in mitochondria, after emergence of rats from a constantly dark environment.

Cortical spreading depression (SD), which produces a "reversible decortication" and which can produce amnesia (Burešová and Bureš, 1969), also inhibits the synthesis of cortical glycoproteins and proteins (e.g. DeLuca *et al.*, 1975). Both dendrites (Van Harreveld, 1958; Ochs and Hunt, 1960; Ochs, 1962) and glial cells (e.g. Sugaya *et al.*, 1975) appear to be involved in SD, the initiation of which may be linked to glutamate release and its subsequent increase of Na^+ entry into various neural elements (e.g. Van Harreveld and Fifkova, 1973, 1975a, b) and to cessation or overload of glial intercellular K^+-buffering function (see, e.g. Sugaya *et al.*, 1975). The swelling of dendritic spines produced by SD could enhance "short-term" memory (e.g. Van Harreveld and Fifkova, 1975a; see Chapter 3, Part B3).

In other studies, Ramirez *et al.* (1974) showed that one hour after intraventricular administration of D-1-^{3}H-glucosamine to rats that had been subjected for 30 days to a complex environment the specific radioactivity of the glycoprotein fraction was increased (with respect to deprived rats) in subcellular fractions (especially in synaptic plasma membrane of both cerebral cortex and hippocampus) (Table 19). However, three hours after injection of precursor, opposite results were obtained, except for incorporation into the hippocampal synaptic membrane fraction which still occurred to a greater extent in "enriched" rats (Table 20; see also Hydén and Lange, 1972; Levitan *et al.*, 1972a, b). This discrepancy might be explained by considering that a difference in the time courses of glycoprotein synthesis existed between these two groups of rats. In a related study in which 1-^{3}H-glucosamine was injected subcutaneously 15 min before training mice for 15 min at a conditioned avoidance "jump-up" task, the mice being sacrificed one hour after injection, Dunn and Hogan (1975) observed no difference in incorporation of radioactivity into a cerebral glycoprotein-enriched fraction between trained and untrained mice, whereas incorporation into gangliosides was significantly greater in trained animals than in "yoked" or quiet controls (Table 21). This difference was not specific to any ganglioside species.

Stimulated by their previous results which indicated that significant increases in the incorporation of precursors of RNA and protein into rat hippocampus and visual cortex occurred during and after acquisition of a brightness discrimination (Pohle and Matthies, 1974a, b; Popov *et al.*, 1975) and that membrane-bound ribosomes of hippocampal pyramidal cells of rats increased after acquisition of this same discrimination (Wenzel *et al.*, 1975), Popov *et al.* (1976) studied the incorporation of L-1-^{3}H-fucose into proteins of rat hippocampus and visual cortex during acquisition of a shock-motivated brightness discrimination. Using polyacrylamide gel electrophoresis, it was shown that solubilized insoluble proteins of three regions of the hippocampus incorporated more fucose during learning, in some discrete, slow-moving, carbohydrate-rich protein bands; such changes did not occur in corresponding bands of visual cortex (Figs 48 and 49). The most significant changes were found mainly in substructures consisting of densely packed neuronal cells. These workers suggested that the acquisition of new behavior probably causes "*relatively specific changes* in gene expression and, consequently, modifications of the synthesized protein patterns within particular brain structures". Glycoproteins appear to play a special role in this process, which supports the contention that the development of permanent memory traces is paralleled by changes in neuronal connectivity.

These data help to clarify the contradictory results of Damstra-Entingh *et al.* (1974) and Routtenberg *et al.* (1974) concerning the

TABLE 19. Effects of environmental complexity on the *in vivo* incorporation of ^{3}H-glucosamine into glycoproteins of subcellular fractions of rat cerebral cortex and hippocampus (1-h labelling time).

Subcellular fraction	Corrected specific radioactivities (counts/min/mg protein)								
	Cortex			Hippocampus			Liver		
	Enriched	Deprived	E/D	Enriched	Deprived	E/D	Enriched	Deprived	E/D
Plasma membrane	2530	2300	1·10	—	—	—	—	—	—
Cytosol	1650	1450	1·14	5750	4620	1·24	367	390	0·94
Nuclei	4180	3060	1·37	—	—	—	—	—	—
Perikaryal mitochondria	2800	2530	1·11	11 530	9000	1·28	—	—	—
Endoplasmic reticulum	5860	5020	1·17	24 200	18 900	1·28	—	—	—
Synaptic plasma membrane	4730	3270	1·45	12 110	12 430	1·78	—	—	—
Synaptosol	3490	2490	1·40	16 520	14 920	1·11	—	—	—
Synaptic vesicles	2500	2220	1·13	—	—	—	—	—	—
Synaptic mitochondria	2980	2380	1·25	—	—	—	—	—	—
Total particulate	—	—	—	—	—	—	915	900	1·02

Sprague-Dawley rats, weaned at 21 days of age, were assigned to "enriched" or "deprived" environments. Thirty days later the rats received 50 *μCi* of ^{3}H-glucosamine intraventricularly, were returned to their cages, and killed 1 h later. Subcellular fractions were prepared separately from cerebral cortices and hippocampi from 8 "enriched" and 8 "deprived" rats. Homologous materials were pooled prior to fractionation to avoid creating artificial differences due to different or delayed handling. No deviation parameters are therefore available, but the results represent the true mean of the group. Missing values in the hippocampus column denote insufficient material to obtain meaningful duplicate values of the glycoprotein specific radioactivity (at least 100 μg of protein per aliquot). (Reproduced with permission from Ramirez *et al.*, 1974.)

TABLE 20. Effects of environmental complexity on the *in vivo* incorporation of ^{3}H-glucosamine into glycoproteins of subcellular fractions of rat cerebral cortex and hippocampus (3-h labelling time).

Subcellular fraction	Corrected specific radioactivities (counts/min/mg protein)								
	Cortex			Hippocampus			Liver		
	Enriched	Deprived	E/D	Enriched	Deprived	E/D	Enriched	Deprived	E/D
Plasma membrane	5270	6440	0·82	11 220	15 200	0·74	—	—	—
Cytosol	3340	4190	0·80	10 800	13 400	0·81	454	470	0·97
Nuclei	7440	8200	0·91	42 150	68 700	0·61	—	—	—
Perikaryal mitochondria	5320	7310	0·73	20 150	22 800	0·88	—	—	—
Endoplasmic reticulum	9980	12 230	0·82	44 300	57 100	0·78	—	—	—
Synaptic plasma membrane	6330	11 250	0·56	66 710	43 000	1·55	—	—	—
Synaptosol	7230	7330	0·99	37 350	61 300	0·61	—	—	—
Synaptic vesicles	6210	8700	0·71	—	—	—	—	—	—
Synaptic mitochondria	6000	6350	0·94	—	—	—	—	—	—
Total particulate	—	—	—	—	—	—	1428	1540	0·93

The experiment was performed as detailed in the legend to Table 19 except that 3 h were allowed between the injection of radioactive glucosamine and the sacrifice of the rats. (Reproduced with permission from Ramirez *et al.*, 1974.)

TABLE 21. Incorporation of 1-^{3}H-glucosamine into chemical fractions of mouse brain: comparison of behaviors.

Comparison	C-M Soluble	C-M Residue	Lipid	Ganglioside	Ganglioside s.r.
Trained/Quiet	1·03 (0·91–1·18)	1·08 (1·03–1·20)	1·05 (0·94–1·30)	1·21* (1·14–1·28)	1·17* (1·12–1·21)
Trained/Yoked	1·05 (0·93–1·10)	1·03 (0·96–1·13)	1·00 (0·95–1·13)	1·12† (1·06–1·21)	1·08 (1·00–1·20)

The mean of the radioactivity in each chemical fraction for each group of animals in each experiment was calculated. The numbers presented are the geometric means for all experiments, of the ratio of the means for each group in each separate experiment. The numbers in parentheses are the ranges of the ratio of the means obtained in the separate experiments. Eight separate experiments involving 41 trained and 41 quiet animals are included in the trained/quiet comparison. Five separate experiments involving 25 trained and 27 yoked animals are included in the trained/yoked comparison. The data on ganglioside specific radioactivity were treated similarly, but only 7 experiments were involved in the trained/quiet comparison and 3 in the trained/yoked comparison. Probabilities were determined by a chi-squared analysis of the probability values obtained by t-comparisons on individual experiments. * $p < 0{\cdot}001$; † $p < 0{\cdot}05$. (Reproduced with permission from Dunn and Hogan, 1975.)

fucosylation of glycoprotein during behavior, but it must be noted that different behavioral paradigms and methods were used. Routtenberg *et al.* (1974) showed, in male albino rats injected intraperitoneally with ^{3}H-L-fucose prior to learning a simple withdrawal from footshock, that a set of glycoproteins of a particular molecular weight range appeared to exhibit altered turnover rates after this experience.

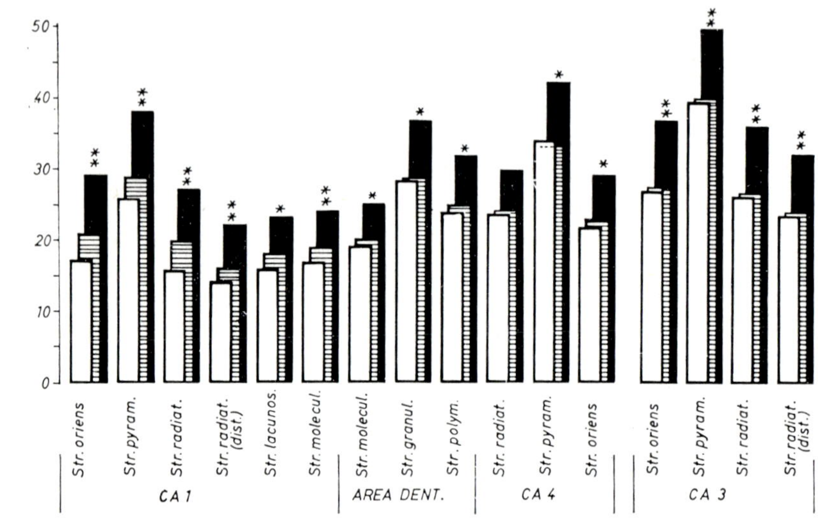

FIG. 48. Increased incorporation of fucose into four regions of the rat hippocampus during learning. Blackening patterns (expressed as absorbance in arbitrary units) using histophotometry of unstained hippocampal slices from each of 9 trained animals (*black columns*), 9 active controls (*shaded columns*) and 9 passive controls (*white columns*). Significance: * = $p < 0{\cdot}05$; ** = $p < 0{\cdot}01$ when comparing trained animals with active and passive controls. (Reproduced with permission from Popov *et al.*, 1976.)

These variations occurred in synaptosomal fractions of the caudate-putamen region and in temporal cortex, but not in other regions studied.

It should be noted that changes in cerebral glycolipid and glycoprotein metabolism can be caused by many other conditions which alter the physiological state of the animal (e.g. induced convulsions; drug and hormone treatments, exposure to hypercapnia; see, e.g. refs in Irwin and Samson, 1971; Irwin, 1974).

These findings which have revealed that changes in the synthesis and turnover of cerebral glycoproteins and glycolipids occur during various experiences indicate that changes in energy utilization had occurred, and that such changes appear to be involved in the process-

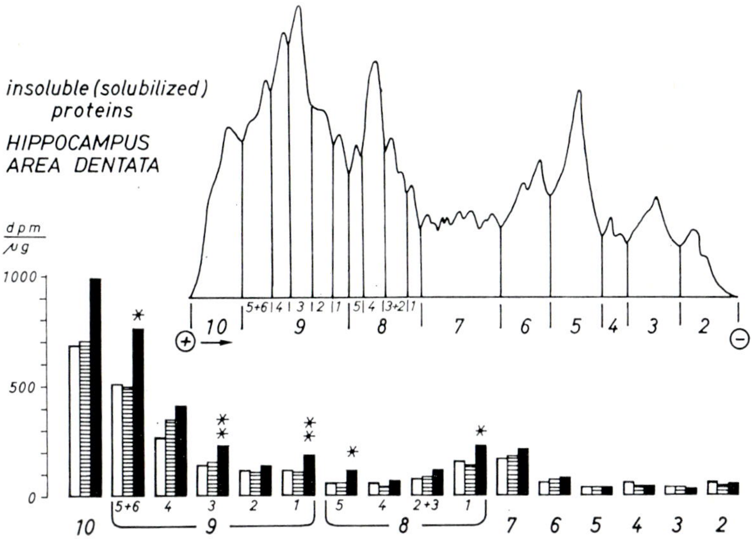

FIG. 49. Increased incorporation of fucose into rat hippocampus during learning. Radioactivities of electrophoretically separated bands from solubilized insoluble proteins obtained from the hippocampal "area dentata" of rats. The upper part shows schematically the densitometric tracing of the 16 gel bands investigated and the lower part shows the incorporation of radioactivity of L-1-^{3}H-fucose into proteins (in dpm/μg protein). *Black columns* = trained animals; *shaded columns* = active controls; *white columns* = passive controls; * and ** indicate, respectively, $p < 0{\cdot}05$ and $p < 0{\cdot}01$ for differences between trained *vs* both active and passive controls. (Reproduced with permission from Popov *et al.*, 1976).

ing of information and in the formation of new synapses (see also Chapters 5 and 7). Morgan *et al.* (1973) have reviewed results concerning the characterization of the various protein and glycoprotein constituents of synaptosomal and synaptic vesicle membranes.

E. Peptides

1. *Introduction*

Roles for peptides in learning and memory have been postulated. Two groups, peptides possibly formed during learning and memory (see, e.g. Ungar, 1975a, b), and peptides possessing hormonal functions (see, e.g. DeWied *et al.*, 1974, 1975b; DeWied, 1974) have been studied. Early claims that such molecules might actually carry "memory traces" have been largely disproven, except in a few cases (see Ungar, 1973a, b; Fjerdingstad, 1973). "Scotophobin", a peptide identified in the brains of rodents trained to avoid the dark is the prototype of the series thought to be formed in response to learning (see Ungar *et al.* 1972). It has been claimed that synthetic scotophobin can induce avoidance of dark compartments by fish and mice (see refs in Malin, 1974). Perhaps the major difficulty in accepting early views about peptides and memory involves the ideas that these peptides (e.g. scotophobin) are absent from the brains of untrained animals and that specific peptides code specific behaviors; i.e. "one peptide, one behavior" (e.g. Ungar, 1975b). Although such a mechanism might exist, it seems more likely that the involvement of peptides in facilitating synaptic transmission is of a more general ("non-specific") nature, as seems to be the case for proteins (see, e.g. Barondes, 1972; Shashoua, 1974; Part C, above). Hormones of the pituitary gland (e.g. adrenocorticotrophic hormone (ACTH); vasopressin) and of the hypothalamus (e.g. the "releasing factors") which play roles in the integration of the nervous and endocrine systems are likely to be the main group of peptides related to behavior.

Although peptides are not generally considered to be macromolecules, they will be discussed here since they involve mechanisms of protein synthesis and have considerable informational content and since they appear to be involved in memory processes and in behavior. Emphasis will be placed on the effects that these "intermediate-sized" molecules exert on behavior and on their roles in mediating neural changes produced by environmental modification.

2. *Pituitary Hormones*

Pituitary hormones or their derivatives can antagonize retrograde amnesia (e.g. DeWied, 1969, 1974; Flexner and Flexner, 1971; Lande *et al.*, 1972; Quinton, 1972; Rigter *et al.*, 1974). Increases in ACTH release following adrenalectomy also attenuated amnesia (Flexner and Flexner, 1970). Since ACTH stimulates adrenal corticosteroid synthesis in addition to exerting a direct effect on the CNS, and since corticosteroids can exert behavioral effects that antagonize those of ACTH (DeWied, 1967; Bohus and Korányi, 1969), most recent studies have been conducted with $ACTH_{4-10}$ which, though lacking significant adrenocorticotrophic activities, affects avoidance behavior similarly to ACTH, itself (DeWied, 1969). It is noteworthy that $ACTH_{4-10}$ also represents a fragment of another anterior pituitary secretion, melanocyte-stimulating hormone (MSH).

DeWied (1969, 1974) showed that normal learning could be restored in hypophysectomized rats by administration of $ACTH_{4-10}$, and Gispen and Schotman (1973) found that the restoration of learning promoted by ACTH was paralleled by restorations to normal of RNA levels and incorporation of amino acids into protein in the brain stem.

$ACTH_{4-8}$ appeared to be fully active, while not causing steroidogenesis in the adrenal gland (DeWied, 1974). DeWied *et al.* (1975b) showed, using extinction of a "pole-jumping" avoidance response in rats as the criterion, that the tetrapeptide $ACTH_{4-7}$ (H-MET-GLU-HIS-PHE-OH) contains the essential elements required for the observed behavioral effects (Table 22).

Rigter and van Riezen (1975a, b) showed that $ACTH_{4-10}$ attenuated amnesia in rats when given prior to retrieval but not when given prior to acquisition by an effect that was independent of both the behavioral task and the nature of the amnestic agent. In a further study aimed at examining the time courses of the anti-amnestic effects of $ACTH_{4-10}$ and desglycinamide-lysine vasopressin (DG-LVP) in rats, Rigter *et al.* (1975) found, using CO_2-induced retrograde amnesia for a one-trial "step-through" passive avoidance response as the criterion, that subcutaneously administered $ACTH_{4-10}$ attenuated amnesia when given within 8 h prior to retrieval (Fig. 50) and that DG-LVP attenuated amnesia when administered either within 2 h of acquisition or within 8 h of retrieval. At the human level, Miller *et al.* (1974) showed that $ACTH_{4-10}$ improved visual memory, decreased anxiety and increased attention, whereas $ACTH_{1-24}$ caused none of these effects.

Reading (1972) found that $ACTH_{4-10}$ caused an increase in ^{14}C-leucine incorporation into protein of rat brain. Using slices of the brain stems of hypophysectomized rats, Reith *et al.* (1975) showed further that the presence of $ACTH_{1-10}$ in the incubation medium stimulated incorporation of ^{14}C-leucine into protein, whereas $ACTH_{1-10}$-7-D-PHE, which acts in the opposite way to the naturally occuring analog on behavior, also caused opposite effects on *in vitro* protein synthesis. It should be noted that ACTH, itself, and β-melanocyte-stimulating hormone (β-MSH) also increase amino acid incorporation into protein (Semiginovsky and Jakoubec, 1971; Rudman *et al.*, 1974).

The novelty of new experience may provide the stress which enhances the secretion of ACTH which then starts the chain of chemo-morphological events leading to memory storage (see Dunn, 1975, and Part C2, above). It should also be remembered that postulated central transmitters (e.g. DA, NE and 5-HT) are very likely to be involved in the secretion of ACTH and other pituitary hormones (e.g. Imura *et al.*, 1973; Chambers and Brown, 1976). Since ACTH also acts on extra-cerebral tissues, its release provides a link between peripheral tissues and the CNS in the explanation of learning, memory and behavior (see also Chapter 8, Part C). It should be realized also that other pituitary hormones (e.g. growth hormone) may possess actions similar to those of ACTH and thus may also be involved in the stress-induced initiation of memory and behavioral adaptation.

Regarding the sites of action of hormones of the pituitary-adrenal

TABLE 22. Effect of progressive shortening of the sequence ACTH 1–10 from the amino- and the carboxyl end on inhibition of extinction of a pole-jumping avoidance response in rats.

	1 2 3 4 5 6 7 8 9 10	Approximate potency
ACTH 1–10	H—Ser—Tyr—Ser—Met—Glu—His—Phe—Arg—Try—Gly—OH	1
ACTH 2–10	H—Tyr—Ser—Met—Glu—His—Phe—Arg—Try—Gly—OH	1
ACTH 3–10	H—Ser—Met—Glu—His—Phe—Arg—Try—Gly—OH	1
ACTH 4–10	H—Met—Glu—His—Phe—Arg—Try—Gly—OH	1
ACTH 5–10	H—Glu—His—Phe—Arg—Try—Gly—OH	0·5
ACTH 6–10	H—His—Phe—Arg—Try—Gly—OH	0·1
ACTH 7–10	H—Phe—Arg—Try—Gly—OH	0·1
ACTH 4–9	H—Met—Glu—His—Phe—Arg—Try—OH	1
ACTH 4–8	H—Met—Glu—His—Phe—Arg—OH	1
ACTH 4–7	H—Met—Glu—His—Phe—OH	1
ACTH 4–6	H—Met—Glu—His—OH	0·3

Note that shortening of the sequence $ACTH_{4-10}$ step by step from the carboxyl end revealed that the tetrapeptide $ACTH_{4-7}$ contains the essential elements required for the behavioral effect of ACTH analogs. (Reproduced with permission from De Wied *et al.*, 1975b.)

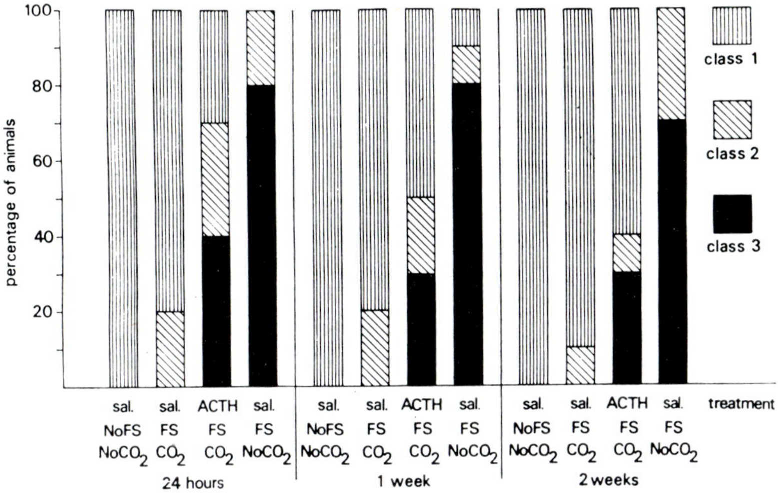

FIG. 50. Permanence of the anti-amnesic effect of $ACTH_{4-10}$ on CO_2-induced amnesia for a one-trial passive avoidance response. The retrieval test was given either 24 h, 1 week or 2 weeks after acquisition. Saline or 100 μg $ACTH_{4-10}$/rat was administered 1 h prior to retrieval. The figure shows the latencies of groups of rats (10/group) at the retrieval test. The scores were divided into 3 classes: (1) 0–10·0 sec (no avoidance); (2) 10·1–299·9 sec (incomplete avoidance); and (3) 300·0 sec (complete avoidance). FS = footshock; No FS = no footshock; CO_2 = CO_2 treatment; $NoCO_2$ = sham amnestic treatment. (Reproduced with permission from Rigter *et al.*, 1975.)

system in the brain, it has been shown that $ACTH_{1-10}$ delayed, whereas $ACTH_{1-10}$-7-D-PHE facilitated extinction of a "pole-jumping" avoidance response when these substances were implanted into rostral mesencephalon, caudal diencephalon (posterior thalamic level) or injected into the CSF (van Wimersma Greidanus and DeWied, 1971). Hence, it appears likely that these agents act on the ascending reticular system, and perhaps more specifically on the nuclei parafascicularis; the latter are non-specific thalamic nuclei which (like the centromedianum complex) are believed to be involved in the maintenance of conditioned avoidance behavior (see Cardo, 1965, 1967; Delacour, 1970). Steroids (e.g. coricosterone, dexamethasone) appear go act on limbic forebrain regions as well as on the ascending reticular system (see van Wimersma Greidanus and DeWied, 1969; DeWied, 1974). Other studies have shown that a decrease in multiunit activity of the reticular formation and in some forebrain structures occurred after administration of ACTH, itself (Korányi *et al.*, 1971a, b). More recently, intravenous administration of $ACTH_{4-10}$ (both the 7-D-PHE and 7-L-PHE analogs) was shown to produce significant decreases in hippocampal "theta" activity of a dog subjected to mild food deprivation

(Urban *et al.*, 1974). This study, together with the findings that lesions of the rostral septal area of rats prevented the inhibitory actions of $ACTH_{4-10}$ and vasopressin on extinction of a "pole-jumping" avoidance response, indicated that the mesencephalic limbic system plays a significant role in mediating the behavioral effects of these peptides (van Wimersma Greidanus *et al.*, 1975a).

Further studies have indicated that puromycin-induced amnesia in mice can be reversed by vasopressin (Lande *et al.*, 1972), and that injections of the hypothalamic releasing factors, thyrotropin-releasing hormone (TRH) and luteinizing hormone-releasing hormone (LHRH), can delay extinction of avoidance behavior, perhaps due to their structural similarity to ACTH analogs (DeWied *et al.*, 1975a, b). The possible relationship between vasopressin and memory processes has been strengthened further by studies which showed that Pitressin, lysine-8-vasopressin (LVP) and/or DG-LVP restored resistance to extinction in posterior lobectomized rats (DeWied, 1965), facilitated learning of passive avoidance tasks (Ader and DeWied, 1972) and increased resistance to extinction of avoidance responses (DeWied and Bohus, 1966; DeWied, 1971; DeWied *et al.*, 1974; Fig. 51). Since an increased resistance to extinction of a "pole-jumping" avoidance response was shown after intra-hypothalamic and intra-ventricular injections of LVP, (DeWied *et al.*, 1974) this action does not appear to be mediated by peripheral horomonal mechanisms. Furthermore, intra-ventricular administration of anti-vasopressin serum, which served to remove vasopressin from the CSF, inhibited memory consolidation in rats (van Wimersma Greidanus *et al.*, 1975b; see Table 23).

Perhaps the most interesting study was conducted recently by DeWied *et al.* (1975a) using a strain of rat (Brattleboro strain) that is afflicted with hereditary hypothalamic diabetes insipidus since it cannot synthesize vasopressin (Valtin and Schroeder, 1964; Valtin, 1967). This study revealed that these vasopressin-deficient rats had serious difficulties in performing a passive avoidance task which were reversed by single subcutaneous injections of either rat arginine vasopressin or DG-LVP (Table 24). Since this enhancement of memory was elicited by DG-LVP, which lacks the C-terminal glycinamide and therefore does not possess most of the classic endocrine activities of the parent hormone, including that of reversing the water diuresis of diabetes insipidus (DeWied *et al.*, 1972), these workers suggested that they had demonstrated the physiological significance of vasopressin in relation to memory. Further studies indicated that these vasopressin-deficient rats displayed an impaired development of tolerance to morphine-induced analgesia that could be reversed by treatment with DG-LVP (DeWied and Gispen, 1976; see also Chapter 8, Part F).

Also, using electrophysiological methods Urban and DeWied (1975a, b) showed that qualitative differences in "rhythmic slow activity" (RSA, or hippocampal "theta" activity), which may be related to memory consolidation processes, exist between homozygous (HO-DI) and heterozygous (HE-DI) rats of this same strain. The major difference in frequency content of the RSA during paradoxical (REM) sleep

H.Cys.Tyr.Phe.Gln.Asn.Cys.Pro.Arg.Gly.NH_2

H.Cys.Tyr.Phe.Gln.Asn.Cys.Pro.Lys.Gly.NH_2

H.Cys.Tyr.Phe.Gln.Asn.Cys.Pro.Lys.OH

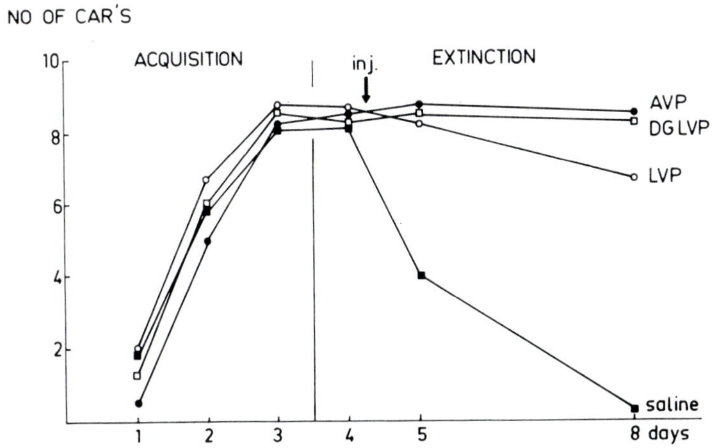

FIG. 51. Influence of a single subcutaneous injection of various vasopressin analogs on extinction of a pole jumping avoidance response. Peptides were injected in 0·5 ml saline immediately after the first extinction session on day 4. AVP (225 I.U./mg), 0·2 μg LVP (60 I.U./mg), 0·6 μg DG-LVP (0·3 μg/ml); 4 rats per group. (Reproduced with permission from De Wied *et al.*, 1974.)

was confined to the band of 5·5 to 10·0 Hz, and the RSA of HO-DI rats contained lower frequencies than that of HE-DI rats. Treatment with DG-AVP, while not affecting the symptoms of diabetes insipidus (DeWied *et al.*, 1972), almost completely normalized the RSA. In this regard, it should be recalled that REM sleep appears to be involved in learning and memory (see, e.g. Pagel *et al.*, 1973 and refs therein). These studies with vasopressin-deficient rats have provided a good example of how a "genetic error" can be used to elucidate an element of the *Behavioral Code*.

TABLE 23. Median latency scores during retention sessions of a passive avoidance response at different time intervals between learning trial and retention session, following intraventricular or intravenous administration of AVP-antiserum.

Treatment	Time interval				
	2 min	1 h	4 h	24 h	48 h
INTRAVENTRICULAR					
1 μl AVP-antiserum	300 (7)	300 (7)	93* (7)	33† (13)	43† (11)
1 μl oxytocin-antiserum	—	—	—	300 (9)	300 (7)
1 μl "normal" rabbit serum	300 (6)	300 (6)	300 (7)	300 (11)	300 (13)
INTRAVENOUS					
100 μl AVP-antiserum	—	—	—	300 (9)	300 (9)
100 μl "normal" rabbit serum	—	—	—	300 (9)	300 (9)

* $p < 0{\cdot}02$; † $p < 0{\cdot}01$; number of animals per group in parentheses; AVP = arginine-8-vasopressin. (Reproduced with permission from Van Wimersma Greidanus *et al.*, 1975b.)

TABLE 24. Passive avoidance behavior in rats with hereditary diabetes insipidus; comparison between homozygous and heterozygous rats. Effect of arginine-8-vasopressin (AVP) and desglycinamide lysine-8-vasopressin (DG-LVP).

				Median avoidance latency in sec					
		Shock		Before	After shock				
Group	N	mA	sec	shock	0 h	3 h	24 h	48 h	120 h
Heterozygous	6	0·00	—	6·0	—	—	8·5	6·0	8·5
	8	0·25	3	7·5	—	—	8·5	21·0	12·0
	8	0·50	3	9·0	—	—	30·5	143·0*	101·5*
	8	1·00	3	8·0	—	—	300·0†	300·0†	300·0†
	6	1·00	10	5·0	—	—	300·0†	300·0†	300·0†
	8	0·00	—	9·0	9·5	—	—	—	—
	8	1·00	3	8·5	300·0†	—	—	—	—
	8	0·00	—	9·0	—	10·5	—	—	—
	8	1·00	3	7·5	—	300·0†	—	—	—
Homozygous	6	0·00	—	4·5	—	—	8·0	8·5	8·0
	6	0·25	3	9·0	—	—	11·5	13·0	20·0
	7	0·50	3	9·0	—	—	34·0	37·5	33·0
	7	1·00	3	10·0	—	—	34·0	42·0	40·0
	6	1·00	10	6·0	—	—	28·5	12·5	9·0
	8	0·00	—	7·5	9·5	—	—	—	—
	6	1·00	3	7·0	287·5†	—	—	—	—
	8	0·00	—	8·0	—	10·0	—	—	—
	6	1·00	3	7·5	—	25·5†	—	—	—
DG-LVP 1 μg[a]	6	0·00	—	8·0	—	—	11·0	11·0	10·0
DG-LVP 1 μg[a]	8	1·00	3	7·5	—	—	253·5†	300·0†	300·0†
AVP 1 μg[a]	7	1·00	3	6·0	—	—	300·0†	300·0†	260·0†

[a] Dose per rat; * $p < 0{\cdot}05$; † $p < 0{\cdot}01$. (Reproduced with permission from De Wied *et al.*, 1975a.)

DeWied and colleagues noted further that vasopressin analogs elicited "long-term" effects on conditioned avoidance behavior, whereas ACTH analogs produced "short-term" effects. Therefore, vasopressin analogs may effect the consolidation of learned behavior, whereas ACTH analogs may increase "arousal" level by acting on mesencephalic limbic structures, thereby increasing the potency of motivational effects of environmental events (see DeWied, 1974; DeWied *et al.*, 1975b). Such "short-term" effects of ACTH could be due to conformational changes in membranes which could enhance cyclic-AMP production (see DeWied *et al.*, 1975b).

It is noteworthy that these hormones do not act selectively on avoidance behaviors. ACTH-like peptides can also induce excessive grooming, yawning and stretching responses, and sexual excitement after their intracranial administration in various mammals (e.g. Ferrari *et al.*, 1963; Baldwin *et al.*, 1974). Thus, these studies on ACTH analogs may serve to define further the role played by "non-specific" stress phenomena in learning, memory and behavioral processes. Many studies have indicated that ACTH and corticosteroids exert direct effects on cerebral excitability (see, e.g. refs in DeWied, 1974). Since ACTH analogs (like ACTH) appear to cause an increase in cerebral protein synthesis (Reading, 1972; Reith *et al.*, 1975; Dunn, 1975), plastic changes in the brain could be triggered by a release of ACTH that is produced by the stress of novel experience (see step 2 of Fig. 80). Recent studies have further substantiated the view that the increase in cerebral protein synthesis caused by behavioral stimulation is mediated by ACTH since: (1) ACTH mimics the effects of behavioral stimulation on the incorporation of radioactive lysine into protein; (2) adrenalectomy does not alter these effects; (3) dexamethasone (a potent glucocorticoid known to produce a strong inhibition of corticotrophic stimulation of the adrenal cortex) blocks the changes; (4) injected corticosterone has no effect; (5) $ACTH_{1-24}$ and non-steroidogenic $ACTH_{4-10}$ both increase lysine incorporation into brain protein, but only the former possesses this action in the liver, in both normal and adrenalectomized mice (A. J. Dunn, personal communication, 1975).

This stress factor has also been revealed in another study in which genetically seizure-resistant C57 Bl/6 mice were induced to develop high susceptibility (about 90%) to audiogenic seizures by prior exposure to auditory stimulation ("acoustic priming") on day 19 of age. It was shown that inhibitors of adrenal corticosteroid synthesis (metopyrone; amino-glutethimide), given 5 h before acoustic priming, or 11-deoxy-17-hydroxycorticosterone (an antagonist of glucocorticoid receptors) prevented the induction of seizure susceptibility; therefore it appears that the release of pituitary hormones and a "non-specific" stress reaction (corticosteroid secretion) are required for the induction of audiogenic seizure susceptibility (Sze, 1970; Sze and Maxson, 1976). Interestingly, both cycloheximide and puromycin pre-treatments (but not post-treatments) antagonized the development of seizure susceptibility, indicating that this phenomenon requires protein synthesis (for $\frac{1}{2}$ to 1 h) post-stimulation (Maxson *et al.*, 1974). From these results it seems clear that susceptibility to audiogenic seizures can depend upon prior auditory input as well as upon genetic differences and that a "non-specific" stress reaction appears to mediate the behavioral effect.

This stress factor has also been related to changes in the cyclic nucleotide content of the brain. Interestingly, the cerebral content of cyclic-GMP, but not that of cyclic-AMP was found to be markedly increased by subjecting rats and mice to agitation or cold stress (Mao *et al.*, 1974; Dinnendahl, 1975; Table 25). In this connection, it is noteworthy that although depolarizing concentrations of K^+ have been shown to elevate the contents of both cyclic-GMP and cyclic-AMP in incubated slices of mouse cerebellum, only the depolarization-induced increase in cyclic-GMP appeared to be dependent on Ca^{2+} (see Ferrendelli *et al.*, 1976). Hence, Ca^{2+} entry into neurones during increased neural activity could trigger an increase in cyclic-GMP which could then lead to a change in protein phosphorylation (see this chapter, Part C3) and/or to an increased synthesis of peptido-amines (see Reichelt *et al.*, 1976, and Section 4, below). Other studies have indicated that cyclic-AMP levels of various brain regions can also be increased by electroshock-induced stress (Goldberg *et al.*, 1972; De la Paz *et al.*, 1975). Recent studies have also revealed that various peptides, including vasopressin and oxytocin, can alter the membrane properties of certain molluscan neurones in a fashion similar to that of cyclic nucleotides, indicating further that cyclic-AMP might mediate the actions of peptides on membranes (Treistman and Levitan, 1976).

These studies on peptides have revealed an early stage of information processing (see step 2 of Fig. 80) which involves the transduction of environmental stimulation into a general stress-mediated change in cerebral (and peripheral; see Knox, 1963) metabolic activity. Therefore, it no longer seems necessary to control experiments for non-specific effects of stress, since these effects, along with mechanisms of "arousal", motivation and attention, appear to represent integral phases of behavioral regulation.

The relationship between another adenohypophyseal peptide, growth hormone (GH, or somatotropin) and behavior has been demonstrated at the human level. Note that GH contains one or two polypeptide chains, depending on the mammalian species considered, and

TABLE 25. Effects of agitation and stress on the cyclic nucleotide contents of mouse brain.

Treatment	c-GMP (p-mol/mg wet wt.)		c-AMP (p-mol/mg wet wt.)	
	Cerebellum	Cerebrum[a]	Cerebellum	Cerebrum[a]
Resting mice	0·280 ± 0·012 (13)	0·056 ± 0·005 (11)	0·86 ± 0·09 (6)	0·64 ± 0·07 (6)
Fighting mice	0·644 ± 0·040 (13)†	0·087 ± 0·006 (13)*	0·88 ± 0·11 (6)	0·57 ± 0·04 (6)
Stressed mice (ice-water; 30 sec)	1·136 ± 0·100 (6)*	0·163 ± 0·017 (13)†	0·80 ± 0·04 (6)	0·51 ± 0·04 (6)
Stressed mice (hot plate; 20 sec)	0·612 ± 0·100 (6)*	0·100 ± 0·020 (6)	0·84 ± 0·07 (6)	0·54 ± 0·07 (6)

[a] This region consisted of cerebral cortex, hippocampus, hypothalamus, thalamus, striatum and midbrain. Values represent means ± S.E.M.; numbers of mice in parentheses; * and † indicate, respectively, $p < 0{\cdot}02$ and $p < 0{\cdot}001$ between control (resting mice) and experimental values. (Reproduced with permission from Dinnendahl, 1975.)

that it serves a regulatory role in many metabolic alterations as well as having growth-promoting and anabolic properties. Insulin-induced hypoglycemia, exercise, fasting, infusion of 2-deoxy-D-glucose or arginine, and the stresses associated with cold exposures, ether and surgery all appear to cause decreases in plasma GH levels in man (e.g. Glick *et al.*, 1965; Greenwood, 1967). Consequently, GH may function as a stress hormone (Schalch, 1967; Meyer and Knobil, 1967), especially since many conditions which enhance its release also increase ACTH secretion (e.g. Liddle *et al.*, 1962). Depressed patients exhibited a lower GH release than normal persons in response to insulin-induced hypoglycemia (P. S. Mueller *et al.*, 1969; Endo, 1970; Sachar *et al.*, 1971). More recently, a deficient GH release was usually observed upon exposure of depressed, psychomotor-retarded patients to high temperature, a finding which indicated that a disturbance of hypothalamic-pituitary function might be involved in mental depression (Schilkrut *et al.*, 1975). Changes in GH release may be controlled by aminergic (NE, DA, 5-HT) mechanisms of the hypothalamus (e.g. Martin, 1973; Brown and Chambers, 1974; Chambers and Brown, 1976). Other studies have revealed that human GH release is significantly increased during slow (synchronized) sleep stages and that this differential release of GH follows the reversal of sleep-wakefulness cycles (e.g. Sassin *et al.*, 1969).

3. *Hypothalamic Peptide Hormones ("Releasing Factors")*

Hypothalamic "releasing factors" affect brain and behavior (e.g. Burgus and Guillemin,1970; Schally *et al.*, 1973; Besser, 1974; Prange *et al.*, 1975; Reichlin *et al.*, 1976), and their cerebral effects may be distinct from those exerted on the anterior pituitary gland (see Prange *et al.*, 1975).

Direct central actions of thyrotropin releasing hormone (TRH) were realized when it was shown that its administration potentiated the behavioral excitement caused by pargyline and L-DOPA in hypophysectomized rats (Plotnikoff *et al.*, 1972, 1974a) and in thyroidectomized rats (Plotnikoff *et al.*, 1974a, b). TRH also antagonized the sedation and temperature loss produced by pentobarbital and ethanol, even in hypophysectomized animals (Breese *et al.*, 1974a, b; Prange *et al.*, 1974) and the actions of other sedative agents (Breese *et al.*, 1975). Biochemical evidence also supports a direct central action of TRH (Horst and Spirit, 1974; Keller *et al.*, 1974), and intravenously administered TRH appears to improve depression in humans (see refs in Prange *et al.*, 1975). Good evidence also exists to support the notion that the behavioral effects of somatotropin release inhibiting factor (SRIF) and luteinizing hormone-releasing hormone (LHRH), which promote, respectively, an inhibition of the release of growth hormone (GH) and the release of luteinizing hormone (LH) from the anterior pituitary, are at least in part centrally mediated (e.g. Siler *et al.*, 1973; Moss and McCann, 1973; Pfaff, 1973; Segal and Mandell, 1974). Recent reviews about the chemistry, physiology and pharmacology of hypothalamic hormones and releasing factors have been published (see, e.g. Motta *et al.*, 1975). Other hypothalamic peptides, such as "Substance P" (see Takahashi *et al.*, 1974) and peptido-amines (see Reichelt *et al.*, 1976) may be involved in behavioral regulation (see below).

4. *Other Peptides*

Other neural peptides may be involved in learning and memory processes (e.g. Flexner *et al.*, 1967; Ungar, 1968, 1975a, b). Peptides, which could be associated with the release of glutamate and aspartate from nerve endings (i.e. γ-glutamyl peptides (see, e.g. Waley, 1957, 1958; Cliffe and Waley, 1958; Sano *et al.*, 1966; Reichelt and Kvamme, 1967; Reichelt, 1970) and N-acetyl-aspartyl peptides (see, e.g. Kvamme *et al.*, 1975) have been described. γ-Glutamyl-peptides appear to be formed by reactions between the γ-glutamyl moiety of glutathione and amino acids, and this mechanism may be involved in cerebral amino acid transport (see Meister, 1973). These acidic peptides could affect membranes and thereby play regulatory roles in CNS function. In support of this contention, the activity of the membrane-bound enzyme, γ-glutamyl *trans*-peptidase, which catalyzes this transfer of the γ-glutamyl moiety of glutathione to amino acids, has been found to be increased in the sera of humans suffering from certain neurologic diseases (Ewen and Griffiths, 1973; Alevizos *et al.*, 1976).

A monoamine-dependent synthesis of N-acetyl-aspartyl-peptides and peptido-amines has been shown to occur in homogenates and extracts of monkey and mouse cerebral cortex and porcine hypothalamus (Kvamme *et al.*, 1975; Reichelt *et al.*, 1976; Fig. 52). Interestingly, this *in vitro* synthesis of peptido-amine complexes appeared to be non-ribosomal since it was stimulated by addition of cyclic-AMP to the medium (Reichelt *et al.*, 1976); cyclic-AMP and monoamines have been shown to inhibit protein synthesis in brain tissues and in polysomes (B. F. Weiss *et al.*, 1974). The findings that histamine and some of its analogs cause an accumulation of cyclic-AMP by brain slices of guinea pig (Rogers *et al.*, 1975; Dismukes *et al.*, 1976) strengthen the contention that cyclic-AMP may be involved in the synthesis of histamine-peptide complexes (or other peptido-amines). An association of this system with nervous impulses is indicated by recent findings that high K^+-induced depolarization of brain tissue produced an increase in its cyclic-AMP content (Ferrendelli *et al.*, 1976). A link with McIlwain's (1974) work, which indicated a role for adenosine in neurotransmission, derives from the finding that histamine-induced accumulation of cyclic-AMP in brain slices is greatly increased by adenosine (Ferrendelli *et al.*. 1976). It should also be recalled that an L-glutamyl-γ-phosphate residue has been isolated in higher amounts from peptides derived from the phosphorylated form of brain Na^+,

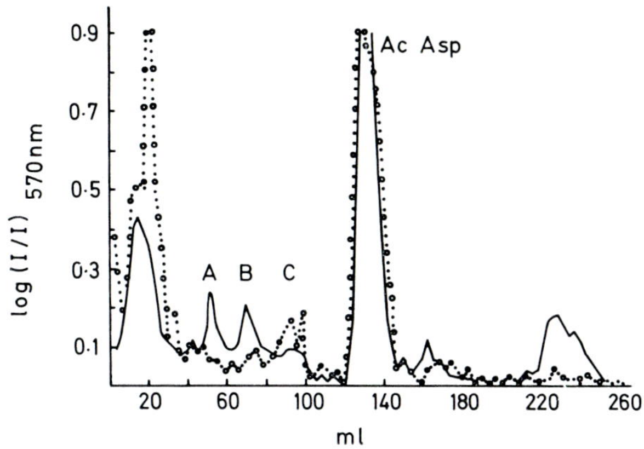

FIG. 52. The effect of norepinephrine (*solid line*) and dopamine (*dotted line*) on the levels of N-substituted amino acid derivatives in monkey brain cortex homogenate. The homogenate was incubated for 20 min at 37°C. An ATP-regenerating system, a monoamine oxidase inhibitor (nialamide), amino acids (acetyl-aspartate, glutamate, serine, glycine, α-alanine, GABA; 0·5 mM of each), and histamine (0·5 mM) or norepinephrine (0·5 mM) were added to the incubation mixture. The N-substituted compounds were eluted with H_2O from a Dowex-50 column, followed by fractionation on a Dowex-1 column (linear formic acid gradient, 0–2 M). The ninhydrin color following hydrolysis was measured. (Reproduced with permission from Reichelt and Edminson, 1974 and from Kvamme *et al.*, 1975.)

K^+-ATPase than from peptides of the non-phosphorylated enzyme (Kahlenberg *et al.*, 1967). The suggested involvement of γ-glutamyl peptides with amino acid transport (see above) may be related to this acyl phosphate "intermediate" of Na^+, K^+-ATPase.

The syntheses of these peptides, many of which had not been previously identified and some of which contained amino acid transmitter candidates (e.g. GABA, taurine, glutamate, glycine), appeared to depend upon the monoamine transmitter candidate (e.g. histamine, norepinephrine, dopamine) present in the incubation medium. These peptides and their monoamine complexes, some of which have been shown to be formed *in vivo*, could be involved in mechanisms of cell-to-cell recognition related to information processing and storage (see Kvamme *et al.*, 1975; Fig. 52; steps 3 and 4 of Fig. 80), and also provide links between transmitter release and protein synthesis as well as between postulated monoamine and amino acid transmitter systems. Like the undecapeptide, "Substance P" (Von Euler and Gaddum, 1931; Chang *et al.*, 1971), these peptides may stimulate cerebral adenyl cyclase activity (Williams *et al.*, 1969; Duffy *et al.*, 1975), and could have a slow, predominantly excitatory activity (as has been demonstrated for "Substance P" using the iontophoretic method) which could facilitate the activity of some central neurones (e.g. Otsuka *et al.*, 1972; Takahashi *et al.*, 1974). Also, histamine, 5-HT and dopamine have all been shown to exert stimulatory effects on cyclic-AMP generating systems in brain slices (see, e.g. Daly, 1973 and refs therein). With further regard to protein synthesis, it is also of interest that ACh, NE, GABA and epinephrine all appear to stimulate, whereas 5-HT appears to inhibit ^{14}C-leucine incorporation into protein in the presence of an energy-generating system *in vitro* (see Baxter, 1976). Other recent experiments have indicated that GABA may play a role in the regulation of cerebral protein synthesis in pre- and/or post-synaptic membranes (see Tapia and Sandoval, 1974; Sandoval and Tapia, 1975). Also, monoamines have been recently implicated as possible mediators of fast axoplasmic flow (Boegman and Wood, 1976).

5. *Conclusions and Comments*

The foregoing discussion leaves little doubt that CNS peptides play significant roles in learning and memory and that these substances are elements of the *Behavioral Code*. However, the roles of peptides are varied; i.e. some appear to act directly on the CNS, whereas others may act by triggering central responses (e.g. by means of increased stress) through their actions on peripheral target organs (e.g. adrenal gland). The "transfer of learning" experiments (e.g. Ungar and Ocequera-Navarro, 1965; Ungar and Irwin, 1967; Ungar *et al.*, 1968; Radcliffe and Shelton, 1974 and refs therein; Ungar, 1975a and refs therein) which involved injections of brain extracts from trained animals into naïve animals, are most likely explained by the presence of various peptides, which by their peripheral and/or central actions, could have produced increases in stress which enhanced learning.

F. Roles of Axoplasmic Flow and Dendritic Transport

1. *Rapid and Slow Processes*

Rapid and slow processes of axoplasmic (axonic) flow exist (see, e.g. Weiss and Hiscoe, 1948; Lasek, 1967, 1968; Grafstein, 1967; McEwen and Grafstein, 1968; Bray and Austin, 1968; Ochs, 1972a, b: Dahlström, 1972; Droz and DiGiamberardino, 1974; Guth, 1974; Cook and Whitlock, 1975; Lubinska, 1975; Cowan and Cuénod, 1975). Rapid (proximo-distal, or cellulifugal) axoplasmic flow occurs at a rate of about 30–400 mm/day, is bi-directional, and appears to account for transport of the major portion of proteins and conjugated proteins (e.g. transmitter-synthesizing enzymes; membrane constituents) from their sites of synthesis in neuronal perikarya to axolemmal membranes and nerve endings (see, e.g. Lubinska, 1964, 1975; Barondes, 1969a; Droz and Barondes, 1969a; Karlsson and Sjöstrand, 1971a, b; Edstrom and Mattsson, 1972; Ochs, 1972b; Droz *et al.*, 1973, 1974; Figs 53, 54, 55) mainly in association with "particulate" fractions (see, e.g. Droz and DiGiamberardino, 1974; Leestma, 1976). For example, convincing evidence supports the contention that all of the enzymes involved in NE synthesis are transported at different rates from the perikaryon to more distal portions of the adrenergic neurone (see Thoenen, 1975, and refs therein).

Ultrastructural autoradiographic studies have shown that the smooth endoplasmic reticulum plays a major role in the rapid transport of proteins, after their synthesis on polysomes and their complexification in the Golgi apparatus (e.g. Droz and DiGiamberardino, 1974). The strict axonic localization of rapidly transported ^{3}H-proline-labelled substances in the sciatic nerve of the toad (*Bufo marinus*) has recently been demonstrated using light microscopic autoradiography (Cook and Whitlock, 1975; Fig. 56).

Rapidly transported proteins and glycoproteins, presumed to participate mainly in the renewal of axolemmal and presynaptic structures and thus to preserve structures essential for impulse transmission and conduction, exhibit rapid turnover (see, e.g. Guth, 1974; Lubinska, 1975). Six classes of glycoprotein appear to be rapidly transported (Barker *et al.*, 1975), but this may differ among various species. Gangliosides and mucopolysaccharides (including heparin-SO_4 and chondroitin-SO_4) also appear to undergo rapid axoplasmic flow (e.g. Holm, 1972; Forman and Ledeen, 1972; Rösner *et al.*, 1973; Elam and Peterson, 1976). Recently, it was shown that a rapid distribution of radioactivity occurred along the chicken optic pathway after intraocular injection of N-^{3}H-acetyl-mannosamine (a specific precursor of NANA), and one of the ganglioside fractions (GD_{1a}) acquired much more radioactivity than other gangliosides of the optic lobe (Rösner, 1975).

Slowly transported proteins, recovered mainly in "soluble" fractions of neural tissues (Grafstein, 1969; Barondes, 1969a), appear to be synthesized on free polysomes and conveyed to the axon either as constituents of the axoplasmic matrix or as constituents of large polymeric structures such as microfilaments or microtubules (e.g. Droz

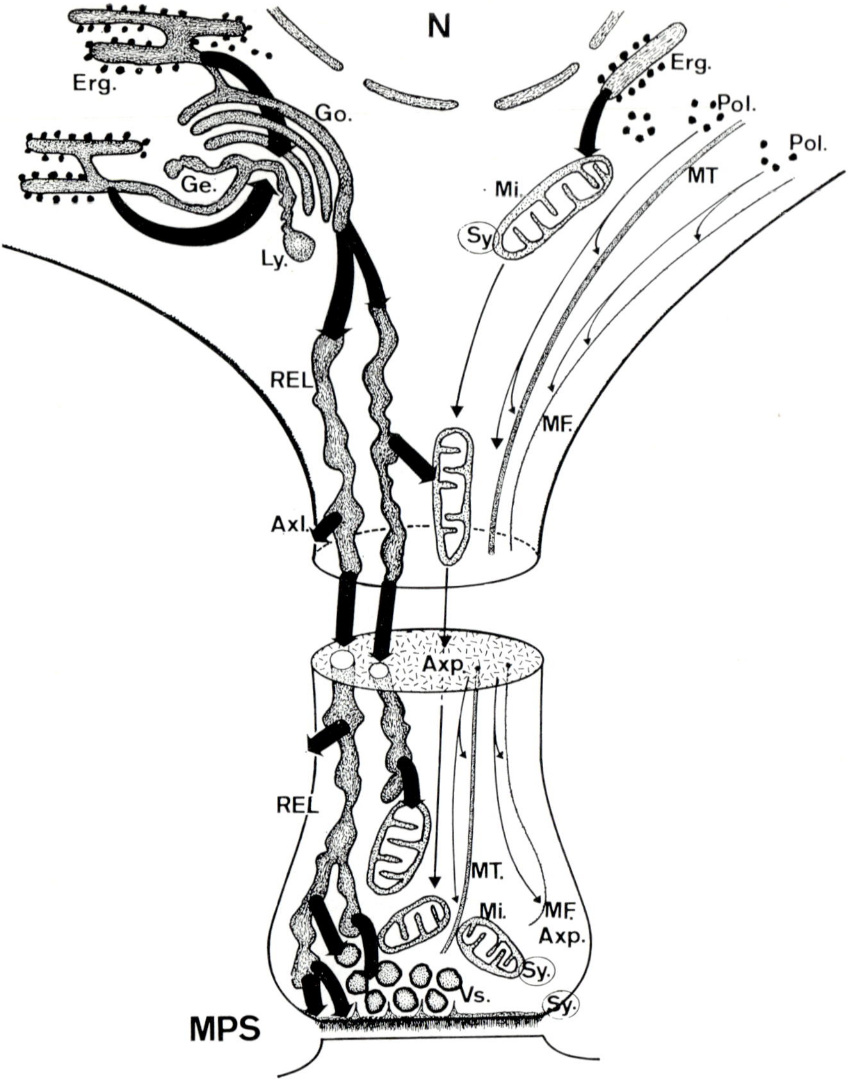

FIG. 53. Diagrammatic representation of the contribution of the axoplasmic migration of macromolecules to the maintenance of nerve cell processes and synapses. On the left side, *thick* arrows indicate the role of rapid transport. Polypeptide chains synthesized in the ergastoplasm of the Nissl substance (Erg.) are transferred to the Golgi apparatus (Go.) and give rise to protein and glycoprotein sequestered in the smooth endoplasmic reticulum (REL). Passing into the axon with the axonal endoplasmic reticulum (REL), they are transported at a high rate and yield new membrane components to the axolemma (Axl.) and mitochondria (Mi.). They accumulate in the terminal part of the axon and ensure the renewal of constitutents of synaptic vesicles (Vs.) and presynaptic plasma membranes (MPS). Thus rapid transport purveys with new membrane components the logistical support of the conduction and

and DiGiamberardino, 1974). Brain-specific 14–3–2 protein is an example of a protein that is synthesized mainly on free polysomes (Zomzely-Neurath *et al.*, 1973) and which appears to be transported by a relatively slow process (Marangos *et al.*, 1975). Most axoplasmic proteins involved in the maintenance of cellular structural integrity,

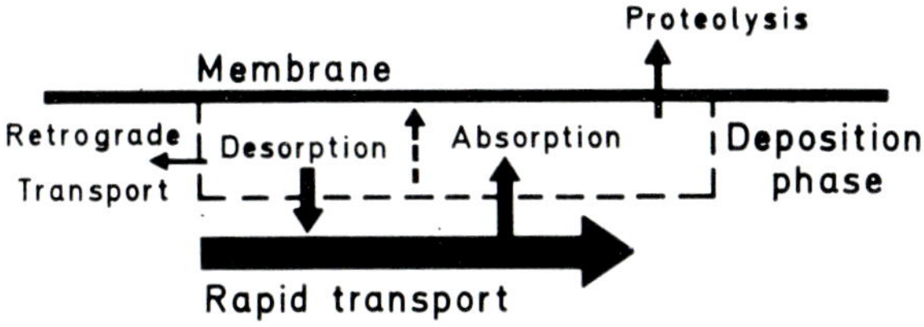

Fig. 54. Mass fluxes required to explain profile variations if they reflect primarily intra-axonal redistributions and if only one rapid transport system exists in these fibers. A deposition phase (stationary and slowly moving axonal constituents) sorbs material from the rapid transport system. The material may then become part of the membrane and organelles, undergo retrograde transport, or re-enter the orthograde transport system. Labelled macromolecules may also be subjected to some proteolysis and are lost from the axon. (Reproduced with permission from Gross and Beidler, 1975.)

including those that make up whole organelles appear to be transferred by slow axoplasmic processes that proceed only proximodistally at rates of about 1·5–12 mm/day, and turn over relatively slowly (e.g. Weiss and Hiscoe, 1948; Weiss and Pillai, 1965; Pomerat *et al.*, 1967; Barondes, 1966; Dahlström and Haggendal, 1966; Jeffrey *et al.*, 1972; Droz, 1973, 1974; Droz *et al.*, 1973; Giorgi *et al.*, 1973; Khan and Ochs, 1975; Elam, 1975).

Although axonal mitochondria in tissue culture appeared to move more slowly than smaller organelles, (Pomerat *et al.*, 1967) they still moved in the range of fast flow (see also Banks *et al.* (1969) who studied the translocation of mitochondrial enzymes in constricted nerves). Lubinska (1975) has suggested that the available evidence indicates fast flow for mitochondria, and that the failure of Jeffrey *et al.* (1972)

mainly of the transmission of nerve impulses. On the right side, *thin* arrows indicate the elements transported with the slow axoplasmic flow. Polypeptide chains, released from free polysomes (Pol.) into the neuronal cytoplasm, migrate slowly into the axoplasm (Axp.), where they may assemble as protein sub-units giving rise to microfilaments (MF.) and microtubules (MT). Mitochondria receive by transfer the great majority of their own proteins; these organelles are displaced along the axon and continue to exhibit a slight synthesis of hydrophobic polypeptides (Sy.). (From Droz, 1974; reproduced with permission from the author and from "Actualités Neurophysiologiques", Vol. X (A. Monnier, ed.), Masson, Paris.)

to observe fast flow when using ^{59}Fe as a mitochondrial marker was related to the slow incorporation of label into mitochondria and the time chosen for analysis, since at such time axonal mitochondria had not yet become heavily labelled. Mitochondrial axonal transport appears to be bi-directional (e.g. see Dahlström, 1972), and since this

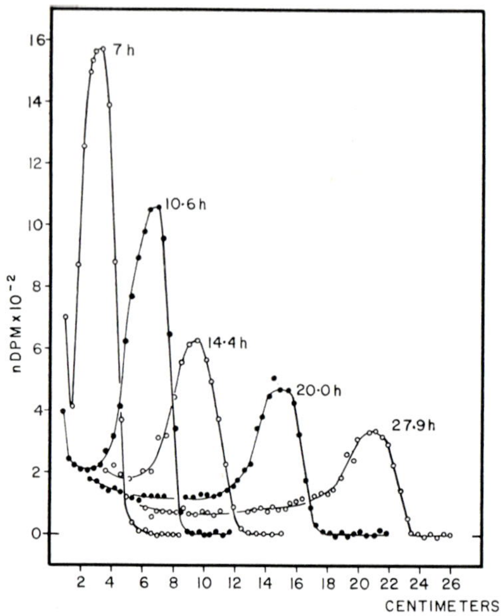

FIG. 55. Normalized isotope concentration distributions obtained at various times after application of labelled leucine to the lamelli of the olfactory mucosa of a garfish. The data are plotted as functions of distance from the olfactory mucosa. Nerves were cut behind the olfactory mucosa 6 h after isotope application. The macromolecule-bound isotope distributions display characteristic changes with distance from the olfactory mucosa. Peak heights decrease exponentially, peaks broaden, and wave-fronts become more shallow. Plateau regions are established behind the peaks reflecting a deposition of material along the axons. Upon arrival at the synaptic region, the peak contains only about 20% of the total activity in the nerve. (Reproduced with permission from Gross and Beidler, 1975.)

process was not prevented by colchicine in rat sciatic nerve (Kreutzberg, 1969) or in cat hypogastric nerves (Banks *et al.*, 1971), it does not appear to depend upon axonal microtubules.

A proximo-distal axoplasmic transport of RNA has been reported (e.g. Miani *et al.*, 1966; Bray and Austin, 1968; Sjöstrand, 1970; Bondy, 1971). After intraocular injection of ^{3}H-uridine, RNA transport in the optic system of the chick occurred at the rather slow rate of about 12 mm/day (Bondy and Madsen, 1973). Such RNA transport

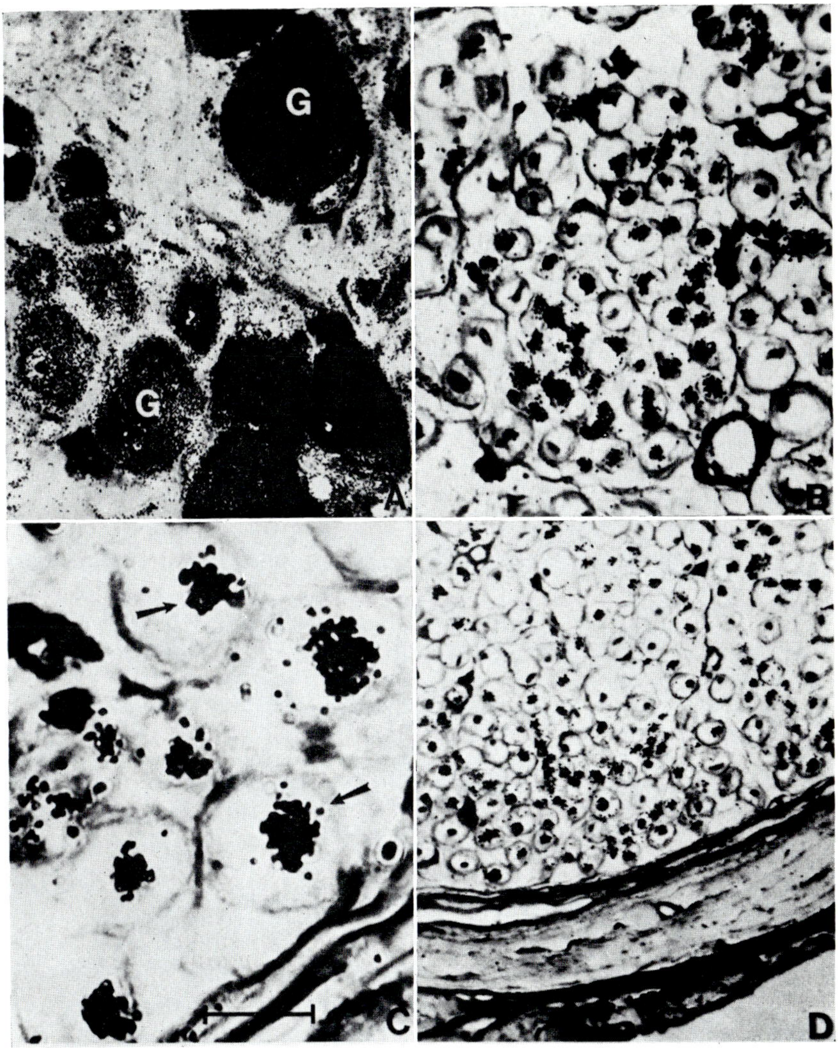

FIG. 56. Axoplasmic transport in the sciatic nerve of the toad, *Bufo marinus*, after hydraulic injection of ^{3}H-proline into a surgically exposed dorsal root ganglion. Light microscopic autoradiographs taken from the dorsal root ganglion and sciatic nerve 10 h after injection. A, Densely labelled ganglion (G) cell bodies. B, transverse section of sciatic nerve showing silver grains overlying several axons in myelinated fibers. C, Higher magnification of nerve showing silver grains (arrows) overlying axons and not the myelin sheaths. Scale bar = 10 μm. D, Transverse section of sciatic nerve showing silver grains overlying axons. No silver grains are seen overlying the endo-, peri-, or epi-neural tissue. (Reproduced with permission from Cook and Whitlock, 1975.)

in the chick optic pathway occurred previous to the onset of electrical activity (Bondy and Marchisio, 1973) and has been correlated with an axoplasmic migration of ribosomes (Bondy and Purdy, 1975). Barondes (1974) has argued that a major pitfall in evaluating such studies involves the separation of RNA transport in axons from the transport of precursors that become incorporated into RNA in periaxonal (glial) cells whose perikarya are abundant in nerves (see also Peterson *et al.*, 1968; Autilio-Gambetti *et al.*, 1973). This pitfall was partially overcome in the recent study of Ingoglia *et al.* (1975) who examined the distribution of labelled RNA and RNA precursors in goldfish optic tecta after intraocular injection of ^{3}H-uridine during various stages of optic nerve regeneration. Autoradiographic results revealed that some of the labelled RNA found in the tectum had originated from RNA transported along optic axons rather than from local synthesis. Thus, large amounts of RNA might be carried with regenerating optic axons as they enter the tectum during the reconnection phase of nerve regeneration. It is also noteworthy that when intracranial injections of ^{3}H-uridine were employed, silver grains were distributed specifically over perikarya during the same stage of regeneration. However, it should be remembered that RNA could have been transported within axonal mitochondria (see Wolburg, 1972).

2. *Retrograde Axoplasmic Flow*

Retrograde (disto-proximal, or cellulipetal) axoplasmic flow of macromolecules has been demonstrated by using protein markers (e.g. horseradish peroxidase; HRP). This process could be involved in the communication between the axon (and/or nerve terminals) and the cell nucleus (see Kristensson and Olsson, 1971, 1974; LaVail and LaVail, 1972, 1974, 1975; LaVail *et al.*, 1973; Kuypers and Maisky, 1975). Figures 57 and 58 provide electron micrographs of cross sections of the optic nerves and the retinae of a chick at 10 h after a relatively restricted tectal injection of HRP (LaVail and LaVail, 1974). It is evident that retrograde axoplasmic flow of HRP occurred only contralateral to the injection, as expected, since in this avian system the optic nerve is totally decussated at the optic chiasma. With such a mechanism the cell nucleus and polysomes could be signalled to synthesize more transmitter-synthesizing enzymes after depletion of presynaptic transmitter stores caused by nervous activity (e.g. Agranoff, 1974), or membranes could be re-utilized (Holtzman *et al.*, 1973). It is also possible that chemical factors from effector cells (e.g. muscle cells, in the case of motor nerves) could be taken up by nerve endings and transported to perikaryal sites at which they could regulate the formation of new synaptic contacts. This latter concept is supported by the recent findings of Cull (1975a) who showed that application of drugs that inhibit retrograde axoplasmic flow (e.g. colchicine, vinblastine) to rat hypoglossal nerves caused hypoglossal neurones to lose boutons (Fig. 59). Although this drug-induced loss of boutons was similar to that caused by axotomy (Cull, 1974, 1975b), neither electrical nerve-muscle contacts or axon numbers were significantly altered (Cull, 1975a). As a control for these findings, it was shown that cuffs containing lignocaine exerted no significant effect on axonal flow or boutons/neurone perimeter (Fig. 59). The existence of intracellular communication from axons (and/or terminals) to the cell nucleus had also been implied from earlier studies on regulatory reactions, such as chromatolysis after axonal injury (Cragg, 1970), retrograde *trans-synaptic* changes (Cowan, 1970), acute glial reactions (Sjöstrand, 1965) and growth regulation according to the size of the peripheral field (Prestige, 1970). Retrograde axoplasmic transport now appears to be the mechanism of this communication.

A good example of a possible role for retrograde flow involves the transport of nerve growth factor (NGF), a substance which appears to be necessary for the normal development and maintenance of function in sensory and sympathetic ganglia and which is taken up rather selectively by adrenergic nerve terminals and transported to their perikarya (Levi-Montalcini and Angeletti, 1968; Hendry and Thoenen, 1974; Hendry *et al.*, 1974). NGF produces a selective induction of tyrosine hydroxylase (TH) and dopamine-β-hydroxylase (key enzyme of the NE synthetic pathway) in the peripheral adrenergic nervous system (Thoenen *et al.*, 1971, 1972b), and the increase in TH activity occurs earlier than the general hypertrophic reaction of NGF (Stöckel *et al.*, 1974). These findings, taken together with the disclosure that NGF is synthesized in adrenergically innervated effector organs, indicate that NGF might transfer trophic information from effector organs to cell bodies (see Thoenen and Stöckel, 1975; Varon, 1975, and refs therein). At the ultrastructural level, subcutaneous injections of NGF (10 μg NGF/g body weight) produce effects on ribosomes (Fig. 60) and on the Golgi apparatus (Fig. 61) of adrenergic neurones of the superior cervical ganglion of young rats as early as 6 h after injection (Schwab and Thoenen, 1975).

A comprehensive study of the growth characteristics of adrenergic nerves in the adult rat has appeared (Olson and Malmfors, 1970). The reader should also recall the earlier studies of Parker (1932) and of Luco and Eyzaguirre (1955) which revealed that the material basis for the trophic influence of effector organs (cells) on neurones appeared to depend upon axonal continuity; after nerve section, trophic support continued only for short periods that were proportional to the length of the degenerating fibers.

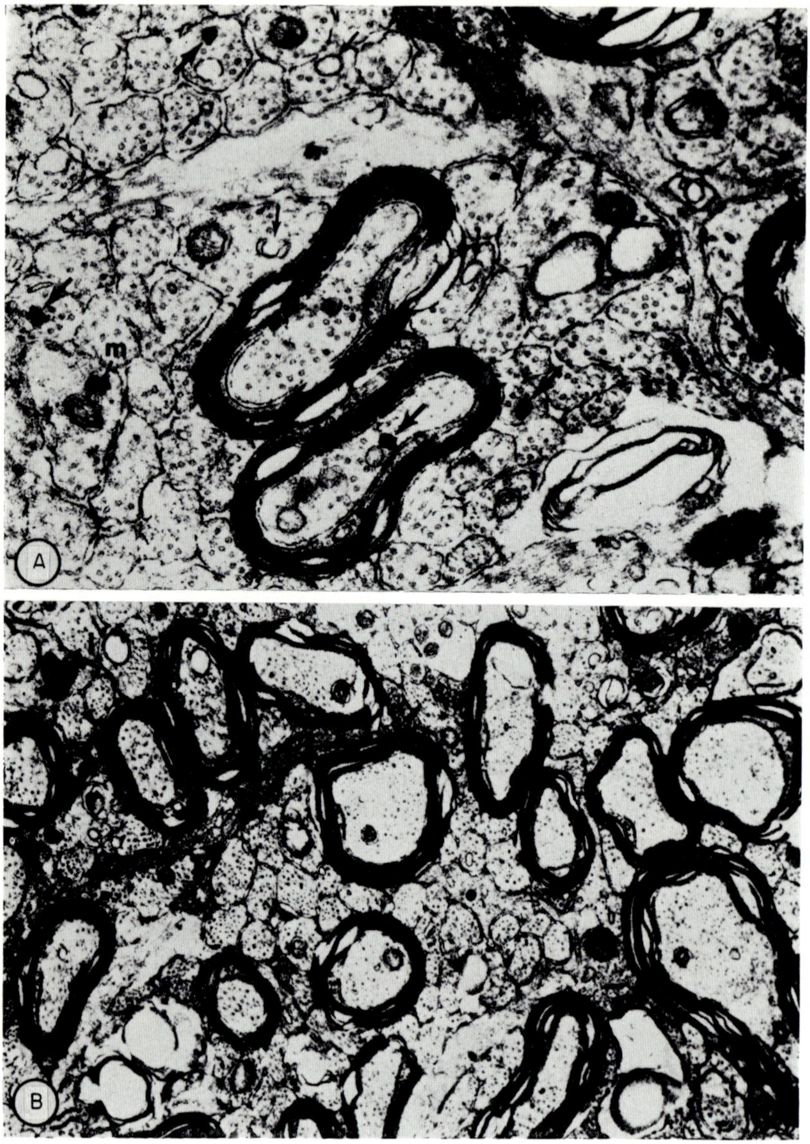

FIG. 57. Retrograde axoplasmic flow. Electron micrographs of cross-sections of the optic nerves 10 h after a relatively restricted tactal injection of horseradish peroxidase (HRP). A, Contralateral to the injection. Large, HRP-containing vesicles (*large arrows*) are present in both myelinated and unmyelinated axons. Vesicles not containing HRP (*small arrows*) are also seen in the axons. No conspicuous extracellular HRP is evident. *m*, multivesicular body. × 41 600. B, Ipsilateral to the injection. No HRP is present extracellularly or in the axons. *l*, dark lysosome in a glial cell process. ×22 200. (Reproduced with permission from LaVail and LaVail, 1974.)

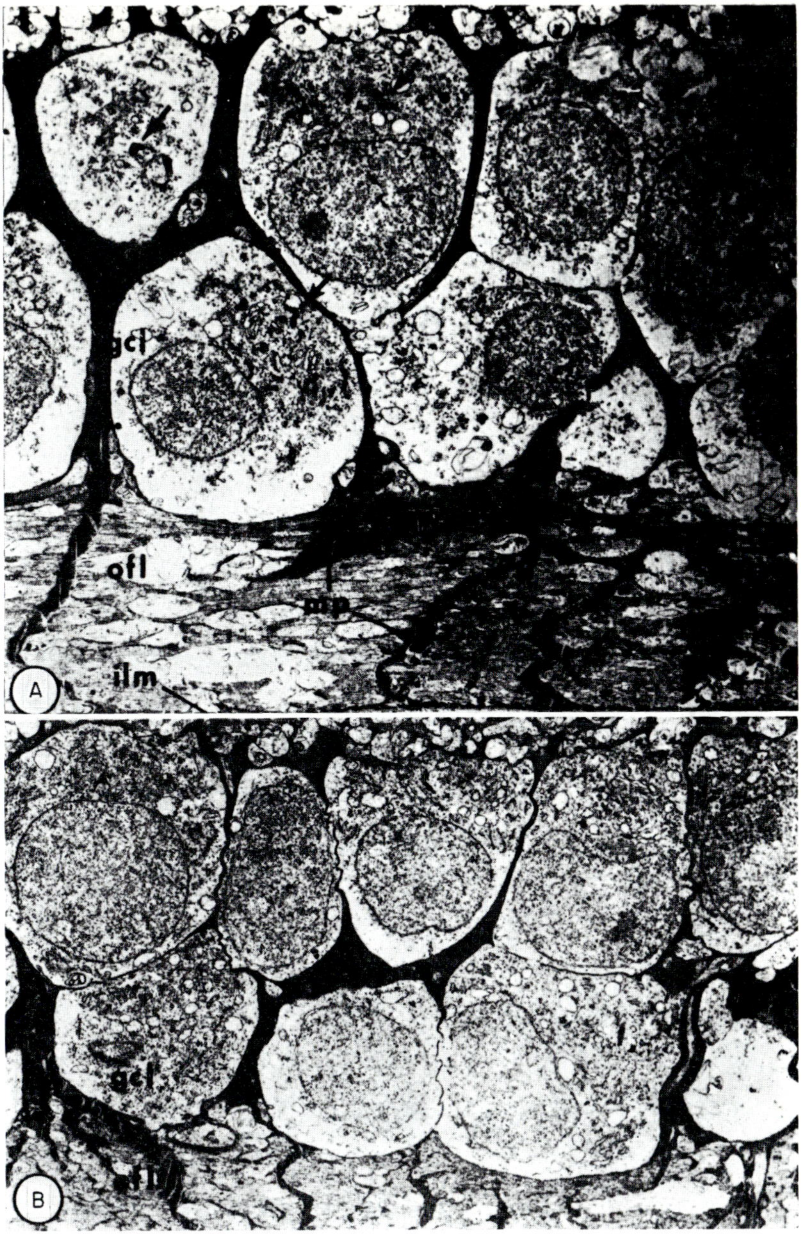

FIG. 58. Retrograde axoplasmic flow. Electron micrographs of parts of the two retinae 10 h after a relatively restricted tectal injection of horseradish peroxidase (HRP). *gcl*, Ganglion cell layer; *ofl*, optic fiber layer; *ilm*, inner limiting membrane; *mp* Müller cell process.
A, Contralateral to the injection. HRP-positive vesicles are present in some of the axons of the optic fiber layer (*small arrows*). The majority of the larger, HRP-containing vesicles (*large arrows*) evident in most of the ganglion cell somas is located in the supranuclear (outer) region. ×3700. B, Ipsilateral to the injection. No HRP is present. A few of the lysosomes are darker than those seen in tissues not reacted cytochemically. ×4200. (Reproduced with permission from LaVail and LaVail, 1974.)

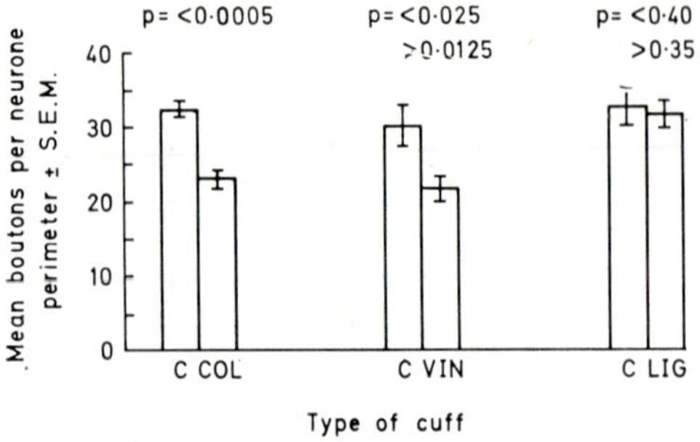

FIG. 59. Mean boutons/neurone perimeter ± S.E.M. Hypoglossal nerves received: C, Control, plain silastic cuff; COL, 0·1% colchicine cuff; VIN, 0·1% vinblastine cuff; LIG, 20% lignocaine cuff. The significant reductions in bouton number caused by colchicine and vinblastine are indicated above the columns. Note that cuffs containing lignocaine did not affect bouton number or axonal flow, whereas those containing colchicine and vinblastine affected both axonal flow and bouton number. (Reproduced with permission from Cull, 1975a.)

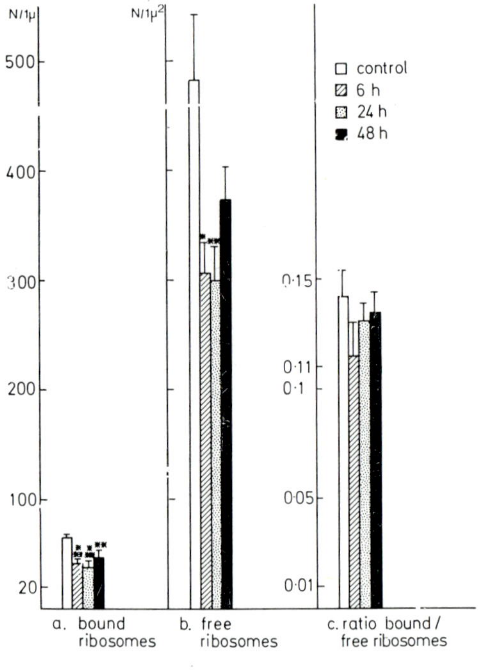

FIG. 60. Numerical density of bound and free ribosomes within a Nissl body after 6 h to 48 h of nerve growth factor (NGF) treatment. (a) Numerical density of bound ribosomes within a defined segment of RER cisterna (N/1 μ); (b) numerical density of free ribosomes within a defined field of cytoplasm in the Nissl body (N/1 μ^2); (c) ratio of numerical densities of bound to free ribosomes. (Reproduced with permission from Schwab and Thoenen, 1975.)

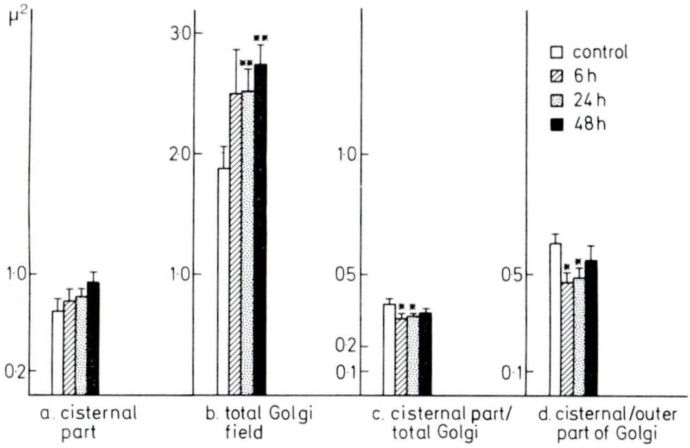

FIG. 61. Golgi field after 6 h to 48 h of nerve growth factor (NGF) treatment. (a) Mean cross sectional area of inner cisternal part of the Golgi apparatus; (b) mean cross sectional area of the total Golgi field; (c) ratio of the cisternal part (a) to the total Golgi field (b); (d) ratio of the cisternal part (a) to the surrounding outer part of the Golgi apparatus (b minus a). (Reproduced with permission from Schwab and Thoenen, 1975.)

3. *Dendritic Transport; Transneuronal Transfer*

Newly formed macromolecules are also transported to dendritic arborizations (e.g. Droz and LeBlond, 1963; Utakojo and Hsu, 1965; Kreutzberg *et al.*, 1973; Kiss, 1975). This process is significant since it supplies postsynaptic sites of neurones; i.e. regions involved in receiving information. Transport of proteins and other substances to dendrites has been shown by iontophoretic injection of substances directly into neurones (e.g. Globus *et al.*, 1968; Lux *et al.*, 1970; Schubert *et al.*, 1971; Kreutzberg *et al.*, 1975a, b). Since colchicine, which destroys neurotubules, blocks dendritic transport (Schubert *et al.*, 1972), this process depends upon a mechanism which is similar to that of axoplasmic transport (see Schubert and Kreutzberg, 1975c). Other evidence obtained using HRP has indicated that dendritic transport, like axoplasmic transport, is a bi-directional process (see Lynch *et al.*, 1975).

AChE has been used as a marker for dendritic transport and for dendritic secretion or transneuronal transport (Kreutzberg *et al.*, 1975b; Fig. 62). Transfer across the dendritic membranes of mammalian neurones has also been demonstrated after intra-cellular injections of radioactive glycoprotein precursors, such as fucose or sulfate (Schubert, 1974; Fig. 63). Evidence for transneuronal transfer has also been obtained with labelled glycine (Globus *et al.*, 1968; Schubert, 1974; Kreutzberg *et al.*, 1975a; Figs 64 and 65), and a transneuronal

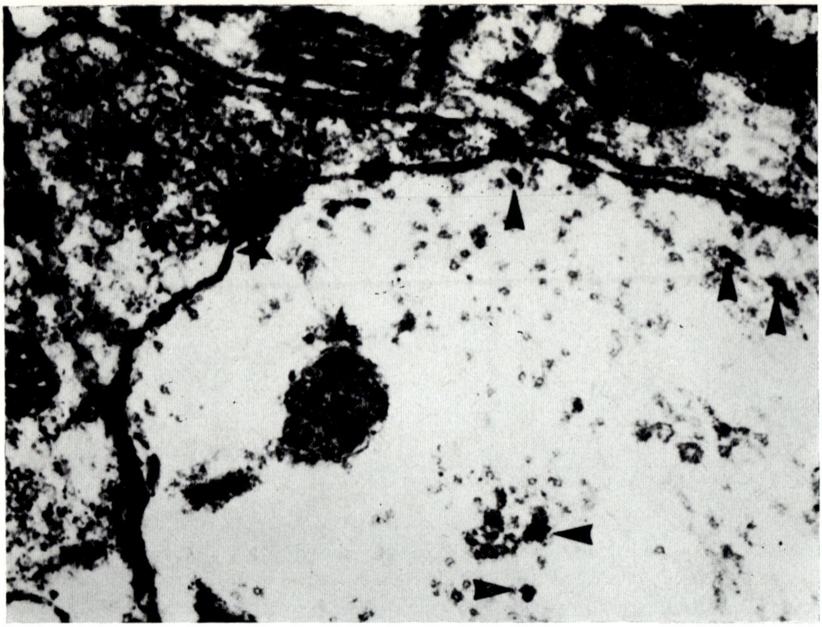

FIG. 62. Dendritic transport of AChE. The electron-dense reaction product demonstrating AChE activity is found in tubulo-cisternal structures inside the dendrite (*arrows*), in the dendritic membrane, and within the synaptic cleft (*star*). EM, ×49 000. (Reproduced with permission from Kreutzberg *et al.*, 1975b.)

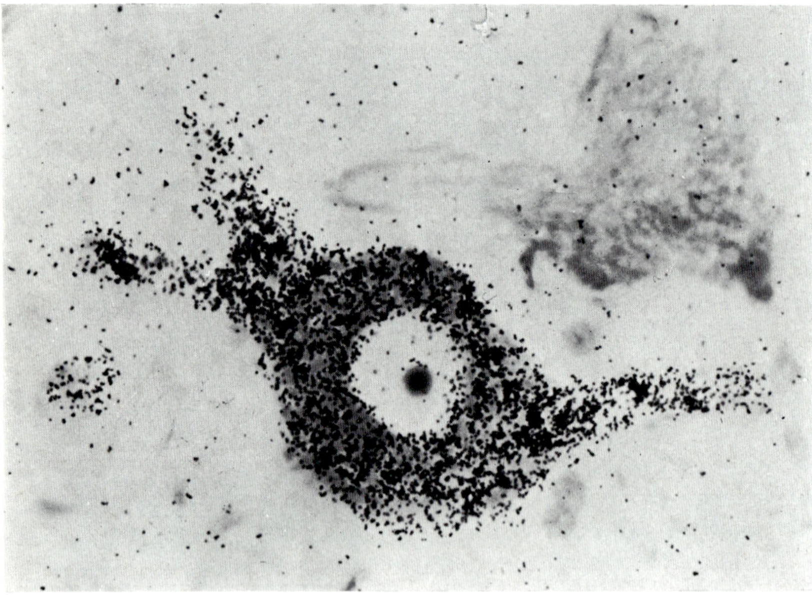

FIG. 63. Incorporation of ^{3}H-fucose in a cat spinal motoneurone, 4 h after intracellular electrophoresis. The patchy labelling pattern of the perikaryon probably indicates the accumulation of fucosyl compounds in the Golgi complexes where glycoproteins are formed. The proximal dendrites are labelled but there is no obvious transport of fucosyl-glycoproteins into the peripheral dendritic branches. Autoradiography, stained by toluidine blue. Magnification, ×640. (Reproduced with permission from Kreutzberg *et al.*, 1973.)

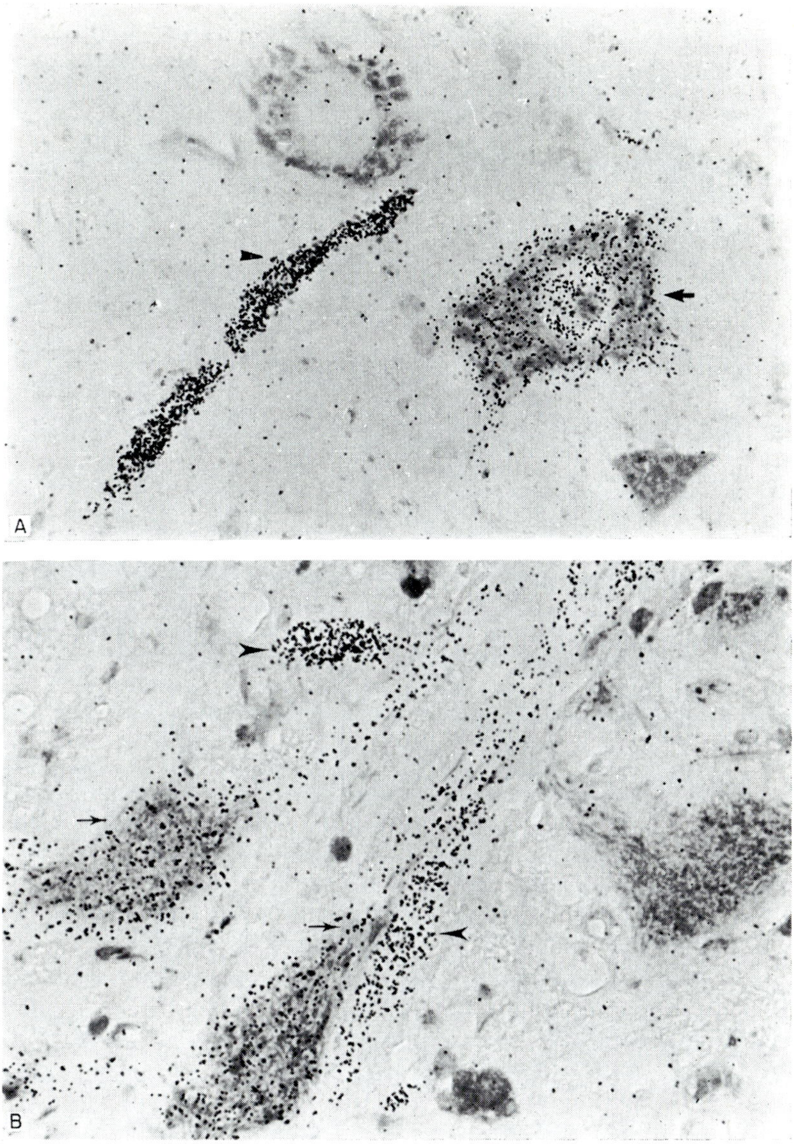

FIG. 64. Neuro-neuronal transfer 5 h after intracellular ^{3}H-glycine injection. A, the intensely labelled dendrite (*arrow head*) belongs to an injected neurone (IN). A non-injected neurone (NIN) also shows radioactivity in the nucleus and cytoplasm (*small arrow*). Transfer seems to be selective since two other neighboring neurones have no activity; ×450. B, two dendrites (*arrow heads*) from an injected neurone are seen passing close by the dendrites of two non-injected motoneurones with a slight amount of label (*small arrows*). A third neurone (right) and the glial cells are unlabelled; ×580. (Reproduced with permission from Kreutzberg and Schubert, 1975 (Part A) and from Schubert, 1975 (Part B).)

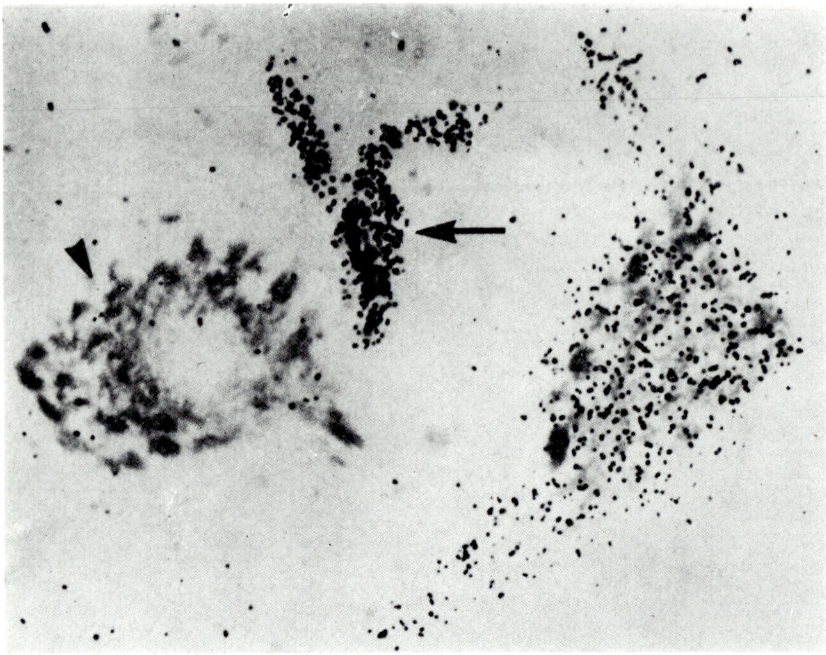

FIG. 65. Non-injected neurone revealing a distinct labelling. It is found close to a more heavily labelled dendrite (←) of the injected neurone which has received ^{3}H-glycine about 5 h before sacrifice. Only single silver grains are seen over the neighboring nerve cell (◂). Autoradiography, stained with toluidene blue. ×750. (Reproduced with permission from Kreutzberg *et al.*, 1975a.)

transfer from axon to postsynaptic neurones has been reported (Grafstein, 1971; Droz *et al.*, 1973; Ingoglia *et al.*, 1973).

Other studies have shown that electrical synapses of the leech CNS mediate transneuronal movements of radioactive amino acids and RNA precursors by a mechanism that does not appear to involve glial cells (Globus *et al.*, 1973a; Rieske *et al.*, 1975). Also, some exchange of precursor ^{3}H-lysine with electrically coupled axons and glial cells has been observed after its intracellular injection into the crayfish giant motoneurone (Droz *et al.*, 1974). Most recently, a transjunctional flux of intracellularly injected ^{3}H-glycine and ^{3}H-glucosamine across the electrotonic synapses of lateral giant axons of the crayfish (*Astacus fluviatilis*) has been shown (Hermann *et al.*, 1975). It was concluded that protein precursors move by diffusion and/or active retrograde axoplasmic transport from injection sites in the axon via gap junctions of electrical synapses to the neuronal soma where they are incorporated into macromolecules. This transneuronal transfer appeared to be

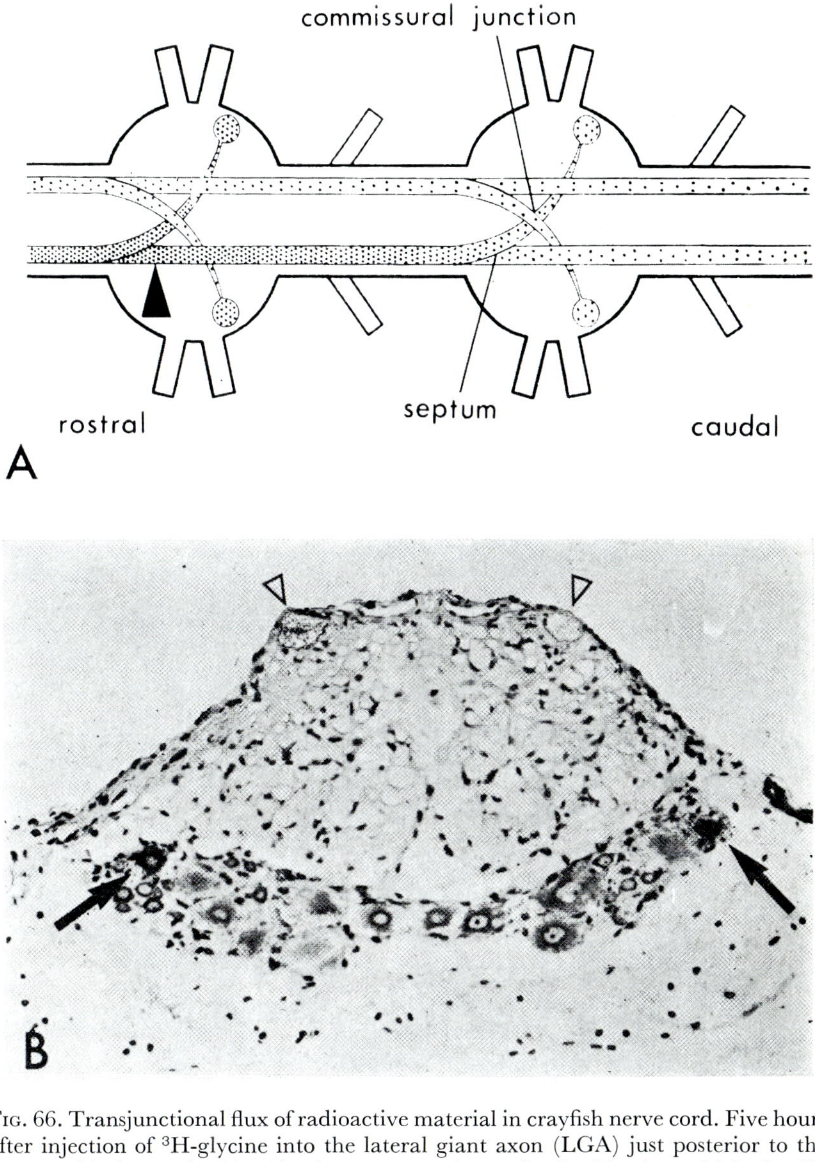

FIG. 66. Transjunctional flux of radioactive material in crayfish nerve cord. Five hours after injection of ^{3}H-glycine into the lateral giant axon (LGA) just posterior to the septum (*black arrowhead* in A), the nerve cord was fixed with glutaraldehyde. To facilitate LGA penetration by electrodes, the dorsal part of the connective tissue sheath was removed. A, schematic reconstruction from serial sections. Silver grains which represent newly synthesized proteins and their soluble precursors are found exclusively in the coupled LGAs at the ipsilateral and contralateral sites indicating a selective material transfer across the electrotonic synapses at the septum and the commissural junction. B, autoradiograph of a cross-section through the third ganglion. *White arrowheads* point to labelled LGAs, and *black arrows* indicate labelled LGA soma. ×95. (Reproduced with permission from Hermann *et al.*, 1975.)

selective to the extent that no other neurones or glia participated in it; i.e. in this preparation there is no alternative pathway through other cells (see Figs 66 and 67).

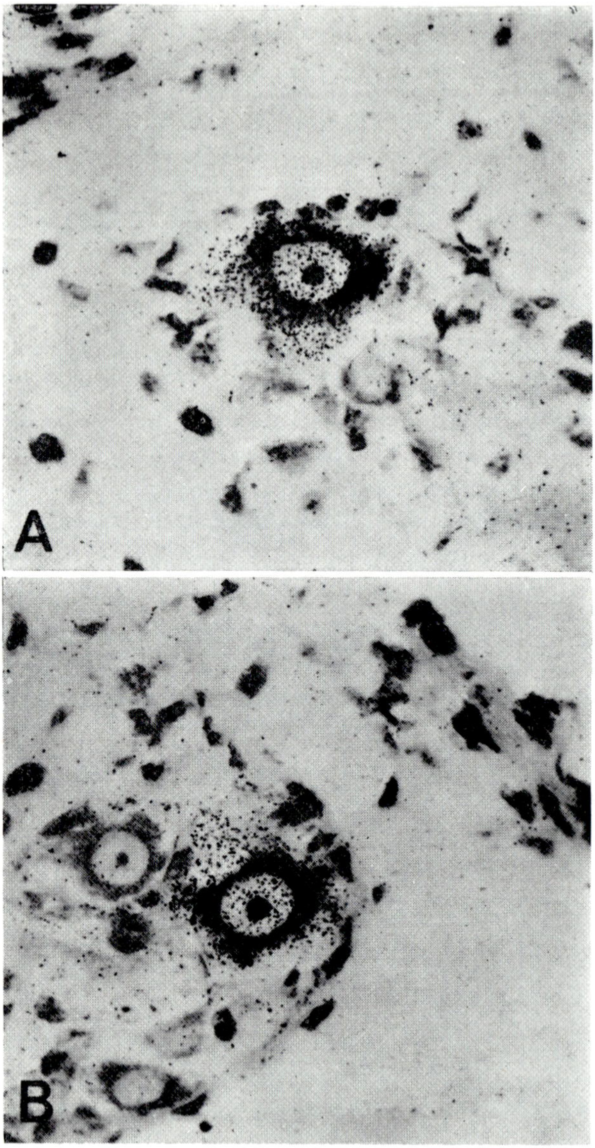

FIG. 67. Labelling of two lateral giant axon (LGA) soma in the third ganglion 2·5 h after injection of ^{3}H-glycine into the LGA belonging to the soma of A; glutaraldehyde fixation. To be available for uptake in the contralateral soma (B) (note the location of the soma near the second root), precursor must have passed through the electrical synapses at the septum and the commissural junction. ×200. (Reproduced with permission from Hermann *et al.*, 1975.)

Both dendritic and axonal transport of nucleoside derivatives occur in cat spinal motoneurones (Kreutzberg and Schubert, 1973; Schubert and Kreutzberg, 1974, 1975a, b). This was a significant finding since these substances are precursors of nucleic acids and of important cellular coenzymes, energy carriers and cyclic nucleotides. Autoradiographs showing the incorporation and dendritic transport of ^{3}H-adenosine and ^{3}H-orotic acid in motoneurones are provided in Figs 68 and 69. Since the release of adenosine and other adenine derivatives from cerebral tissues can be enhanced by electrical stimulation *in vitro* (Pull and McIlwain, 1972; McIlwain, 1974; Sun *et al.*, 1976), their release from dendrites and axons could enable neurones to regulate the diameters of their adjacent blood vessels according to changes in functional demand which, in turn, could cause localized increases in blood flow as have been observed in functionally active regions of the brain (e.g. Ingvar, 1973).

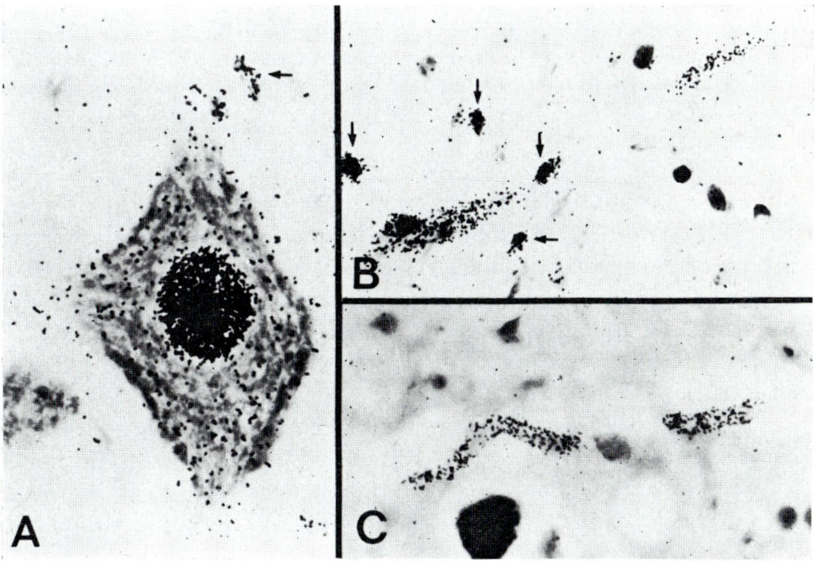

FIG. 68. Incorporation and transport of ^{3}H-adenosine derivatives, 4 h after intracellular injection. A, the ratio of somal to nuclear labelling is considerably greater than after intracellular injection of ^{3}H-orotic acid (see Fig. 69), indicating the presence of nucleotides in addition to RNA; $\times 600$. B, transport in dendrites. Some of the ^{3}H-adenosine derivatives have been obviously released from dendrites as indicated by the strong labelling of neighboring glial cells (*arrows*); $\times 400$. C, intra-axonal transport. Within 4 h, a large amount of nucleotide derivatives has reached the level of ventral root formation, 3 mm distant from the injected nerve cell soma. Autoradiography, counterstained with toluidine blue; $\times 300$. (Reproduced with permission from Schubert and Kreutzberg, 1975a, b.)

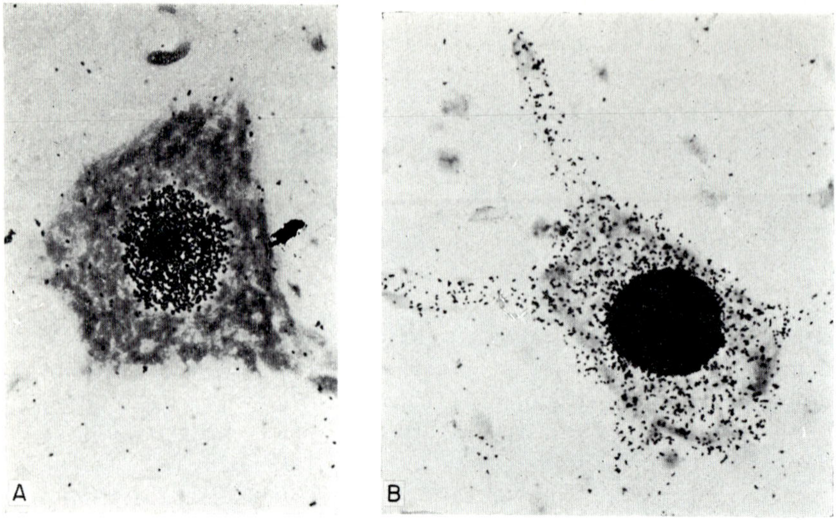

FIG. 69. Incorporation and transport of the RNA precursor ^{3}H-orotic acid. A, 15 min after orotic acid injection into the nerve cell, incorporation in the nucleus and nucleolus can be demonstrated. B, 4 h later radioactive RNA is also present in the perikaryon and the stem dendrites; $\times 700$. (Reproduced with permission from Kreutzberg and Schubert, 1973.)

Other recent studies have led to the notion that a local dendritic release of dopamine (DA) might alter the excitability of dopaminergic neurones of the substantia nigra (Groves *et al.*, 1975; Korf *et al.*, 1976; Geffen *et al.*, 1976). Such a contention gains additional impact when it is considered that atypical synapses in various regions of the CNS of higher vertebrates appear to possess vescicle-containing processes than can be defined as dendrites, and hence, that dendrites can be "presynaptic" (see, e.g. Ralston and Herman, 1969; Morest, 1971; reviews by Hamori *et al.*, 1975; Ramón-Moliner, 1975). The axonal release of active substances (e.g. glutamate) which had been previously shown (Wheeler *et al.*, 1966; DeFeudis, 1969, 1971a) could also participate in such processes.

Mechanisms of intercellular and transneuronal transport could play integral roles in the *Behavioral Code* since they could be involved in informational transfer and storage by affecting neuronal plasticity and the formation of new synaptic connections. The studies of DeRobertis (1964) which revealed the existence of filamentous connections between presynaptic and postsynaptic elements provide morphological support for this view. Such mechanisms of transfer, which could be stimulated by increased neuronal activity, could control the synthesis of synaptic macromolecules, and hence provide a good example of the association between environmental influences and gene expression at the subcellular level (see P. M. Milner, 1970). Nucleic acid mechanisms *need not be changed* under these conditions; the only requirement is that they be *activated.*

4. *Axoplasmic Flow, Dendritic Transport and Behavior*

Axoplasmic flow may play a significant role in memory processes and behavior since the syntheses of RNA and protein scarcely occur in nerve-endings, if they occur at all. Although synaptosomal RNA could account for some local synthesis of protein (Barondes, 1966, 1968a, b, 1969b; Balázs and Cocks, 1967; Austin and Morgan, 1967; Morgan, 1970; Appel, 1972), the limited number of molecular species formed (Ramirez *et al.*, 1972; Gilbert, 1972) could replace only $< 2\%$ of the rapidly renewed proteins (Droz, 1973). Therefore, changes in synaptic protein produced by environmental modification (see, e.g. Levitan *et al.*, 1972a, b; Hydén, 1973a; Conde and DeFeudis, 1977) probably reflect changes mainly in the *amount* of newly synthesized protein that is transported from cell bodies to nerve endings. Changes in the *rate* of axoplasmic transport could also be involved. This latter contention is based on a recent study which indicated that modification of the *internal* environment by drugs which modify cerebral monoamine metabolism can alter the rate of rapid axoplasmic transport of ^{3}H-leucine-labelled material in rat sciatic nerve (Boegman and Wood, 1976).

The findings that rapid axoplasmic flow requires mitochondrial synthesis of ATP locally all along axons (Banks *et al.*, 1969), that it is sensitive to metabolic inhibitors (e.g. Ochs, 1972b; Gross, 1973; Lubinska, 1975), and that its energy supply depends mainly on glycolysis, with increasing dependency occurring at later developmental stages (Arese *et al.*, 1975), indicate that this phenomenon may be related primarily to earlier stages of memory formation (see step 3 of Fig. 80). This conception gains some support from recent results concerning the temperature-dependency of memory formation in goldfish: (1) a 10°C decrease in temperature subsequent to training prolonged the period of susceptibility of memory to disruption by ECS and puromycin; (2) a 10°C increase immediately after training decreased the period of susceptibility of memory to consolidation; (3) cooling the fish for 24 h after training caused a pronounced decrease in protein synthesis, but did not selectively inhibit memory (Neale *et al.*, 1973a; Agranoff, 1974). On the basis of these studies, Agranoff (1974) has suggested that metabolic processes must be functioning during the period of decreased macromolecular synthesis in order for memory to be disrupted. Such a mechanism also adds support to the argument that "short-" and "long-term" memory processes might be best explained as being parts of a continuum (see also Chapter 7 and Chapter 8, Part E).

It should also be noted that rapid axoplasmic flow is not directly dependent upon protein synthesis since it has been shown to continue after separation of the axon from its perikaryon (see Lubinska, 1975) and since inhibitors of protein synthesis do not hinder this process (e.g. McEwen and Grafstein, 1968; Barondes, 1969a). However, cycloheximide treatments (10 mg/kg, s.c.) in rats produced, in 4–6 h, a depletion of a minor, but rapidly-transported compartment of dopamine-β-hydroxylase in sciatic nerve. Other recent studies by Ambron *et al.* (1975) have shown that brief exposure of the cholinergic giant neurone of the abdominal ganglion of the marine mollusc, *Aplysia californica*, to anisomycin (another protein synthesis inhibitor that blocks memory; see Part C4 above) almost completely blocked the appearance of newly synthesized ^{3}H-glycoprotein in the axon after intrasomatic injection of L-^{3}H-fucose. However, in support of the above contention, these workers showed further that though anisomycin inhibited both fucosylation of glycoproteins and their export from cell body to axon, it did not affect the transport of glycoproteins once they were in the axon. Therefore, it was suggested that the drug had interfered with the synthesis of proteinaceous organelles (e.g. microtubules) required for export of glycoproteins from the cell body. Furthermore, colchicine, which blocks axoplasmic transport and which also blocks memory in the goldfish (Cronly-Dillon, 1973), does not appear to interfere with protein synthesis (e.g. Karlsson and Sjöstrand, 1968; Chang, 1972). The above mentioned findings indicate that a relationship between memory and rapid axoplasmic transport might exist at the level of oxidative energy metabolism (step 3 of Fig. 80) rather than at the level of protein synthesis.

The foregoing discussion indicates that major emphasis should be placed at the level of macromolecular synthesis and transport to the synaptic environment. Proteins, conjugated proteins and gangliosides, situated near synaptic junctions, which require axoplasmic flow, appear to be basic elements of the *Behavioral Code*. However, some rapidly synthesized macromolecules (e.g. some glycoproteins and gangliosides) localized at postsynaptic regions may not require axoplasmic or dendritic transport. It has been claimed that a local synthesis of glycoproteins can occur in synaptosomal fractions *in vitro* (Barondes, 1968b; Festoff *et al.*, 1971; Appel, 1972). However, such data, when acquired by using radioactive glucosamine as the precursor substance, are subject to the criticism of Ramirez (1974); i.e. that the presumed incorporation of glucosamine into glycoproteins could be due to a non-specific "binding" of glucosamine to macromolecules.

Local synthesis of gangliosides is indicated by the presence of glycosyltransferase enzymes in synaptosomal fractions (e.g. Den *et al.*, 1970; DiCesare and Dain, 1972). In any case, the synaptic plasticity related to "short-term" memory could be a postsynaptic phenomenon that depends upon local macromolecular synthesis, whereas that related to "long-term" memory could be presynaptic, requiring axoplasmic flow.

A role for axoplasmic flow in memory processes is supported further by the recent studies that have correlated abnormal protein metabolism in the brains of humans suffering from presenile-senile dementia with abnormal neurofibrillary structures (Iqbal *et al.*, 1974b, 1975; see also Chapter 8, Part B). In this connection, it is noteworthy that colchicine and vinblastine have been shown to cause a proliferation of neurofilaments in animals (Shelansky and Taylor, 1970; Wisniewski and Terry, 1970; Wisniewski *et al.*, 1970), and perhaps more importantly, that colchicine appears to interact with the tubulin of microtubules and to disassemble microtubules (Borisy and Taylor, 1967; Wilson and Meza, 1973; see also Part C, above and Chapter 8, Part B). On the basis of the studies by Roisen *et al.* (1972a, b), one could speculate that memory deficits might be reversed by producing an increase in cyclic-AMP since this agent might reverse colchicine effects by stimulating the assembly of microtubules from the pre-existing subunit pool (see also Lubinska, 1975). Also, nerve growth factor could promote such an effect by its possible stimulatory actions in the synthesis of neurotubule subunit protein (Hier *et al.*, 1972) and/or on the growth of axons (Yamada and Wessells, 1971).

It should be noted further that colchicine binds to other membrane proteins besides tubulin (see, e.g. Dahl *et al.*, 1970). Also, Chang (1972) found that doses of colchicine that reduced considerably the number of microtubules in axons of cultured chick sensory ganglia only slightly diminished the rate of movement of particles. In contrast, other studies have shown that colchicine can prevent rapid axoplasmic transport while not affecting microtubules (Fernandez *et al.*, 1971; Karlsson *et al.*, 1971; Flament-Durand and Dustin, 1972). However, since microtubules appear to be involved in the transport of NE (Banks *et al.*, 1971), it can only be concluded that the transport of some organelles (e.g. storage granules of NE) involve microtubules whereas others perhaps involve microfilaments or other types of mechanisms. In any case, longitudinal fibrous elements do appear to be involved in axoplasmic flow (see Lubinska, 1975). The role of axoplasmic flow in maintaining central synaptic connections has been recently emphasized (Krygier-Brévart *et al.*, 1974; Cull, 1975a; see above and Table 26). It should be noted further that axoplasmic transport of several classes of compounds appears to be

TABLE 26. Subcellular fractionation of the lateral geniculate nucleus 7 h and 24 h after ^{3}H-leucine application to the striate cortex.

Fraction	Sucrose (M)*	7 h experiment: % protein†	7 h experiment: % counts/min†	7 h experiment: RSA	24 h experiment: % protein†	24 h experiment: % counts/min†	24 h experiment: RSA
(A) DIFFERENTIAL CENTRIFUGATION							
homogenate		100‡	100‡	1	100‡	100‡	1
crude nuclear	P_1	38·1	16·9	0·44	36·3	13·1	0·36
crude mitochondrial	S_1	63·3	84·6	1·34	64·0	94·5	1·48
shocked crude mitochondrial	P_2	37·3	54·5	1·46	37·9	59·5	1·55
microsomes and soluble	S_2	26·0	30·2	1·16	23·4	25·2	1·08
(B) GRADIENT I[a] (separated: S_1)							
soluble supernatant	0·32	4·93	6·28	1·27	5·78	7·09	1·24
myelin and microsomes	0·55	18·37	15·28	0·83	17·63	13·28	0·76
synaptosomes	1·00	11·69	31·31	2·67	13·58	32·30	2·39
mitochondria	1·40	10·28	10·45	1·01	10·05	12·79	1·28

Fraction	Sucrose (M)*	7 h experiment % protein†	7 h experiment % counts/min†	7 h experiment RSA	24 h experiment % protein†	24 h experiment % counts/min†	24 h experiment RSA
(C) GRADIENT II[a] (separated: shocked P_2)							
1 soluble	0·05	8·54	5·58	0·65	8·15	4·59	0·56
2 soluble	0·10	2·09	1·66	0·80	3·58	2·01	0·56
3	0·18	1·34	0·77	0·58	1·54	1·21	0·79
4 synaptic vesicles	0·28	0·85	0·97	1·14	—[b]	—[b]	—[b]
5	0·40	1·18	1·78	1·51	1·57	2·09	1·32
6 myelin	0·57	4·60	3·83	0·83	2·99	2·98	0·99
7	0·77	4·08	10·39	2·55	3·62	11·10	3·03
8 external synaptosomal membranes	0·90	2·30	11·24	4·89	2·73	11·57	4·20
9	1·00	1·99	7·23	3·63	2·99	7·42	2·46
10 partially disrupted synaptosomes	1·09	2·73	5·89	2·16	2·86	7·48	2·60
11	1·20	2·51	2·99	1·19	3·16	3·46	1·08
13 mitochondria	pellet	4·48	1·24	0·28	3·88	3·63	0·93

A typical 7 h and 24 h experiment was chosen to present the distribution of protein, radioactivity, and relative specific activity (RSA; % counts/min/percent mg protein) of the subcellular fractions. The morphological description of the subfractions after differential centrifugation (A), was made according to Whittaker (1965), while those of the gradient fractions (B and C), were identified by electron microscopic inspection.

* Average sucrose molarities of the harvested fractions as determined by refractometry.

† Percentage of each fraction based on the homogenate being 100%.

‡ Absolute amounts were: 5·05 mg protein and 10 800 counts/min (7 h experiment) and 4·80 mg protein and 20 500 counts/min (24 h experiment).

[a] Gradients were calculated as though the whole of fraction S_1 or P_2, respectively, had been applied to the gradient, although in every experiment fraction S_1 was divided and both gradients simultaneously run (see Methods). [b] Amounts were too small for reliable assay or ratio determination. Note that radioactivity in the lateral geniculate nucleus represented transported proteins and that RSA was highest in synaptosomal fractions. (Reproduced with permission from Krygier-Brévart *et al.*, 1974.)

more rapid in the young animal and to decrease during development (see Bondy and Madsen (1974) and refs therein), a finding which also indicates that it is a behaviorally sensitive process that can be affected by both maturation and environmental modification.

The further study of changes in axoplasmic and dendritic transport and in neurotransmitter uptake mechanisms which may occur during various induced, or normal, function states in animals should greatly augment our approach toward understanding the structural basis of learning, memory and behavior.

5. A Change in Energy Metabolism

A. Introduction

A proper setting for this chapter is provided by the following statements of Quastel and Quastel (1961):

> The integrative activity of the central nervous system, whatever the physiological mechanisms by which it is mediated, depends ultimately upon its endogenous energy supply. Disturbances at this level are often implicated in cerebral dysfunction or alteration.

These statements bring to mind Cannon's (1932) notion that bodily integration is secured not merely by the nervous system but still more by the blood and tissue fluids, which supply the glucose and oxygen (O_2) required to maintain cerebral energy metabolism. However, it should be understood that although blood and tissue fluids are components of the *internal* (cellular) environment, the O_2 requirement for aerobic energy metabolism reveals the intimate relationship between organism and *external* environment. Haldane (1917) stated, with regard to his own question of when O_2 becomes part of the organism as opposed to the environment, that "An organism is one with its environment both internal and external." Actually, O_2 passes from the *external* to the *internal* environment of the cells when it penetrates the alveolar membrane. White (1931) has defined the organism as only *a specialized form of certain aspects of the environment*, a contention that is easy to accept when one considers the elasmobranch in which tissue fluids are really little more than walled-off sea water (see Dunbar, 1954). Glucose, the principal substrate of cerebral energy metabolism is also required in constant supply from the external environment. The coupling of neural activity to O_2 consumption and glucose utilization and its involvement with membrane ionic fluxes and Na^+, K^+-ATPase systems is well known. From these considerations, it becomes obvious that neural energy metabolism and the organism's environment cannot be separated.

B. Carbohydrate Metabolism

Under normal *in vivo* conditions and in controlled *in vitro* experiments, D-glucose is the major energy-yielding oxidative substrate of brain (see, e.g. Quastel and Quastel, 1961; Quastel, 1969; McIlwain and Bachelard, 1971). However, during starvation, at least in adult man, β-hydroxybutyrate and acetoacetate appear to equal glucose as the brain's major fuel (see, e.g. Owen *et al.*, 1967). Also, since brains perfused *in situ* can survive for relatively long periods by metabolizing non-carbohydrate compounds (Geiger, 1958) and since hypoxia and hypercapnia can cause changes in cerebral ganglioside content and structure (Lowden and Wolfe, 1964; Prokhorova *et al.*, 1964, 1967), glycolipids and glycoproteins appear to be linked to cerebral energy metabolism (see also Chapter 4, Part E). By merely considering that glucose and mannose are residues of glycoproteins and glycolipids, one arrives at the conclusion that changes in these substances can reflect changes in cerebral energy metabolism.

However, it should be noted that increases in the cerebral utilization of amino acids, proteins and other substances (e.g. nucleotide sugars) also occur when the cerebral glucose supply is decreased (e.g. in insulin-induced hypoglycemia; see, e.g. Cravioto *et al.*, 1951; Dawson, 1953; Massieu *et al.*, 1962), or during increased functional activity of the CNS (see, e.g. Jakoubec and Semiginovsky, 1970; Talwar and Singh, 1971; Lovell, 1971), and that an adequate O_2 supply is necessary for continued normal function of many central neurones (see, e.g. refs in Krnjević, 1975; Meier-Ruge, 1976). The effects of various psychoactive agents on cerebral energy metabolism have been well documented (see, e.g. reviews by Quastel and Quastel, 1961; McIlwain, 1966; Quastel, 1969, 1975). These observations have provided significant information about the involvement of energy metabolism in the *Behavioral Code* since they have served to relate behaviorally induced changes in energy metabolism to changes both in structural constituents (e.g. glycoproteins and glycolipids) of nerve ending membranes and in neurotransmitter enzymes. It seems likely that changes in energy metabolism (step 3 of Fig. 80) might lead to alterations in structural components of nerve endings which could be involved in the storage of experiential information in the nervous system. In this regard, it is also noteworthy that animals which are maintained on diets rich in glucose exhibit many changes, especially in liver and other peripheral tissues; e.g. repression of the induction of enzymes involved in gluconeogenesis; stimulation of the induction of enzymes involved in glucose utilization; decreased RNA synthesis; repression of glucocorticoid activity; changes in morphology of cell nuclei (see, e.g. Goldberg,

1975). Cyclic-AMP and cyclic-GMP appear to participate in this "glucose effect".

Carbohydrate metabolism of the CNS can be dramatically altered by subjecting mice to environmental changes (e.g. DeFeudis, 1971b; DeFeudis and Black, 1972; DeFeudis, 1975a; Table 27). Since the

TABLE 27. Incorporation of radioactive carbon atoms of glucose, fructose, pyruvate and inulin into the brains of differentially housed, male mice.

Days of differential housing	Substance injected	Radioactivity in brain dpm/g ($\times 10^{-3}$) Homogenate	Extract
	"Isolated" mice		
27	U-^{14}C-D-glucose	304 ± 7 (5)*	306 ± 6 (5)*
27	U-^{14}C-D-fructose	273 ± 7·5 (5)*	273 ± 5·5 (5)*
35	1-^{14}C-pyruvate	30 ± 2 (8)*	17 ± 0·6 (7)*
27–35	U-^{14}C-inulin	11 ± 0·2 (4)	12 ± 1·8 (4)
	"Aggregated" mice		
27	U-^{14}C-D-glucose	474 ± 17 (6)	468 ± 16 (6)
27	U-^{14}C-D-fructose	394 ± 11 (6)	395 ± 12 (6)
35	1-^{14}C-pyruvate	49 ± 1·7 (6)	35 ± 0·6 (6)
27–35	U-^{14}C-inulin	12 ± 1·7 (4)	13 ± 2·2 (4)

Male C-57 Black mice were fasted 12–18 h before experiments, and were sacrificed either by immersion in liquid N_2 or by decapitation 30 min after subcutaneous injections of "tracer" amounts of radioactive compounds. Values are presented for homogenates of brain (rostral to the level of the inferior colliculi; excluding the cerebellum) and for 80% ethanolic (v/v) extracts of these same homogenates. Means ± S.E.M.; numbers of mice in parentheses; * indicates $p < 0{\cdot}001$ when these values are compared with corresponding values for "aggregated" mice; Student's *t*-test; two-tailed. (Reproduced with permission from DeFeudis, 1971b; DeFeudis and Paolino, 1972.)

decreased entry of carbon atoms of energy substrates (e.g. glucose, mannose) into the brains of mice caused by "long-term" individual housing was reversed by "short-term" (1·5 h) exposures of the animals to new environmental conditions (DeFeudis, 1973), this type of change may not involve cerebral morphology. Other experiments have revealed that slices and homogenates from the cerebral cortices of rats submitted to 25 min of exploratory activity ("active rats") had a lower endogenous O_2 uptake than identical preparations from inactive controls (Eterović *et al.*, 1973; Table 28). Also, the O_2 uptake of homogenates from active rats, but not that of inactive rats, was increased in a Ca^{2+}-free medium.

TABLE 28. Oxygen uptake by cortical slices from rats that explored and controls.

Substrate in medium	No. of pairs	Oxygen uptake (μl/g/h) Experimental	Control	% Difference	p
None	24	1391 ± 34	1497 ± 44	−7·1 ± 2·3	0·01
D-glucose (3·3 mM)	24	1822 ± 24	1809 ± 30	0·7 ± 1·3	N.S.

Means ± S.E.M. Note that no difference between experimental and control animals existed when glucose was present in the incubation medium. (Reproduced with permission from Eterović *et al.*, 1973.)

Such environmentally induced changes in energy metabolism may "trigger" changes in RNA and protein synthesis (see step 3 of Fig. 80). Ochs (1972b) has emphasized the dependency of rapid axoplasmic transport on oxidative metabolism (see also Ochs and Ranish, 1970; Ochs *et al.*, 1970), and his calculations have indicated that the effects of cerebral asphyxiation on synaptic transmission and behavior might be explained by a blocking action on this transport process. Also, the relationship that has been shown to exist between oxidative metabolism and macromolecular synthesis in isolated synaptosomes indicates that local mechanisms related to carbohydrate metabolism play significant roles in synaptic function (Autilio *et al.*, 1968; Appel, 1972). A tight coupling among oxidative metabolism, membrane function, ion fluxes, and amino acid incorporation into membrane proteins appears to exist (Appel *et al.*, 1969; Festoff *et al.*, 1970; Appel, 1972).

Animals subjected to restricted environments exhibit lower metabolic rates and have higher body weights than animals kept under "enriched" conditions (see, e.g. Welch, 1965; Krech *et al.*, 1966; DeFeudis and Black, 1972). These findings, together with those discussed above, tend to support the observations that carbohydrate metabolism of psychotic patients may be more sensitive than that of control subjects to a variety of abnormal conditions (e.g. injections of insulin and epinephrine; see, e.g. Altschule, 1970), and that many emotional and mental disturbances in man may be accompanied by disturbed carbohydrate metabolism (e.g. Henneman *et al.*, 1954; Altschule, 1970; Shopsin *et al.*, 1972; Sacks, 1973; Bogoch, 1973). All of these findings add support to the pioneering studies of J. H. Quastel and colleagues which revealed that glucose is the major fuel of the brain *in vitro* (e.g. Quastel and Wheatley, 1932a) and that some human mental disorders were associated with an increased susceptibility to barbiturate toxicity that was caused partly by an interference with cerebral carbohydrate metabolism (Quastel and Wheatley, 1932b; see also Quastel and Quastel, 1961; Quastel, 1974). Much further experimental evidence now supports the notion that barbiturates and other central depressant drugs, with the possible exception of morphine and related substances (see Takemori, 1971), depress cerebral carbohydrate metabolism both *in vitro* and *in vivo* (e.g. Gatfield *et al.*, 1966; Folbergrova *et al.*, 1970; Quastel, 1975).

Also, prolonged barbiturate anesthesia (6 h of phenobarbital-Na^+; 250 mg/kg) induced glycogen accumulation in the brains of mice, the primary site being the cytoplasm of astrocytes in regions of high synaptic density (Phelps, 1972; see Fig. 70). Furthermore, chronic administration of phenobarbital to neonatal rats can retard the growth of the brain and some other organs, by a mechanism which involves reduced food intake and possibly an inhibitory effect of the drug on cerebral energy metabolism (Schain and Watanabe, 1975, 1976; see also McIlwain and Bachelard, 1971). Retardation of brain growth by undernutrition is well documented (e.g. Winick, 1969; see also Chapter 1, Part F), but differs from that caused by phenobarbital since this drug affects other organs similarly to brain, whereas undernutrition relatively spares brain growth (Wigglesworth, 1964). Administration of other behavioral depressants (e.g. reserpine, chlorpromazine and haloperidol; see, e.g. Shimizu and Ishii, 1964; Koizumi and Shiraishi, 1970a, b), traumatic experiences (e.g. Hager *et al.*, 1967; Guth and Watson, 1968), and X-irradiation (e.g. Miquel and Haymaker, 1965) also caused glycogen accumulation in astrocytes. It is also noteworthy that alterations in neuronal structure produced by insulin shock or by anoxic-ischaemic insult involve swelling and disorganization of mitochondria which are caused by the induced hypoglycemia and reduced cerebral blood flow (e.g. McGee-Russell *et al.*, 1970; Brierley *et al.*, 1971a, b). Austin (1972) has reviewed evidence that links disorders in the metabolism of glycogen (e.g. glycogen storage diseases) and related macromolecules to nervous dysfunctions.

The cerebral syntheses of "active" (transmitter-candidate) amino acids, acetylcholine (ACh), and some other environmentally sensitive substances are also intimately related to glucose utilization and to general metabolic rate (see, e.g. DeFeudis, 1975a). Perhaps the most significant finding regarding environmentally induced changes in carbohydrate metabolism is that such changes are "non-specific", i.e. they affect many, if not all, regions of the CNS as well as the liver and some other peripheral tissues (see Table 29). Such findings offer support, at the metabolic level, to the postulates of Lashley (e.g. 1950, 1958) which suggested that integrative activities of large regions of the nervous system are involved in information processing and in memory formation (see also Chapter 1, Part C and Chapter 2, Part A).

Many other stressors that affect behavior (e.g. convulsions, hypercapnia), as well as the ageing process, affect cerebral carbohydrate

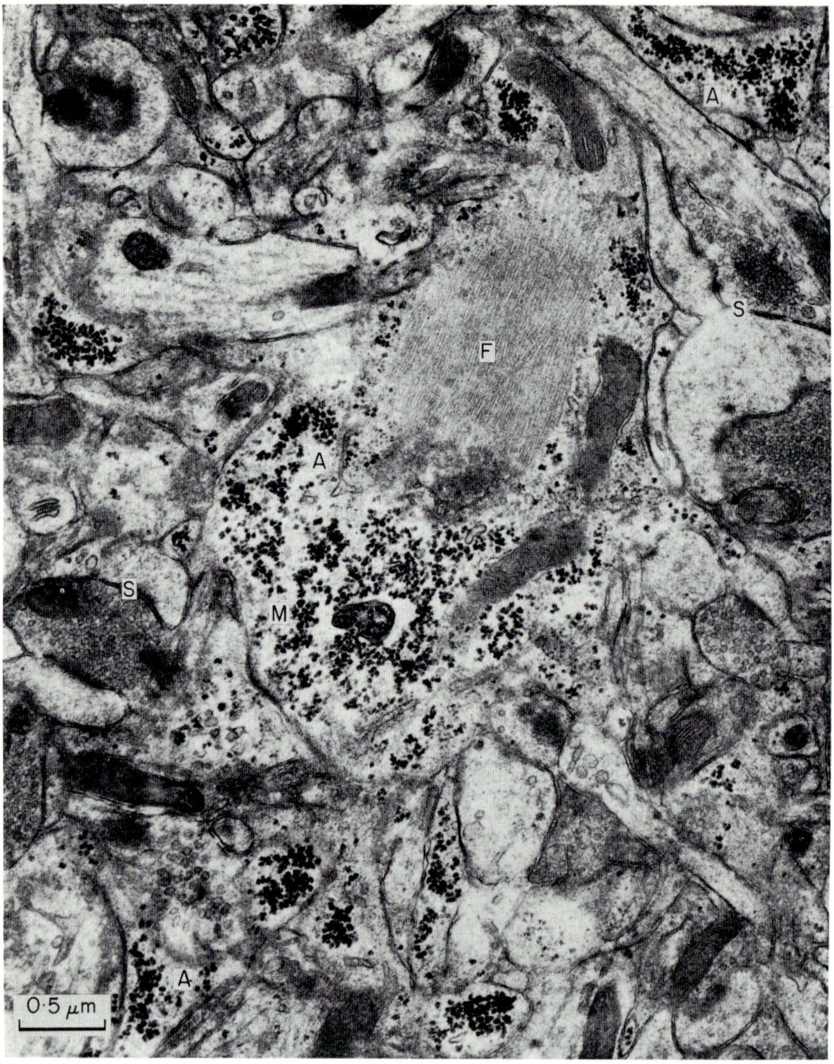

Fig. 70. Electron microscopic evidence for barbiturate-induced glycogen accumulation in the brain. Synaptic neuropile in the area dentata of the hippocampus of a mouse treated with sodium phenobarbital for 6 h; glycogen-laden astrocytic processes (A) are interspersed among axo-dendritic synapses (S). In the large astrocytic processes in the center is a collection of gliofilaments and tubules (F) and a membrane-bound body containing glycogen (M). (Reproduced with permission from Phelps, 1972.)

metabolism (e.g. Kety, 1955; Quastel and Quastel, 1961; Quastel, 1969; Balázs, 1970; Maker and Lehrer, 1972; Folbergrova *et al.*, 1975), and therefore this metabolic response appears to have universal applicability to the *Behavioral Code* (see step 3 of Fig. 80). By using the

TABLE 29. Incorporation of carbon atoms of U-^{14}C-D-glucose into various tissues of differentially housed mice.

Tissue	Radioactivity in tissue; dpm/g fresh wt. ($\times 10^{-3}$) "Isolated" mice		"Aggregated" mice	
	Homogenate	Extract	Homogenate	Extract
Brain	400·8	395·2	702·4	690·8
Spinal cord	312·0	297·0	516·4	588·6
Liver	329·7	300·6	393·1	404·2
Heart	221·3	228·6	164·5	176·0
Spleen	141·7	130·3	139·0	129·1
Kidney	240·8	226·9	245·2	220·5
Adipose tissue	42·4	39·9	179·1	151·7
Lung	184·8	176·4	214·2	201·4
Testis	85·4	79·3	73·1	72·6
Skeletal muscle	227·8	223·2	214·7	224·3
Stomach	235·5	185·5	242·5	191·9
Small intestine	267·2	220·3	269·0	212·6
Large intestine	205·5	154·1	368·3	281·3
Bone (femur)	91·2	91·6	106·3	100·4

Male C-57 Black mice were "isolated" or "aggregated" for 15 days. Means of closely agreeing values obtained from two mice are shown. U-^{14}C-D-glucose was administered subcutaneously (0·2 μ*Ci*/g, mouse), 30 min before sacrifice. (Reproduced with permission from DeFeudis, 1971b.)

recently-developed radioautographic technique, based on ^{14}C-2-deoxy-D-glucose, for measuring local changes in cerebral metabolic rate (Sokoloff *et al.*, 1974; Kennedy *et al.*, 1975; Sharp *et al.*, 1975; Sokoloff, 1975; see Figs 71 and 72, and Table 30), regional specificity of cerebral glucose utilization may be shown to be related to environmentally induced alterations in information processing and behavior. This new methodology employs low doses of 2-deoxy-D-glucose so that no appreciable effect on tissue utilization of glucose and no subsequent hyperglycemia, cytoglycopenia and central hypoglycemia are expected to occur (see, e.g. Tower, 1958). It should be recalled that Sols and Crane (1954) first showed that 2-deoxyglucose is phosphorylated at a rate comparable to that of glucose by cerebral hexokinase, but that the phosphate ester formed is not a substrate for either glucose-6-phosphate dehydrogenase or phosphohexoseisomerase. From the above discussion, it is easy to conceive that information processing (e.g. "short-term" memory) can be disrupted by alterations in energy metabolism. Changes in metabolic rate are likely to be caused by "non-specific" effects of stress that increase neuronal and glial activity, thus promoting an increased demand for ATP which is required to maintain membrane ionic gradients. Such effects of stress are likely to be mediated by changes in cerebral blood flow and O_2 consumption.

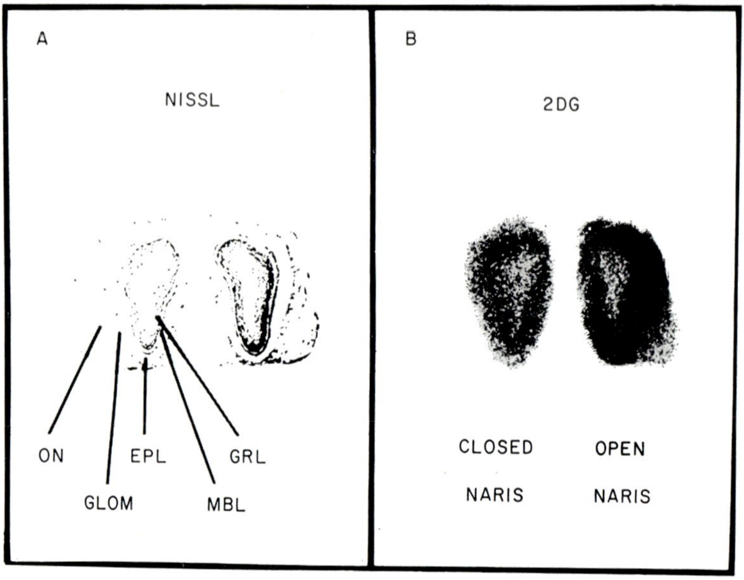

FIG. 71. A, thionin-stained Nissl section of the olfactory bulb, used to obtain the autoradiograph in B. ON, olfactory nerve layer; GLOM, glomerular layer; EPL, external plexiform layer; MBL, mitral body layer; GRL, granule cell layer. B, X-ray film autoradiograph of the olfactory bulbs of a rat which had one naris sutured closed 24 h prior to the administration of ^{14}C-2-deoxy-D-glucose. The bulb which was ipsilateral to the operated naris is labelled "closed naris", and that ipsilateral to the unoperated side "open naris". Note lighter density of the bulb on the operated side. ×15 (Reproduced with permission from Sharp *et al.*, 1975.)

C. Metabolic Dispositions of Psycho-active Agents

Studies on the metabolic dispositions of psycho-active agents (Baumel *et al.*, 1969; Lal *et al.*, 1972; DeFeudis and Marks, 1973a, b; Table 31 and Fig. 73) have provided further evidence that alterations of metabolic rate are caused by modification of the *external* environment. Such findings can be extended by applying recently developed methods for monitoring changes in glucose metabolism in discrete regions of the brains of awake animals (DeFeudis *et al.*, 1969, 1970) and for monitoring changes in regional glucose metabolism caused by functional changes (see above).

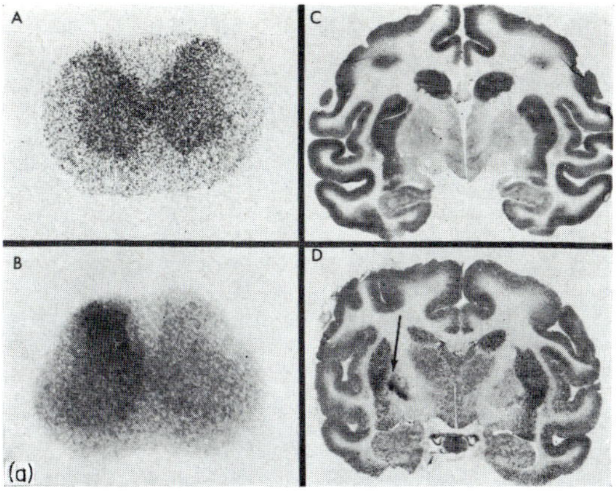

FIG. 72a. Effects of localized increases in functional activity on uptake of ^{14}C-deoxyglucose into neural tissues. A and B, Autoradiographs of sections of the lumbar spinal cord of a control rat (A) and of a rat with unilateral electrical stimulation of the sciatic nerve (B). Note the bilateral symmetry of optical densities in control rat and the asymmetrical increase in optical density in the dorsal horn ipsilateral to stimulation in the animal with sciatic stimulation. C and D, Autoradiographs of comparable brain sections from a normal rhesus monkey (C) and from a monkey in which seizures were induced by application of penicillin to the motor cortex of one side (D). Note the unilateral increases in optical density in discrete areas of the putamen and globus pallidus ipsilateral to the locus of penicillin application in (D). In autoradiographs of other brain sections from the animal with seizures (not shown), areas of increased uptake of ^{14}C-deoxyglucose were seen in the motor cortex adjacent to the site of penicillin application and in parts of the caudate nucleus and thalamus.

D. Other Metabolic Changes

Long-term "isolation" of rats altered *in vitro* steroidogenesis by the brain, and decreased the aromatization of testosterone to 17-β-estradiol by a mechanism that was age dependent, but not sex dependent (Lupo and Dessi-Fulgheri, 1975). The aromatization of testosterone appears to be associated with its central action (see Lupo and Dessi-Fulgheri, 1975). These findings, together with others which have indicated that long-term "isolation" of rats caused an increase in *in vitro* gonadal androgen production (Dessi-Fulgheri *et al.*, 1975), revealed that sex steroids are markedly affected by environmental changes. Interestingly, long-term sexual deprivation of rats, like social isolation, affected plasma levels of sex steroids (Dessi-Fulgheri and Lupo di Prisco, 1975) as well as the *in vivo* conversion of ^{3}H-corticosterone to its metabolites (Lupo *et al.*, 1975).

Other studies have revealed that an increased metabolic demand existed in the rat's hippocampus during training at reversed paw

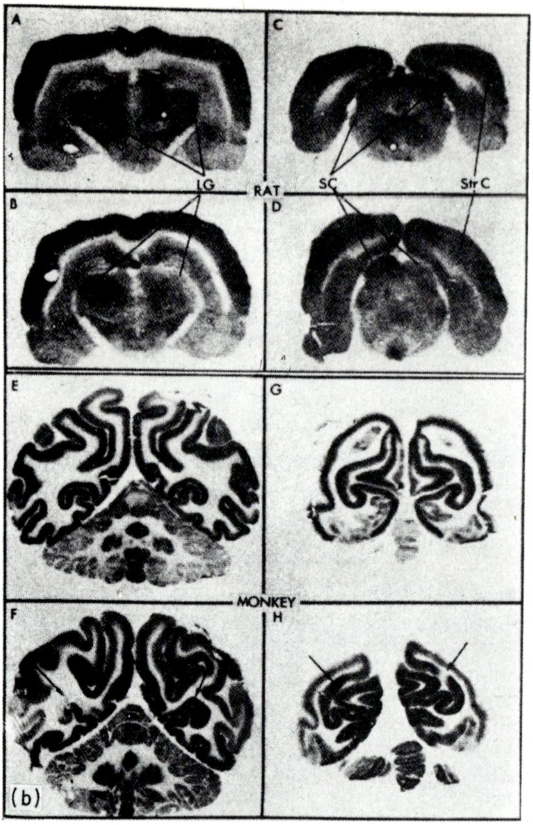

FIG. 72b. Effects of unilateral enucleation on ^{14}C-deoxyglucose uptake in components of the visual system in the rat (A to D) and rhesus monkey (E to H). In the normal rat with both eyes intact the uptakes in the lateral geniculate bodies (LG), superior colliculi (SC), and striate cortex (Str C) are approximately equal on both sides (A and C). In the unilaterally enucleated rat, there are marked decreases in optical density in the areas corresponding to these structures on the side contralateral to the enucleation (B and D). In the unilaterally enucleated monkey, areas of asymmetry are seen only in regions with monocular input, namely, the rostral portion of the calcarine cortex (F) and the loci corresponding to the blind spots of the visual fields (H). Autoradiographs from corresponding sections of brain from a control monkey with both eyes intact exhibited no asymmetry in these regions (E and G). The section in (H) is only one of many serial sections demonstrating the loci of the blind spots, which appear to extend approximately 3·4 mm from front to back; the gross appearance of the section in (G) is somewhat different from that of the section in (H) because of a slight difference in the plane of sectioning, but it includes the portions of cortex containing the loci of the blind spots. (Reproduced with permission from Kennedy *et al.*, 1975.)

preference (e.g. Levitan *et al.*, 1972c) and that an increase in NE metabolism in the brain stem-diencephalon region of rats could be produced by a lever-pressing response for water reward (Lewy and

TABLE 30. Some representative values for local cerebral glucose utilization in the albino rat during normal consciousness and during thiopental anesthesia.

Structure	Mean glucose utilization (μmol/100 g/min) Unanesthetized (8 rats)	Anesthetized* (3 rats)	% Difference
Visual cortex	95	65	−31
Auditory cortex	148	80	−45
Thalamus	106	54	−49
Medial geniculate body	119	63	−47
Lateral geniculate body	98	47	−51
Hypothalamus	67	37	−45
Inferior colliculus	185	129	−30
Superior colliculus	91	57	−37
Pontine gray matter	60	42	−30
Cerebral white matter	44	26	−43

* All decreases caused by anesthesia were statistically significant ($p < 0{\cdot}05$). (Reproduced with permission from Sokoloff, 1975.)

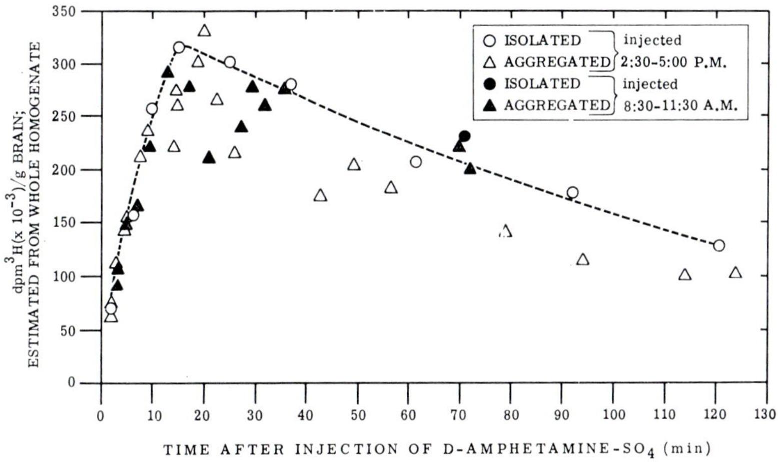

FIG. 73. Time course of accumulation of ^{3}H by the brains of "isolated" and "aggregated" male C-57 Black mice after intraperitoneal injection of ^{3}H-*d*-amphetamine-SO$_4$ (50 mg/kg; 0·1 $\mu Ci/g$). Experiments were conducted in both mornings and afternoons, as indicated. The estimated curve for "isolated" mice lies above most of the points for "aggregated" mice in its declining phase. Weanling mice were housed either individually or in groups of 16–25 for 5–9 weeks. (Reproduced with permission from DeFeudis and Marks, 1973a.)

TABLE 31. Effect of differential housing of mice on hexobarbital metabolism and hexobarbital concentration at awakening.

Sex	n-mol Hexobarbital/g body wt.		p
	Grouped	Isolated[a]	
	15 min after injection[b]		
Male	341 ± 13 (8)	299 ± 17 (6)	<0·05
Female	441 ± 23 (8)	363 ± 17 (8)	<0·01
	At awakening		
Male	170 ± 14 (8)	214 ± 9 (10)	<0·01
Female	221 ± 17 (8)	277 ± 23 (9)	<0·05

[a] Mice housed singly for 5 weeks; [b] 120 mg/kg (550 n-mol/g) injected i.p.
Means ± S.E.M.; numbers of mice in parentheses.
Note the more rapid disappearance of hexobarbital from the whole body and the higher drug concentration at awakening in the "isolated" mice. (Reproduced with permission from Baumel *et al.*, 1970.)

Seiden, 1972; see also Chapters 4 and 7). Such environmentally-triggered cerebral metabolic responses appear to be transient; i.e. the brain appears to undergo metabolic adaptation when the animal is subjected to prolonged exposure to a given stimulus. This concept has been well demonstrated in the study of Maccioni *et al.* (1971), who showed that the increased incorporation of 6-^{3}H-glucosamine into brain gangliosides of rats, caused by exposing them to light or sound stimuli, tended to disappear when the animals were kept under experimental conditions for periods greater than 1 hour.

All of these environmentally induced metabolic changes, in addition to the changes that occur in cerebral protein, nucleic acids, enzyme activities and in brain weight, itself, are likely to be reflections of changes in the energy-linked metabolism of glucose.

6. Changes in Possible Neurotransmitters and in Some Other Small Molecules

A. Introduction

Catecholamines, such as norepinephrine (NE) and dopamine (DA), serotonin (5-HT; an indolealkylamine), histamine (an imidazole), along with acetylcholine (ACh), the "pharmacologically-active" amino acids (e.g. GABA, glutamate, taurine) and some other small molecules, fulfil most of the criteria necessary of neurotransmitters (e.g. Florey, 1960; Eccles, 1964; Werman, 1966; Monnier *et al.*, 1970; McLennan, 1970; Phillis, 1970; Blaschko and Muscholl, 1972; Krnjević, 1974; Curtis and Johnston, 1974; DeFeudis, 1975c; Roberts *et al.*, 1976). These substances have been implicated in many types of animal behavior (e.g. aggressiveness; sleep-wakefulness; convulsions), in the central responses of animals to many psycho-active agents, and in certain neurological and behavioral disorders of man (see, e.g. Blaschko and Muscholl, 1972; Jouvet, 1972; Vogt, 1973; DeFeudis, 1974b; Costa *et al.*, 1974a, b; Lahti, 1975; Pradhan, 1975; Friedhoff, 1975a, b; Mandell, 1975a; Gazzaniga and Blakemore, 1975; Waser, 1975; Domino and Davis, 1975; Berl *et al.*, 1975; Roberts *et al.*, 1976; Goldberg and Hanin, 1976; Huxtable and Barbeau, 1976) and are therefore important elements of the *Behavioral Code*.

Some neurotransmitter suspects and other biologically-active small molecules have been discussed in other chapters of this monograph (e.g. Chapters 3, 5, 7 and 8). It is noteworthy that studies of postulated neurotransmitters have led to the monoamine theory of affective disorders (see, e.g. Schildkraut, 1965; Schildkraut and Kety, 1967; Himwich, 1970; Aillon, 1971; Axelrod, 1972) as well as to transmitter-based theories of the etiologies of other neurological and behavioral disorders in man (see also Chapter 8). A schematic diagram of some of the events which characterize the chemical theory of synaptic transmission is provided in Fig. 74. Excellent reviews on synaptic transmission (e.g. Eccles, 1964; McLennan, 1970; Phillis, 1970; Curtis and Johnston, 1974; Krnjević, 1974) and on the possible roles of endogenous small molecules and ions as regulators of enzymes involved in neurotransmitter synthesis and action (e.g. Mandell, 1975a; Krnjević, 1975)

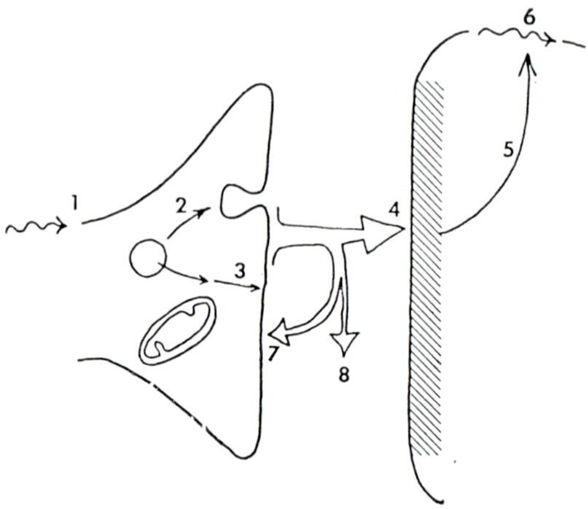

FIG. 74. Chemical theory of neurotransmission. In the diagram, it is shown that a wave of depolarization (nervous impulse) causes a Ca^{2+}-dependent release of neurotransmitter from storage sites in synaptic vesicles either by exocytosis (2) or by a transmembrane mechanism (3). Released transmitter combines with a specific receptor (4) on the postsynaptic membrane, leading to a transduction, perhaps by a second messenger (5), into re-initiation of the electrical signal (6). With this theory, transmitter release and receptor interaction are required for signal transmission. Secreted transmitter can be resorbed presynaptically (7) or eliminated by its enzymatic destruction or by its diffusion away from active postsynaptic sites (8). The actions of almost all known neuropharmacological agents can be explained by their effects on certain steps of transmitter function. (Reproduced with permission from Agranoff, 1975.)

are available. It should be noted that the stimulus-induced release of neurotransmitters appears to be dependent upon Ca^{2+} entry into neurones (e.g. Katz, 1966; Hubbard, 1970, 1973; Llinás and Nicholson, 1975).

Concepts involved in the trans-synaptic regulation of macromolecular (protein) synthesis are perhaps even more significant with regard to formulating the *Behavioral Code* (see Chapters 3 (Part D), 4 and 7; Thoenen, 1975; Mandell, 1975b). Transmitter-induced effects on cell membranes, which appear to control the syntheses of macromolecular cell constituents, including transmitter-synthesizing enzymes, represent well defined cases of the general theme of this monograph; i.e. the reaction of a cell to changes in its environment.

B. Putative Neurotransmitter Systems

1. *Introduction*

"Active" amino acids and ACh are synthesized from glucose in the nervous system, and are the most widely distributed postulated transmitters of the vertebrate CNS (see, e.g. Krnjević and Silver, 1965; Shute and Lewis, 1967; Krnjević, 1974; Curtis and Johnston, 1974; DeFeudis, 1974b, 1975c). Cerebral glucose (or pyruvate) serves as the major precursor for the acetyl groups required for ACh synthesis. After injection of radioactive precursors of acetyl-Coenzyme A into mammals, radioactive ACh appears rapidly in the brain (e.g. Cheney *et al.*, 1969; Tucek and Cheng, 1970), and ACh is readily formed from glucose or pyruvate by brain slices *in vitro* (e.g. Nakamura *et al.*, 1970; Lefresne *et al.*, 1973). Carbon atoms of systemically-injected U-^{14}C-D-glucose are also rapidly incorporated into "active" amino acids of the brain (e.g. Vrba, 1962), and these "non-essential" amino acids can be both synthesized in, and released from, various structures of the brains of monkeys *in vivo* (DeFeudis *et al.*, 1969, 1970). Since ACh and "active" amino acids are formed from glucose, approaches employing injections of labelled glucose with subsequent analysis of newly formed transmitter-suspects should provide results which may permit liaisons to be made between cerebral neurotransmitter mechanisms, energy metabolism, and various environmentally induced behavioral modifications. However, it should be understood that cerebral amino acids arise from two sources; some are synthesized solely within the brain (mainly "non-essential" ones), and some pass from the blood into the brain by processes which depend upon the level of maturation of the animal (mainly the "essential" type).

Although it seems reasonable to suspect that a greater depth of information about the environmental regulation of behavior will be obtained by studying ACh and "active", non-essential amino acids than by studying transmitter suspects that are derived from "essential" amino acids (e.g. NE, DA, 5-HT), the latter group of substances has received more attention to date (see, e.g. reviews by Welch, 1965; Valzelli, 1967, 1971, 1973a, b; Welch and Welch, 1969a, 1971; Garattini *et al.*, 1969; Bennett and Rosenzweig, 1971; Essman, 1972; DeFeudis, 1975a). However, it should be remembered that many hereditary metabolic disorders (e.g. human aminoacidopathics, such as phenylketonuria, Hartnup disease, homocystinuria), characterized by mental retardation and other neurologic dysfunctions, have been associated with defective utilization of one or several related "essential" amino acids, some of which serve as precursors to putative neurotransmitters (see, e.g. Jagenburg, 1965; Nyhan, 1967; Rodnight, 1968; Rosenberg, 1969; Gaull, 1972; Wiltse and Menkes, 1972). Such disorders will not be discussed in detail herein, except when necessary for the development of specific concepts, since their mechanisms do not relate to normal changes in behavior, but it should be noted that they provide exaggerated examples of the genetic control of behavior and that they reveal the value of biochemical methods in the diagnosis of neuro-behavioral disorders. For further information about aminoacidopathies the reader is referred to the above mentioned references, and for further information about heredodegenerative disorders involving lipid, nucleic acid and carbohydrate metabolism, see Wilson (1968), Hsia, (1972) and Brady (1973).

2. *Cholinergic Systems; Effects of Environment*

The first studies on the changes in cerebral acetylcholinesterase (AChE) and pseudo-cholinesterase (ChE) activities caused by environmental modification were carried out by Krech and co-workers (see, e.g. Krech *et al.*, 1962a, b, 1966; Rosenzweig, 1967; Bennett and Rosenzweig, 1971; see also Chapters 2 and 4). With regard to these enzymes that degrade ACh, it was shown that the *total* amounts of both AChE and ChE of the cerebral cortices (especially occipital cortex) of young and adult rats of the S_1 strain were increased by subjecting the animals to "enriched" (EC) environments. However, when these results were expressed on a unit weight basis (see, e.g. Bennett *et al.*, 1964) it was found that the activity of AChE was decreased by environmental "enrichment", whereas that of ChE was increased. These findings, together with others which revealed that increases in cerebral cortical weight also occurred in young and adult rats subjected to environmental "enrichment" (see also Chapter 2), led these workers to state that:

> The occurrence of such cerebral effects among adults supports the argument that these effects, rather than being consequences of accelerated early development, are residuals of experience. (Bennett *et al.*, 1964, p. 616)

The decrease in AChE activity, the persistence of which was later shown to depend upon the time of exposure to EC conditions (see, e.g. Bennett *et al.*, 1974; Fig. 43), indicated that the increased cortical growth caused by "enriched" experience could have occurred mainly in elements which are low in AChE activity (e.g. *glia*; *non-cholinergic neurones*). In this connection, it is known that glial cells are relatively rich in pseudo-ChE and that proliferation of "glia-like" elements is enhanced by complex experience (Diamond *et al.*, 1966; Altman, 1967; Rosenzweig *et al.*, 1969; Walsh *et al.*, 1969; see also Chapter 2 and Chapter 3, Part C).

Environmentally induced effects on cerebral AChE do not occur exclusively in rats. The occipital cortices of A_1 and C-57 strains of mice and of gerbils also contained less AChE per unit weight when these animals were reared under "enriched" conditions; i.e. in groups of 12 in large cages furnished with objects (Rosenzweig and Bennett, 1969; Table 32). However, Goldberg and Welch (1972) found that neither ChAc nor ChE activity was altered in the brains of socially stimulated ("pre-isolated") mice. With further regard to ChAc, Geller and Yuwiler (1968) reported that its activity was higher in some regions (e.g. hypothalamus) of "impoverished" (IC) rats, whereas Consolo and Valzelli (1970) could show no difference in the cerebral activity of this enzyme between "isolated" and "aggregated" mice. However, these results are not strictly comparable due to the different experimental designs that were employed.

Using differentially housed CF-1 mice, Essman (1971a) found that both AChE and BuChE (pseudo-ChE) activities were lower in the brains of "aggregated" mice, and that environmental stimulation of "isolated" mice caused significant elevations in the activities of both enzymes in the somatosensory cortex. Essman (1971a) showed also that the *direction* of the effect of environmental stimulation was contingent upon the endogenous condition caused by differential housing; in somatosensory cortex, decreases in cholinesterase were shown in "aggregated" mice subjected to environmental stimulation, whereas increases were shown in "isolated" mice which had received this same stimulation. In these studies, 21- to 23-day-old mice were housed for 28 days, either singly or in groups of five (see also Essman, 1972).

Other studies involving possible environmentally induced differences in cholinesterases have been performed in rats using the ChE-inhibitor, eserine (physostigmine). Greenough *et al.* (1973) reported that post-trial intraperitoneal administration of eserine facilitated performance in a Lashley III maze to a greater extent in EC rats than in IC rats. This greater sensitivity of EC rats to the drug indicated that the association between their central cholinergic systems and memory had been altered by the environment.

In "isolated" male mice, the contents of "bound" ACh of cerebral cortical homogenates and synaptosomes were increased, a change which might be correlated with their decreased neuronal activity. Also, isolation of male mice for 165 days significantly decreased the "bound"/"free" ratio of ACh in cerebral cortex; prior treatment of these animals with nicotine caused an increase in this ratio (Essman, 1971b). ACh "binding" to subcellular fractions (at 0°C) did not differ appreciably

TABLE 32. Percentage EC-IC differences in brain weights and brain enzymes of rats, mice, and gerbils.

	S_1 Rat		Mouse				Gerbil	
			A strain		C57 strain			
	(N = 29 per group) % difference	P	(N = 12 per group) % difference	P	(N = 12 per group) % difference	P	(N = 23 per group) % difference	P
Weight								
Occipital cortex	10·3	<0·001	19·6	<0·01	5·8	NS	4·0	0·10
Total cortex	5·8	<0·001	10·7	<0·001	8·2	<0·001	4·4	<0·001
Subcortex	−0·1	NS	2·7	NS	2·0	NS	1·7	0·10
Cortex/Rest	5·9	<0·001	7·6	<0·01	6·0	<0·01	2·6	<0·05
AChE/wt.								
Occipital cortex	−4·8	<0·001	−7·8	<0·01	−5·0	0·10	−5·2	<0·001
Total cortex	−3·2	<0·01	−5·1	NS	−1·5	NS	−3·2	<0·05
Subcortex	2·5	<0·05	0·9	NS1	0·6	NS	−1·4	0·10
Cortex/Rest	−5·5	<0·001	−5·6	NS	−2·1	NS	−1·8	NS
ChE/wt.								
Occipital cortex	1·8	=0·10	−14·5	<0·01	−1·2	NS	2·8	0·10
Total cortex	1·2	NS	−0·8	NS	2·7	NS	−0·7	NS
Subcortex	1·2	0·10	−0·3	NS	−3·2	NS	−1·0	NS
Cortex/Rest	0·0	NS	−0·4	NS	5·3	NS	0·5	NS
ChE/AChE								
Occipital cortex	6·9	<0·001	−7·0	0·10	3·0	NS	8·8	<0·001
Total cortex	4·7	<0·001	4·6	NS	4·1	NS	2·5	NS
Subcortex	−1·2	NS	−1·1	NS	−3·8	NS	0·4	NS
Cortex/Rest	6·0	<0·001	6·2	NS	7·3	NS	2·4	NS

(Reproduced with permission from Rosenzweig and Bennett, 1969.)

between "isolated" and "aggregated" male C-57 Black mice (DeFeudis, 1972d). Recent results have also revealed that the cholinolytic drug, scopolamine, increased ambulation and rearing of rats tested in an open field but decreased these responses when the animals were in complex environments (Stewart, 1975). These environmentally induced changes in the central cholinergic system may be useful in explaining the altered "arousal" level which exists in "isolated" animals (see, e.g. Welch, 1965; DeFeudis, 1974b, 1975a).

Evidence indicating that ACh acts as a "first messenger" in the trans-synaptic induction of tyrosine hydroxylase (TH) and dopamine-β-oxidase in the peripheral sympathetic nervous system has been presented (see Thoenen, 1975, and Chapters 5 and 8). Recent studies have indicated further that the activation of muscarinic and nicotinic ACh receptors exert opposing influences on the induction of TH synthesis in sympathetic ganglia (Hanbauer and Costa, 1976). Even though many observations have made it doubtful that cyclic-AMP serves as the "second messenger" in this system (see Thoenen, 1975), this action of ACh can serve as a model system for "long-term" memory formation which may also apply to the CNS, and has served to link the roles of cholinergic and adrenergic systems in neural function.

Comprehensive accounts have been written about the relationship between central cholinergic systems and behavior and about the regional distribution of the elements of the cholinergic system (e.g. Shute and Lewis, 1967; Silver, 1967; Pradhan and Dutta, 1971; Deutsch, 1971, 1972, 1973; DeFeudis, 1974b; Allikmets, 1974; Waser, 1975).

3. *Amino Acid Systems*

a. Introduction

Most of the "pharmacologically-active" amino acids (e.g. GABA, glutamate, aspartate) are of the "non-essential" variety, are present in high concentrations in various regions of the CNS (e.g. Tallan, 1962; Tews and Stone, 1965; Himwich and Agrawal, 1969; refs in Baxter, 1970; Van den Berg, 1970; Robinson and Wells, 1973), and are likely to be involved in the regulation of diverse physiological functions associated with behavior (see, e.g. Krnjević, 1974; DeFeudis, 1975c; Roberts *et al.*, 1976). Changes in the cerebral levels of these amino acids and their derivatives and in their associated uptake or "binding" processes and enzyme systems, which occur during maturation of various mammalian species in the absence or in the presence of diverse environmental insults, have been described (e.g. Sisken *et al.*, 1961; Oja, 1966; Agrawal *et al.*, 1966/7; Oja *et al.*, 1968a; Perry *et al.*, 1968;

Davis *et al.*, 1970; Sims and Pitts, 1970; Hudson *et al.*, 1970; Levi, 1970; Levi and Morisi, 1971; DeFeudis, 1972d, 1975a, c; DeFeudis *et al.*, 1976a). Roles for glutamate, GABA, aspartate and taurine in convulsive activity have been postulated (e.g. Tews and Stone, 1965; Watkins, 1967; Tapia and Massieu, 1969; Lovell, 1971; Tapia, 1975), and other evidence has indicated that these substances (especially GABA) are associated with the mechanisms of anesthesia and hypothermia (e.g. Baxter, 1966; Oja *et al.*, 1968b; Patel and Lal, 1973), with the actions of depressant, anti-convulsant (Hakkinen and Kulonen, 1959; Piha *et al.*, 1962; Cerda and Israel, 1969; Sytinsky *et al.*, 1975; Weinberger *et al.*, 1976) and anti-depressant (Patel *et al.*, 1975) drugs, with the "sleep-wakefulness" cycle (e.g. Godin and Mandel, 1965), with sleep deprivation (Karadzić *et al.*, 1971) and with various neurologic infections (Buryakova and Sytinsky, 1975). However, it must be understood that such involvements are not restricted to amino acid systems; e.g. central cholinergic and monoamine systems also appear to be involved in most (or all) of the above mentioned phenomena, including events of the sleep-wakefulness cycle (see, e.g. Jouvet, 1972; DeFeudis, 1974b).

Changes in hormonal balance also influence postulated amino acid transmitter systems. As examples, thyroidectomy and adrenalectomy caused decreases in cerebral GABA levels (Nishioka, 1960; Rindi and Ventura, 1961), whereas ovariectomy and thyroid hormone treatment increased cerebral GABA levels (Sklenovsky, 1967). Most recently, Ghazal *et al.* (1976) showed that treatment of virgin female rats with the estrogenic substance, mestranol, or with mestranol plus the progestogenic substance, lynestrenol (both components of the oral contraceptive, "Lyndiol") increased the GABA level and decreased the glutamate:GABA ratio of the brain. Such changes could explain the mental depression which accompanies the use of these agents by some women (see, e.g. Kutner and Brown, 1972).

It has also been shown recently that afferent electrical stimulation of the cat brachial plexus caused a reversible reduction of the GABA content in the corresponding (contralateral) cerebral cortex (Lichtshtein and Dobkin, 1976). Furthermore, diurnal rhythmicities appear to affect the CNS content (e.g. of glycine; Piepho and Friedman, 1971) and incorporation into protein (e.g. lysine; Richardson and Rose, 1971) of some amino acids.

b. Effects of Environment

The effects of early weaning of rats and of "isolation-rearing" of dogs on cerebral amino acid content have been described (Himwich *et al.*, 1968; Himwich, 1969; Fig. 75). It has been shown also that "isolation"

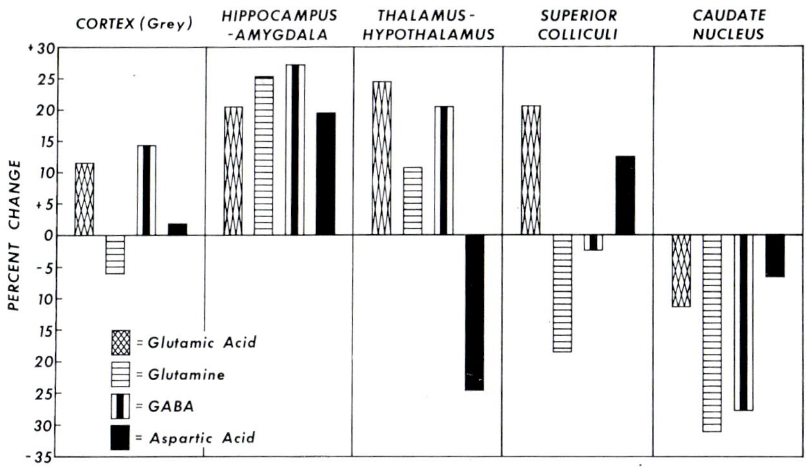

FIG. 75. Effect of isolation on the amino acids of the brain in 5-week-old dogs. For experimental details, see Agrawal *et al.* (1967). (Reproduced with permission from Himwich, 1969.)

of male mice for 21–30 days caused a significant decrease (about 20%) in the N-acetylaspartate contents of several brain regions studied, except for the corpora quadrigemina (Marcucci *et al.*, 1968; Garattini *et al.*, 1969). Such a change was not found in female mice which did not become aggressive by isolation-rearing (Marcucci *et al.*, 1968).

In male, C-57 Black mice, differentially housed at 21 days of age for 35–37 days, less GABA was bound to cerebral subcellular fractions containing nerve endings when mice were kept "isolated" than when they were kept "aggregated" (DeFeudis, 1972d; Table 33; see also Chapter 3, Part D). This finding agreed with the suggestion of Bennett *et al.* (1974) that cerebral changes caused by experience might be reflected mainly in non-cholinergic elements and glial cells; i.e. in elements which are rich in GABA uptake sites. Because the "percentage GABA space" (pellet/supernatant distribution ratio $\times$ 100) of the brains of "isolated" mice was reduced in the "P_2" fraction when homogenizing fluids contained low salt concentrations and in the "P_1" fraction with high salt concentrations, the "binding" of GABA appeared to provide a measure of the number or size of (or development of) cerebral nerve endings which can be affected by chronic environmental changes (see legend to Table 33). Thus, animals exposed to "isolation", by having a lower capacity to bind GABA (an inhibitory substance), may possess less, or smaller-sized inhibitory nerve endings, which could be related to the behavioral changes caused by "isolation" (see also Chapter 3, Part D). These findings were in agreement with those of Lal *et al.* (1972) who found that male, Swiss

TABLE 33. Comparison of ^{3}H-acetylcholine and ^{14}C-GABA spaces in subcellular fractions of the brains of differentially housed mice; effects of Li^+.

Salts in homogenizing fluid	% ^{3}H-ACh space		% ^{14}C-GABA space		% ACh space minus % GABA space	
	P_1	P_2	P_1	P_2	P_1	P_2
"Isolated" mice						
4 mM NaCl	71·1 ± 4·08	105·8 ± 1·46[d]	62·6 ± 2·88	99.8 ± 3·06[e]	8·6 ± 1·41	6·0 ± 1·61
4 mM NaCl + 4 mM LiCl	78·0 ± 3·51	99·2 ± 0·96[b]	72·1 ± 3·84	98·2 ± 1·45[f]	5·9 ± 0·72	1·1 ± 1·52
40 mM NaCl	94·4 ± 1·40	63·5 ± 3·80	141·6 ± 3·17[e]	96·1 ± 4·29	−47·2 ± 3·67[d]	−32·6 ± 2·14
40 mM NaCl + 40 mM LiCl	85·0 ± 1·83[b]	51·2 ± 2·24[a]	132·9 ± 2·41[a]	66·5 ± 1·95[c]	−47·9 ± 0·96[g]	−15·3 ± 0·84[c]
"Aggregated" mice						
4 mM NaCl	74·7 ± 3·78	111·2 ± 1·79	68·5 ± 2·52	110·5 ± 1·53	6·2 ± 2·12	0·72 ± 3·15
4 mM NaCl + 4 mM LiCl	80·2 ± 2·33	102·2 ± 3·01[a]	75·3 ± 3·06	108·7 ± 2·13	4·9 ± 1·72	−6·6 ± 3·45
40 mM NaCl	96·5 ± 1·12	62·6 ± 1·48	159·4 ± 5·20	94·0 ± 3·45	−62·9 ± 4·96	−31·4 ± 2·30
40 mM NaCl + 40 mM LiCl	87·1 ± 0·70[c]	52·3 ± 1·45[b]	151·0 ± 2·59	65·6 ± 1·36[c]	−63·9 ± 2·12	−13·3 ± 1·16[c]

Means ± standard errors; matching values from 5 mice in all cases; [a], [b] and [c] indicate $p < 0·05$, $p < 0·01$ and $p < 0·001$, respectively, for comparisons within "isolated" or "aggregated" groups (i.e. effects of Li^+); [d], [e], [f] and [g] indicate $p < 0·05$, $p < 0·02$, $p < 0·01$ and $p < 0·001$, respectively, for comparisons between fractions from the brains of "isolated" and "aggregated" mice which were homogenized under identical conditions (Student's *t*-test; two-tailed).

Weanling, male C-57 Black mice were housed either singly, or in groups of 25, for 35–37 days; all animals were fasted 19–21 h before sacrifice. Brains (above the level of the inferior colliculi; excluding the cerebellum) were homogenized in 10 vol. (0·32 M) sucrose solutions containing the various concentrations of electrolytes plus "tracer" amounts of ^{3}H-ACh, ^{14}C-GABA and 5×10^{-5} M eserine-SO_4. Aliquots (2·5 ml) of homogenates were centrifuged at 1000 × g, 10 min, 0°C to obtain P_1 and S_1 fractions, and 1·0 or 1·5 ml aliquots of S_1 fractions were re-centrifuged at 17 000 × g, 55 min, 0°C to obtain P_2 fractions; due to "clumping" effects caused by the higher salt concentrations, many nerve ending particles which are normally present in P_2 fractions were sedimented in P_1 fractions (Gray and Whittaker, 1962).

$$\text{"\% Space"} = \left(\frac{\text{dpm/g pellet}}{\text{dpm/g supernatant}}\right)(100).$$

Note that the "% GABA space" of the brains of "isolated" mice was smaller in P_2 fractions in the presence of low salt concentrations, and that it was smaller in P_1 fractions in the presence of high salt concentrations which caused "clumping" of the particles. (Reproduced with permission from DeFeudis, 1972d.)

albino mice that had been "isolated" for 6 or 20 weeks contained less cerebral GABA than those kept in groups of ten. Regarding strain differences, the incorporation of radioactivity of U-^{14}C-D-glucose into GABA and glutamate of the cerebral cortex (*in vivo* or *in vitro*) differed between A2-G and C-57 Black mice (Al-Ani *et al.*, 1970).

Another line of research related to the morphological basis of the *Behavioral Code* is concerned with the effects of injections of large amounts of transmitter-candidate amino acids on brain and behavior. Monosodium glutamate, administered either orally or parenterally to infant animals of several species, caused acute necrosis of neurones of the retina and hypothalamus, changes that were associated with endocrine disturbances and obesity as these animals attained adulthood (Olney, 1969; Olney and Sharpe, 1969; Fig. 76). Interestingly, treatment with other acidic amino acids, (aspartate, cysteine), but not

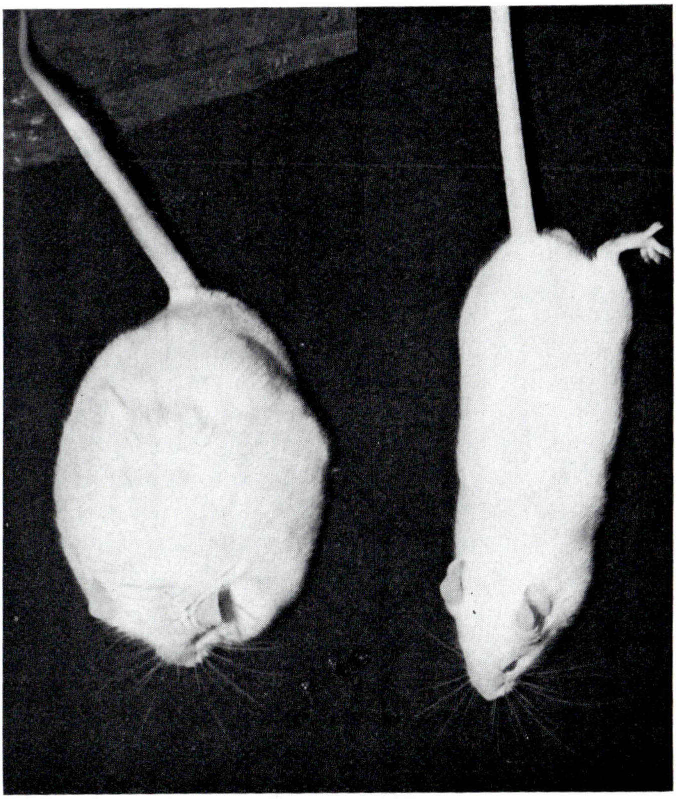

Fig. 76. A 9-month-old Swiss albino mouse (left) which was treated, as a newborn, with monosodium glutamate is shown beside the heaviest untreated male (right) from the same litter. The experimental animal weighed 84 g compared to 44 g for the control. In addition, the treated animal is shorter than the control and its fur is not as sleek as that of the control. (Reproduced with permission from Olney, 1969.)

neutral or basic amino acids, also produced degenerative changes in the maturing nervous system (see Olney, 1971). Such effects of high doses of monosodium glutamate (which has also been associated with the "Chinese restaurant syndrome" in man (e.g. Schaumberg *et al.*, 1969); i.e. with the symptoms of "burning", facial pressure and headache caused by ingestion of more than 3 g of this substance) could be mediated by its inhibition of cerebral glucose uptake. Such an effect has been shown in mice by Creasey and Malawista (1971). It has also been shown that diets supplemented with GABA or glycine can increase brain and body weights of mice, whereas those supplemented with glutamate caused the opposite effect (van Gelder, 1971).

4. *Norepinephrine* (*NE*), *Dopamine* (*DA*) *and Serotonin* (*5-HT*)

a. *Introduction*

Catecholamines and 5-HT appear to be involved in the regulation of many behaviors, including aggressiveness, stereotypy, sleep states and learning ability, as well as in certain affective states of man (see, e.g. Ungerstedt, 1971; Reis, 1971; Aillon, 1971; Papeschi, 1972; Robins and Hartman, 1972; Vogt, 1973). Bourgault *et al.* (1963) found that the aggressive SC-1 strain of mouse had lower cerebral concentrations of NE and 5-HT than the less aggressive C-57 Black strain, Maas (1962) found C-57 Black mice to be more aggressive and to have lower levels of brain stem 5-HT than BALB/c mice, and Lagerspetz *et al.* (1967) found that aggressive strains had lower concentrations of 5-HT in the forebrain and higher concentrations of NE in the brain stem than less aggressive strains. Further studies revealed that most mouse strains which possess lower levels of cerebral catecholamines and 5-HT, though more aggressive, displayed less stereotyped behavior. These observations led Karczmar and Scudder (1967) to suggest that catecholamines and 5-HT might be involved in the inhibitory circuitry of the forebrain and limbic system.

Other studies have indicated that aggressivity may be caused by an increase of cerebral NE, whereas stereotypy may be due to an increase in cerebral DA (Scheel-Kruger and Randrup, 1967; Randrup and Munkvad, 1968, 1970). Aggressive behaviors exhibited by rats and cats also appear to involve increases in cerebral NE (Hernández-Peón *et al.*, 1963; Gunne, 1969; Bliss and Aillon, 1969; Reis and Fuxe, 1969; Fog *et al.*, 1970). In relation to this, it has been shown also that the brain stems of rats subjected to the stress of mild electric foot-shock exhibited a preferential utilization of newly formed NE, as compared with the stored amine (Thierry *et al.*, 1970).

b. Cerebral Concentrations of NE, DA and 5-HT

Comprehensive mapping studies have revealed that certain central pathways appear to be enriched in NE, DA and 5-HT (see, e.g. Dahlström and Fuxe, 1964; Olsen and Fuxe, 1972; Andén *et al.*, 1969; Fuxe *et al.*, 1970; Ungerstedt, 1971; Jouvet, 1972; Holzbauer and Sharman, 1972; Vogt, 1973; Dahlström *et al.*, 1973).

Welch (1965) showed that male DUB/ICR mice, kept in groups of 20 for a week, possessed lower cerebral NE concentrations than mice subjected to individual housing for the same period of time. However, further studies by Welch and Welch (1968a, c) indicated that brain NE was higher in "grouped" than in "isolated" male mice. In other experiments, Valzelli (1967) found that cerebral catecholamine concentrations of male mice were not significantly modified by isolation, and Essman (1971b) showed that cerebral NE concentration did not differ between CF-1 mice which had been kept "isolated" or "grouped" for 28 days, commencing at 21 days of age. However, when differential housing was carried out for 34 days, the brains of "isolated" mice contained less NE than those of "grouped" mice (Essman and Smith 1967). Geller *et al.* (1965) found that "isolated" (IC) rats (weaned at 18 days of age) possessed significantly higher contents of cerebral NE than "group-enriched" (EC) rats, an increase which appeared to be localized to the caudate nucleus (see Fig. 77); tyrosine hydroxylase

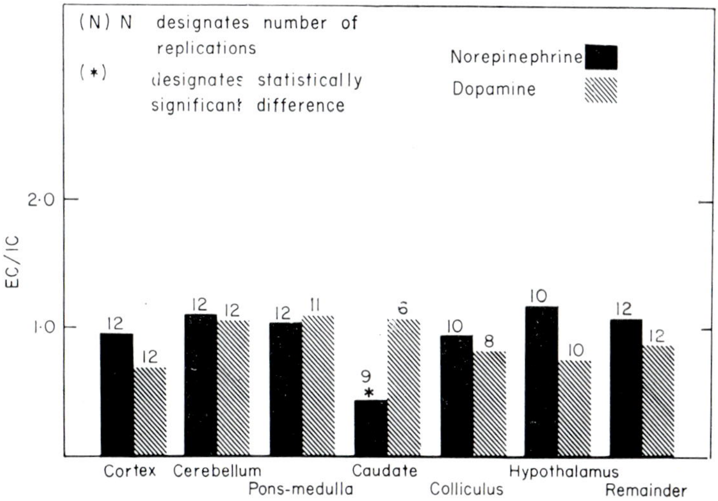

FIG. 77. Differences in norepinephrine or dopamine in various regions of the brains of rats reared under "enriched" (EC) or "impoverished" conditions (IC), expressed as EC/IC. (Reproduced with permission from Geller and Yuwiler, 1968.)

activity was higher, and MAO activity was lower in the cerebral cortices and colliculi of IC animals, as compared with EC animals (see Geller and Yuwiler, 1968).

Garattini *et al.* (1969) found no change in the 5-HT content of whole brain or discrete cerebral structures in differentially housed mice (see also Valzelli, 1967). However, the 5-HT contents of telencephalon and diencephalon-mesencephalon were higher in "grouped" than in "isolated" mice (Welch and Welch, 1968a, c). Isolation of mice reduced cerebral 5-hydroxyindoleacetic acid (5-HIAA) content within one day, and this change persisted for at least a month (Giacalone *et al.*, 1968). Welch and Welch (1968a) also found the levels of 5-HIAA and the 5-HIAA/5-HT ratio to be lower in "isolated" mice (see also Modigh, 1974). In albino CF-1 mice, "isolated" for 34 days (commencing at 21 days of age), Essman and Smith (1967) found no change in either whole brain 5-HT or 5-HIAA, but subsequent results from this same laboratory (Essman, 1968, 1969) indicated that isolation of the same strain of mice for 24 days produced an increase in whole brain and regional 5-HT and 5-HIAA. The discrepancies which exist among these studies appear to be due mainly to variations in experimental design (e.g. differences in the duration of differential housing; differences in the time of onset of isolation, strain differences).

Using subcellular fractionation techniques, Essman (1969) showed that: (1) a greater proportion of "free" 5-HT existed in "isolated" mouse brain; (2) "free" 5-HT could be decreased by environmental stimulation in both "aggregated" and "isolated" mice; (3) a rather stable level of "bound" 5-HT was present in the brain under all conditions (see also Essman, 1971a). Essman (1971a) suggested that a "free" pool of 5-HT might be regulated by certain environmental conditions. Although 5-HIAA levels of the brains of "isolated" mice were found to be significantly decreased (Giacalone *et al.*, 1968), no difference in MAO was found either in whole brain or in various cerebral structures (Consolo and Valzelli, 1970).

c. *Cerebral "Turnover" of NE, DA and 5-HT*

"Turnover" is a term which includes both synthesis and release of the substance considered. Thus, the turnover of a substance is very difficult to approach biochemically, especially since metabolic inhibitors which possess many non-specific actions are generally used for this purpose. However, because the basal metabolic rate appears to be lower in "isolated" than in "grouped" male mice (Welch, 1965), it seems quite likely that the cerebral metabolism (and hence, turnover) of catecholamines and 5-HT (as well as many other substances) can be affected by differential housing.

In this regard, it has been shown that "isolated" male mice had lower rates of synthesis, release and turnover of cerebral catecholamines than "grouped" (8 to 10 per cage) mice, as indicated by a greater decrease of catecholamines after administration of *dl*-α-methyltyrosine (α-MT) in "grouped" mice (Welch, 1967; Welch and Welch, 1968c; Modigh, 1974; Table 34). Such impairment of catecholamine biosynthesis by α-MT markedly decreased aggressiveness in "isolated"

TABLE 34. Effects of tyrosine hydroxylase inhibition on brain and adrenal catecholamines in differentially housed male, Swiss albino mice.

	Isolated (± S.E.)	Group of 10 (± S.E.)	p<
Number	30	28	
Body weight (g)	38·1 ± 0·8	33·0 ± 0·8	0·01
Left adrenal weight (mg)	2·08 ± 0·09	2·44 ± 0·07	0·01[a]
Brain norepinephrine			
Saline control (ng/g)	402 ± 14·8	456 ± 16·7	0·025[a]
α-MT decrease (% control)	58·4 ± 1·38	64·4 ± 1·96	0·01
Brain dopamine			
Saline control (ng/g)	788 ± 30·9	748 ± 29·1	NS
α-MT decrease (% control)	53·8 ± 0·99	62·0 ± 0·89	0·001
Adrenal epinephrine			
Saline control (ng/left adrenal)	3·66 ± 0·25	4·01 ± 0·33	NS[a]
α-MT decrease (% control)	20·9 ± 3·25	33·3 ± 4·16	0·01
drenal norepinephrine			
Saline control (ng/left adrenal)	0·44 ± 0·04	0·42 ± 0·10	NS
α-MT decrease (% control)	11·4 ± 10·28	14·3 ± 9·84	NS

Mice were group- or individually-housed for 14 weeks subsequent to weaning at the age of 4 weeks.

[a] Differences are enhanced when values are adjusted to body weight by co-variance analysis; on this basis, the adrenals of grouped mice contained significantly more epinephrine ($p < 0{\cdot}001$). (Reproduced with permission from Welch and Welch, 1968c.)

mice (see Welch and Welch, 1969a). Also, cerebral 5-HT and catecholamines usually accumulated more slowly after treating "isolated" mice with MAO-inhibitors (MAOI) (Welch and Welch, 1966; Garattini *et al.*, 1967; Giacalone *et al.*, 1968; Valzelli and Garattini, 1968; Welch and Welch, 1971; Modigh, 1974). The differences in rate of NE and DA turnover in brain appeared to be graded with respect to level of environmental stimulation (group size), and graded differences in the accumulation of these cerebral amines were shown within a few hours after injecting mice with the fast-acting MAOI, pargyline (Welch and Welch, 1971). Valzelli (1971) showed a minor reduction of NE turnover and an increase in DA turnover in the brains of "isolated" mice (see Table 35). These disparities among the data obtained in different laboratories perhaps reflect technical differences as well as strain and age differences of the animals used. A recent study has indicated that NE turnover in the brain can also be increased by prenatal environmental stress (Huttunen, 1971).

Welch and Welch (1968a) showed that cerebral NE and DA contents were decreased by *para*-chlorophenylalanine (p-CPA) to about the same small extent as 5-HT, and that this change occurred to a greater extent in "grouped" than in "isolated" mice. These results indicated a lack of specificity of p-CPA as an inhibitor of 5-HT synthesis.

Recently, the influence of group size on the CNS metabolism of NE and some peripheral parameters was investigated in NMRI mice. NE turnover decreased in basal ganglia, diencephalon and mesencephalon-pons from animals housed singly to groups of 16/cage and then increased again; no changes were found in cerebral cortex and cerebellum (Frank and Raab, 1975). Perhaps the most interesting finding was that changes in plasma corticoid levels were correlated with these density-induced changes in NE metabolism of the basal ganglia in both male and female mice (Fig. 78).

Brain 5-HT was increased more in "grouped" than in "isolated" mice after administration of pargyline (Welch and Welch, 1971) or tranylcypromine (Valzelli, 1967; Garattini *et al.*, 1967). The results of Giacalone *et al.* (1968), showed a decreased 5-HT turnover rate in brain and a concomitant increased 5-HT turnover time for "isolated" mice when compared with their "grouped" (10 per cage) counterparts. These changes were also evident in specific brain regions, the greatest difference occurring in the corpora quadrigemina in which the rate was only two-thirds as fast in "isolated" as in "grouped" mice. Differentially-housed female mice did not show differences in cerebral 5-HT turnover.

Another study with CF-1 and Swiss-Webster mice revealed that brain 5-HT turnover was increased in "isolated" mice, as compared to "grouped" (5 per cage) mice (Essman and Smith, 1967; Essman, 1968). These results differed from those reported by Welch's and Garattini's groups, perhaps due to the earlier age (21 days) of initiation of differential housing. Also, since Essman and co-workers took brain samples immediately after mice had been tested for behavioral activity in a novel situation, the basal rate of metabolism could not be determined. In general, the results of Welch's and Garattini's groups are in agreement in that 5-HT turnover occurs more rapidly in "grouped" mice and that the rate of 5-HT synthesis can be increased

TABLE 35. Cerebral serotonin (5-HT), norepinephrine (NE) and dopamine (DA) "turnover" in normal and "isolated" mice.

Type of mice	5-HT "turnover"		NE "turnover"		DA "turnover"	
	Rate (μg/g/h)	Time (h)	Rate (μg/g/h)	Time (h)	Rate (μg/g/h)	Time (h)
Normal	0·46	1·10′	0·05	8·32	0·08	6·10′
"Isolated-aggressive"	2·26[a]	2·20′	0·03[a]	10·15′	0·13[a]	3·54′

Each value represents the mean of 8 determinations; [a] indicates $p < 0{\cdot}01$ with respect to normal animals. (Reproduced with permission from Valzelli, 1971.)

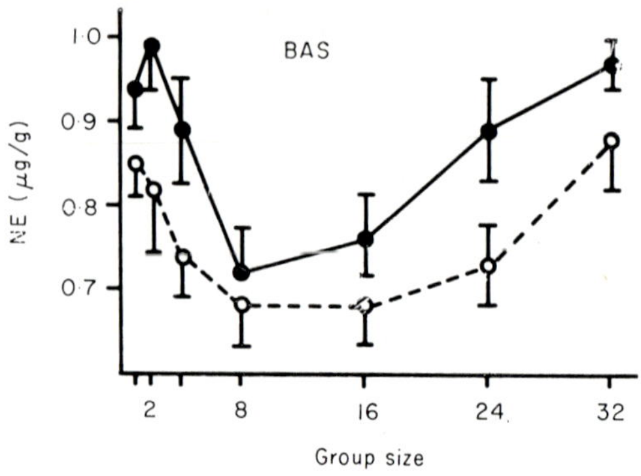

FIG. 78. Concentration of NE (μg/g tissue wet wt.) in the basal ganglia of mice from different groups one hour after injection of pargyline. Each point represents the mean of 6–8 males (continuous line, filled circles) or females (dotted line, open circles) with the standard error of the mean. The significance of apparent differences between groups was assessed by Student's *t*-test and indicated as follows: e.g. ♂♂ 1:16, p = 0·05; this means that singly housed males are significantly (p = 0·05) different from males housed in groups of 16 per cage.

			p		p		p
Statistical evaluation:	♂♂	1: 8	0·03	2: 8	0·01	8:32	0·01
		1:16	0·05	2:16	0·01	16:32	0·01
	♀♀	1:16	0·05	2:16	0·01	16:32	0·05
		2: 8	0·03	8:32	0·05		

(Reproduced with permission from Frank and Raab, 1975.)

by tranylcypromine administration (Garattini *et al.*, 1967; Giacalone *et al.*, 1968). This decrease in 5-HT turnover appears to represent the most typical metabolic change that occurs in cerebral monoamine systems of "isolated" mice (see Valzelli, 1973a; Table 35). However, it should be noted that in a recent study in which rates were designated as "high-aggressive" *vs.* "low-aggressive", on the basis of a food competition test, it was shown that "high-aggressive" rats displayed a more rapid 5-HT turnover than "low-aggressive" rats in two gross brain regions: (1) all structures, minus olfactory bulbs, rostral to the level of the anterior commissure; (2) brain stem; i.e. hypothalamus-midbrain-medulla (Daruna and Kent, 1976).

Induction or potentiation of aggressive behavior or fighting in rats, produced by the dopamine receptor antagonist, apomorphine (Maj *et al.*, 1973; Scheel-Kruger and Hasselager, 1974), by foot-shock coupled with REM sleep deprivation (Stern *et al.*, 1971), by increased temperature (Weiss and Aghajanian, 1971) or by 6-hydroxydopamine administration (Blondaux *et al.*, 1973) were also correlated with increases in cerebral 5-HT turnover (see also Morden *et al.*, 1968; McKenzie, 1971; Berry and Jack, 1971; Thoa *et al.*, 1972). Also, a decrease in foot-shock-induced aggression has been produced by chlordiazepoxide (Quenzer *et al.*, 1974), a substance which decreases 5-HT turnover (Dominic, 1973).

Other studies on regional 5-HT turnover in the CNS revealed that turnover time was lower in "isolated" mice; e.g. in olfactory bulbs and cerebral cortex (Essman, 1971c, 1972). In a related study, Raab and Deisz (1975) showed, after examining six brain regions, that only the septal area of both male and female mice showed increases in 5-HT turnover that were correlated with increases in group size (see Fig. 79). This finding was in accord with the previous observations of Valzelli and Garattini (1968), Garattini *et al.* (1969) and Welch and Welch (1971). Tryptophan hydroxylase activity of the septal area of rats can also be increased by "crowding", indicating an enhancement of 5-HT turnover (Segal *et al.*, 1973), and septal 5-HT turnover in tree

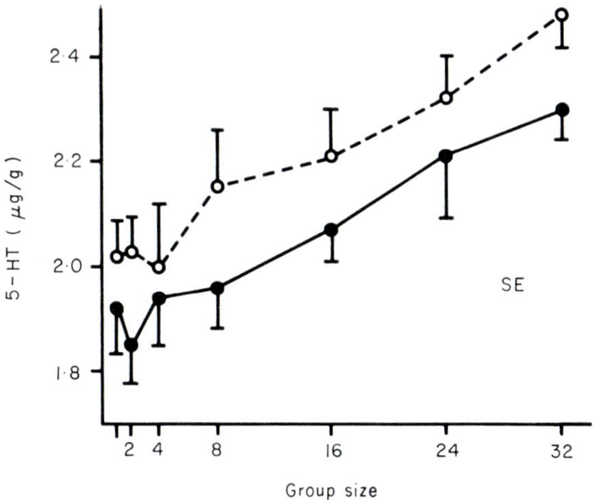

FIG. 79. Concentration of 5-HT (μg/g tissue wet wt.) in the septa of mice one hour after application of parygline. See legend to Fig. 78.

		p		p
Statistical evaluation:	♀♀1:32	0·01	♂♂1:32	0·03
	1:24	0·05	2:32	0·01
	2:32	0·013	4:32	0·05
	2:24	0·05		
	4:32	0·03		

(Reproduced with permission from Raab and Deisz, 1975.)

shrews (*Tupaia belangeri*) was increased by psycho-social stress (Raab, 1971). Such data support the notion that the septum is involved in social adaptation (see Fried, 1972; Lubar and Numan, 1973).

The different results obtained by these various groups of investigators with regard to 5-HT turnover and metabolism indicated that such results depend on many factors (see above). Perhaps the most important are those concerning the entry into, and emergence from, the physiological state caused by "isolation".

d. Other Factors Affecting Cerebral Levels, Metabolism and Turnover of NE, DA and 5-HT

As expected, many factors can modify the metabolism of these possible central neurotransmitters. As examples, increases in cerebral NE turnover have been shown in rats submitted to diverse stressors such as electric foot-shock (Thierry *et al.*, 1968; Bliss *et al.*, 1968), prolonged immobilization (Corrodi *et al.*, 1968) or intense muscular exercise (Gordon *et al.*, 1966). Also, spontaneously (genetically) hypertensive rats possess lower brain stem and hypothalamic concentrations of NE than control strains (Yamori *et al.*, 1970). Also, elevations in 5-HT and reductions in NE occurred in the brains of rats subjected to cold or swimming stress, the latter of which also causes behavioral depression (e.g. Stone, 1970). Even avoidance and escape responding can elevate rat brain NE (Weiss *et al.*, 1970, 1974). Most recently, it has been shown that rats implanted in the area ventralis tegmenti (AVT) and trained to self-stimulate for three hours per day for seven days possessed significant increases in the NE contents of the nucleus accumbens, olfactory tubercle, nucleus amygdloideus centralis and dorsal bundle and in the DA content of the olfactory tubercle (St. Laurent *et al.*, 1975). These changes indicated that NE and DA of specific cerebral regions may have a particular functional relationship with the "reward" mechanism.*

* *Reward mechanism:* The "rewarding" effect of brain stimulation was first demonstrated by Olds and Milner (1954) who trained rats to press a bar to receive short bursts of electrical stimulation (self-stimulation) in the septal area. Subsequent experiments showed that stimulation was rewarding in many regions of the limbic system, and in particular, in the hypothalamus (J. Olds, 1956; M. E. Olds and J. Olds, 1963). Significant rewarding effects have also been elicited from the cingulate cortex, the mamillothalamic tract and the tegmentum (see P. M. Milner, 1970). In almost every structure in the rat some points have been found which are at least mildly rewarding. The best sites found in all mammals were the medial forebrain bundle and the lateral and posterior hypothalamus (see P. M. Milner, 1970). However, "punishing" points have also been found next to rewarding ones in some structures and some "non-rewarding" points were found even in those structures possessing the most consistently rewarding points (see P. M. Milner, 1970).

Although such factors make it more difficult to control experiments aimed at quantitating changes in these monoamine systems that are caused by environmental modification, they have also been useful for sensitizing the methods used. A good example of this has been provided by the results obtained when using drugs that block the rate-limiting enzymes of cerebral catecholamine and 5-HT synthesis (α-MT and p-CPA; see above). Administration of these agents to mice increased the differences in cerebral metabolism of these proposed transmitters which were caused by differential housing. In some cases in which equivocal results were obtained in the absence of drugs (e.g. for cerebral levels of NE, DA and 5-HT), the use of the drugs, or restraint stress, or fighting activity accentuated the differences caused by differential housing, thereby making these easier to demonstrate (see below).

Welch and Welch (1969b), using DUB/ICR male mice, made aggressive by individual housing subsequent to weaning, showed that daily fighting (for 5 to 10 min) for either 10 or 14 days caused a significant increase (about 10%) in whole brain DA levels; cerebral NE was elevated to the same extent but only after 14 days, whereas no change occurred in cerebral 5-HT (see also Modigh, 1974). These studies provided evidence that sustained effects on cerebral catecholamines could be caused by brief periods of stimulation, a finding which agreed with other reports which indicated that cerebral NE content was increased in rats subjected to electroconvulsive shock twice a day for 7 days (Kety *et al.*, 1967), and that rat brain NE was elevated 24 hours after three days of repetitive stress (Thierry *et al.*, 1968).

These sustained effects may differ (in direction) from acute effects. For example, the acute effects of intense fighting include a decrease of brain NE and DA, at least in the brain stem (Welch and Welch, 1968a, d). Hence, the sustained effects (i.e. elevated levels of catecholamines) perhaps reflect adaptive mechanisms. Elevations in cerebral NE and DA were found to be about $\frac{1}{3}$ to $\frac{1}{2}$ as great as those

---

Stimulation of the extrapyramidal system is also rewarding. Self-stimulation from the caudate nucleus of cats (Brady *et al.*, 1957; Justesen *et al.*, 1963), from the globus pallidus in monkeys (Brady, 1960), from the substantia nigra in gerbils (Routtenberg and Kramis, 1967), and from the region of the brachium conjuctivum in rats (Routtenberg and Malsbury, 1969) have been reported. Also, sites from which self-stimulation may be elicited are the same as those from which other species- and drive-related behaviors (e.g. eating, drinking, copulating and grooming) can be elicited. However, the thresholds for eliciting these different responses may differ (see P. M. Milner, 1970). An interlinked system of catecholaminergic neurones stretching from the locus coeruleus rostral to the nigro-striatal region has been associated with self-stimulation (see Crow, 1972, 1973). For recent reviews about the reward mechanism, see Gallistel (1973) and Rolls (1975).

caused by administration of a MAOI (pargyline) for a similar period of time to mice which were then allowed to fight. However, since the effects of pargyline and fighting were not additive, it was suggested that fighting, itself, involves cerebral mechanisms other than MAO inhibition (Welch and Welch, 1971). Although "isolated" female mice of the BALB/c strain which were grouped similarly to male mice (groups of 4) did not fight, small (insignificant) increases in cerebral NE, DA and 5-HT did occur (Welch and Welch, 1971).

Welch and Welch (1969b) found that the reduction of cerebral catecholamines by fighting activity occurred first in brain stem, and that 5-HT tended to be elevated in mesencephalon. Dopamine was reduced to 59% of its control value in the brain stems of mice that fought in groups of four for 150 minutes, while no change occurred in the telencephalon (see Welch and Welch, 1971). Studies on "isolated" mice which were exposed to fighting activity by suspending them in a cage over 4 to 6 other "isolated" mice revealed that merely witnessing fighting for 75 min could markedly reduce the NE content of the pons and medulla (Welch and Welch, 1968b; Table 36). Hence, it seems

TABLE 36. Reduction of brain norepinephrine in male mice that had witnessed fighting among other male mice made aggressive by long-term isolation.[a]

Brain region	Norepinephrine (ng/g)		
	Undisturbed controls	Strange cage	Witness fighting
Telencephalon	568 ± 17·4	557 ± 11·4	531 ± 17·0
Diencephalon and mesencephalon	524 ± 25·2	583 ± 20·5	522 ± 26·6
Pons and medulla oblongata	472 ± 24·9	421 ± 22·4	319 ± 23·4[b]

[a] Values are expressed as mean ± S.E.M.; N = 20 in all cases.

[b] Differs $p < 0{\cdot}001$ from controls, $p < 0{\cdot}002$ from strange cage (Student's *t*-test; two-tailed). Mice were placed in a feed basket above an empty strange cage or above a strange cage that contained fighting mice for 75 minutes. All brain parts were analyzed separately for individual animals. (Reproduced with permission from Welch and Welch, 1968b.)

evident that a purely psychological stimulus can enhance the central release of NE. The marked decreases found in pons-medulla were of interest since cell bodies enriched in these amines are found in this area (e.g. Dahlström and Fuxe, 1964, 1965; Hillarp *et al.*, 1966).

In more recent experiments, Hendley *et al.* (1973) showed that significant increases in the apparent K_m for NE uptake by cerebral cortical homogenates and in the K_i for *d*-amphetamine-induced inhibition of this uptake occurred in "grouped" mice which had been living under chronic attack from aggressive cage-mates (Table 9; see also Chapter 3, Part D). Also, significant increases in apparent K_m and V_{max} for NE uptake by particles of cerebral cortex were observed 18 to 20 hours after the last of a series of short, intense, daily fights between male mice that had been made aggressive by prior "long-term" individual housing (see footnote, p. 60). In male mice, caged individually from weaning to 5 to 9 months of age and then allowed to fight for 10 to 15 min/day for 14 consecutive days, the apparent K_m was increased by about 62% and the V_{max} by about 79% in fighters, as compared with undisturbed "isolated" animals (Table 9). These changes appeared to reflect adaptive mechanisms since no significant changes in K_m and V_{max} could be demonstrated in 2- to 3-month-old "isolated" mice. The elevated V_{max} caused by fighting indicated that an adaptive increase in the number of NE uptake sites had occurred in nerve terminal membranes, and the elevation of K_m indicated a decrease in affinity of NE for its uptake sites in the cerebral cortex. This study and that of DeFeudis (1972d), which showed that a decrease in the "binding" of GABA to nerve ending fractions of brain could be caused by individual housing of mice (Table 33), revealed that cerebral transmitter uptake mechanisms may be adaptable and plastic.

Another study which indicated that monoamines are involved in fighting behavior was conducted by Eleftheriou and Boehlke (1967) who showed that central MAO levels could be modified by exposing C-57 mice to trained fighter mice so that they experienced brief daily periods of fighting and defeat. The greatest changes occurred in hypothalamus; MAO initially increased by nearly 50% after one or two days of fighting and then fell to about half of its original level. The effects of fighting behavior on NE, DA and 5-HT have also been examined in other tissues. An excellent review about the roles of biogenic amines in fighting behavior has been written (Welch and Welch, 1971).

Welch and Welch (1968c) studied the effects of restraint stress in littermate, male, ICR Swiss albino mice (weaned at 4 weeks and housed either individually or in groups for 8 to 12 weeks), by taping the animals lightly to the rounded edge of an inverted cage for 7 min or for 2·5 h. Half of the stressed and half of the non-stressed mice were injected with α-MT. Cerebral NE and DA contents were decreased more during the 2·5 h following α-MT in "grouped" than in "isolated" mice. Hence, when the syntheses of NE and DA were inhibited, stress appeared to accelerate the depletion in "isolated" mice and to retard it in "grouped" mice. Note that such restraint stress for 7 min or 2·5 h significantly elevated cerebral levels of NE, DA and 5-HT. Furthermore, after pargyline, both NE and DA accumulated more rapidly in the brains of subordinate than in those of dominant mice (Welch and Welch, 1971). This finding is of interest since individual housing appears to predispose mice to positions of social dominance. The different response to pargyline showed some similarity to the differences obtained between "grouped" and "isolated" mice.

Increased aggressiveness has also been reported in rats (Sheard, 1969) and cats (Ferguson *et al.*, 1970) treated with p-chlorophenylalanine (p-CPA), an agent that inhibits 5-HT synthesis (Koe and Weissman, 1966), and p-CPA attenuated irritability in rats bearing septal lesions (Dominguez and Longo, 1969). Also, p-CPA has been shown to increase mouse killing (muricidal) behavior in "isolated" rats and to enhance this behavior in bulbectomized (bilateral removal of cortex of olfactory bulb and anterior olfactory nucleus) or bulbotomized (cutting of both olfactory stalks) rats (DiChiara *et al.*, 1971). Although these studies supported the notion that both the anterior olfactory area and 5-HT might exert inhibitory influences on muricidal behavior (see also Vergnes and Karli, 1965), bulbectomy and p-CPA may act by different mechanisms (see DiChiara *et al.*, 1971).

Diurnal and seasonal rhythms affect the cerebral metabolism of these possible neurotransmitters. A phase relationship between rat brain NE and 5-HT levels has been noted: 5-HT peaked during the light phase and was associated with periods of inactivity and sleep, whereas NE peaked during the dark phase and was associated with increased motor activity (Friedman and Walker, 1972). The NE daily rhythm was 180° out of phase with the 5-HT daily rhythm. In mice, both diurnal and seasonal rhythms have been observed for cerebral 5-HT and 5-HIAA (Garattini and Valzelli, 1965); daily rhythm peaks occurred during light hours whereas seasonal variation peaks occurred during June, July and August. Recent findings which revealed that amphetamine toxicity in mice is influenced by time of day as well as by group size might be related to fluctuation in the endogenous catecholamine content of the brain (Walker, 1972).

Variations of exposure to light and dark periods can also affect the metabolism of animals which, in turn, can affect cerebral levels of biogenic amines (e.g. in pineal gland and epiphysis; see Quay, 1963, 1966; Snyder and Axelrod, 1965; Wurtman *et al.*, 1968; Vogel *et al.*, 1969; Klein and Weller, 1970). Also, epiphyseal MAO activity was decreased in squirrel monkeys which were constantly exposed to light (Manocha and Perachio, 1973). Further functional significance may be attached to the above findings by studying recent reviews about the regional distributions and other properties of these monoamine systems (see, e.g. refs in Weil-Malherbe, 1970; refs in de la Torre, 1972; Blaschko and Muscholl, 1972).

Strain differences, extent of handling, age at sacrifice, age of weaning, an especially the sex of animals can influence their levels of cerebral biogenic amines and associated enzymes, as well as other substances, and the responses of animals to psycho-active drugs. For example, Schlesinger *et al.*, (1970) have shown that a correlation exists between endogenous cerebral 5-HT and NE contents and susceptibility to audiogenic seizures in inbred mouse strains (see refs in Schlesinger *et al.*, 1970 for further differences in monoamine systems between strains of mice differing in audiogenic seizure susceptibility; see also Chapter 1, Part D). Developmental differences in mechanisms of monoamine synthesis and/or storage could also contribute to deficient amounts of monoamine transmitter release in the brains of mice susceptible to audiogenic seizures (see Kellogg, 1976). Recent ultrastructural studies on "hypocerebellar" mutant strains of mice have revealed that the molecular layer of the cerebellar cortices of these animals possessed about five times more NE-containing boutons than a control strain (Landis and Bloom, 1975). Also, the study of Weaver and Kerley (1962) clearly revealed strain differences in the response of mice to *d*-amphetamine.

C. General Comments

From the foregoing discussion it is evident that neurotransmitter suspects are integral elements of the *Behavioral Code.* Other small molecules involved in behavioral regulation have been discussed in other chapters of this monograph (see especially Chapters 5, 7 and 8). The major constituent of the nervous system, water, is deserving of mention, and for studies concerning its regional distribution and its roles in neural processes the reader is referred to the following works: Agrawal *et al.* (1968), Pappius (1969), Reynolds (1970), Van Harreveld (1972), Gadea-Ciria *et al.* (1975). Examples of the possible implications of water in behavior, learning and memory have been discussed in relation to swelling of neuronal elements (see, e.g. Chapter 3, Parts B3 and B5).

Nucleotides represent another important series of small molecules associated with behavioral regulation. Roles for cyclic nucleotides as elements of the *Behavioral Code* have been discussed in Chapters 4 and 7. For recent comprehensive studies and reviews of the roles of nucleotides, especially cyclic nucleotides, in neural function, the reader is referred to the following works: Rall and Gilman (1970), Greengard and Costa (1970); Greengard *et al.* (1972, 1973), Greengard and Kebabian (1974), Greengard (1975, 1976); see also Chapter 4, Part C3. For reviews of the roles of vitamins in neural function, see Sourkes (1962), Holtz and Palm (1964), Roberts *et al.* (1964), Dreyfus and Geel (1972), and Ebadi and Costa (1972); for the roles of inorganic ions in relation to nervous conduction, synaptic transmission, cellular metabolism and transport systems, see, e.g. Brink (1954), Eccles (1964), Hodgkin (1964), Huxley (1964), Van Harreveld (1966), Talwar and Singh (1971); Weight (1971), articles in Shamoo (1975), and Hille (1976).

7. Elements of the "Behavioral Code"

An hypothetical scheme has been derived by analyzing environmentally induced effects on cerebral chemistry, physiology and morphology which provides a tentative sequence of events involved in the *Behavioral Code.* It should be noted that in this scheme "specific" (learning and memory) and "non-specific" (stress, exercise) elements which affect cerebral chemistry, physiology and morphology and thus alter behavior have been combined. Figure 80 shows that:

(1) The initial stage of learning, memory and behavioral processes is reception of sensory input from the environment, which can be either *specific* or *non-specific,* which depends upon the frequency and/or strength of stimuli (Bogoch, 1974), and which can elicit either *adaptive* (+) or *non-adaptive* (—) behavioral responses.

(2) This experiential event (initial environmental stress) stimulates primary sensory modalities, thereby initiating changes in the ionic permeabilities of neuronal membranes which lead to nervous impulses (action potentials). An increased release of hormones (e.g. ACTH, vasopressin) also occurs. (It should be noted that "stress", as used herein, refers to the "initiator" of mechanisms of adaptation and biological survival, rather than to the maladaptive consequences of the stressor (see Yuwiler *et al.*, 1973).)

(3) Nervous impulses promote changes in cellular energy-yielding processes (e.g. oxidative metabolism of glucose) since their recovery processes (i.e. restoration of ionic gradients across neuronal membranes) require energy-linked ATP-ase systems. Changes in energy metabolism are not localized to specific regions of the nervous system. They also occur in some extra-cerebral tissues (e.g. liver). At this stage, "short-term" memory could be encoded in nervous pathways by processes involving *conformational changes* in membrane macromolecules (or changes in the structures of macromolecules that contain monosaccharide, phosphate, or other residues; such as gangliosides and phos-

(1) Environmental stimulation; specific or non-specific (+ or −).

↓

(2) Effects on primary sensory modalities; initiation of nerve impulses and transmitter release; release of ACTH and other peptides which cause a stress-induced increase of "arousal" level and a limbic system response.

↓

(3) Effects on O_2 uptake, energy metabolism, and ATP-ase systems; further transmitter release; alterations in the conformations of membrane macromolecules; key stage of "short-term" memory (about a few sec post-stimulation).

↓

(4) Effects on rates of synthesis of cyclic nucleotides, m-RNA and protein.

↓

(5) Effects on the amount of protein transported to nerve endings by axoplasmic flow, and on the amount of protein present in other membranes; changes in amounts of transmitter enzymes, receptor and uptake sites, or in the number of synaptic boutons; key stage of "long-term" memory (about 3 to 20 min post-stimulation).

↓

(6) Changes in synaptic efficacy and in functional activity of the CNS.

↓

(7) Changes in learning, memory, and adaptive behavior.

FIG. 80. Elements of the "Behavioral Code".

phoproteins). Such conformational changes could be associated with a change in the rate of recovery of the extracellular K^+ level and/or with a change in the duration of membrane hyperpolarization (i.e. with a change in membrane resistance) produced by the increased afferent input and neuronal activity (see, e.g. Krnjević and Morris, 1975a, b). Ca^{2+} entry into neurones could be involved in the regulation of such changes in K^+ conductance (Krnjević *et al.*, 1975; see also Chapter 3, Part B3).

(4) Changes occur in the rates of neurotransmitter release from presynaptic terminals which could affect the enzymes that control the syntheses of various substances (e.g. cyclic nucleotides; RNA; protein; conjugated proteins, such as phospho-, nucleo- or glyco-proteins). As examples, ACh release could induce the synthesis of AChE and ChAc (Smith, 1962) or the syntheses of tyrosine hydroxylase and dopamine-β-hydroxylase in terminal adrenergic neurones and adrenal chromaffin cells (Molinoff and Axelrod, 1971; Thoenen, 1975; Mandell, 1975b; Stolk *et al.*, 1975), or the release of monoamines (or other possible transmitters) could stimulate the formation of γ-glutamyl-peptides (e.g. Reichelt and Kvamme, 1967; Reichelt, 1970) or N-acetyl-aspartyl-peptides (see Kvamme *et al.*, 1975). Such mechanisms of enzyme induction would be preceded by a neurotransmitter-induced activation of adenyl cyclase and protein kinase (see, e.g. Glassman *et al.*, 1973) which could increase RNA synthesis (e.g. Doemin, 1968; Mueller *et al.*, 1969a, b; Thoenen *et al.*, 1970; see also Fillenz, 1972; Walsh and Cummins, 1975). Also in this connection, a correlation has been shown between a stress-induced increase in cyclic-GMP and the GABA concentration of the rat cerebellum (Mao *et al.*, 1974). The first involvement of nucleic acids could be activation of RNA polymerase which could enhance the rate of transcription of genetic information, and RNA may be the first substance synthesized which is associated with the structural change (see, e.g. Roberts and Matthysse, 1970; Bennett, 1973). This synthesis, which could involve all types of RNA, could occur via genome activation involving a derepression due to an interaction of the released transmitter (or other substance) and a repressor (Walsh and Cummins, 1975). Of particular interest in this connection is the recent finding that brain-specific S-100 protein(s) stimulates the activity of RNA polymerase I in isolated brain nuclei by an effect that appears to be selective to these nuclei (Miani *et al.*, 1973; Michetti *et al.*, 1976). Since S-100 protein(s) is found in non-histone fractions of brain nuclei (Michetti *et al.*, 1974), it may be involved in tissue-specific gene activation (see Stein *et al.*, 1974; for comprehensive coverage of gene expression and its mechanisms of regulation, the reader is referred to Kenney *et al.*, 1973, and Lewin, 1974a, b).

(5) Changes occur in the *amount of protein* synthesized and transported to presynaptic nerve endings and dendritic arborizations; protein constituents of postsynaptic membranes of initial portions of dendrites that do not require axoplasmic flow would also be affected by this change in protein synthesis (see Chapter 4). This stage represents an early phase of "long-term" memory (which could be linked to step 3, above), the consolidation of which likely depends upon the synthesis of membrane macromolecules.

(6) Adaptive changes in synaptic efficacy and in functional activity of the nervous system occur, which involve changes in synaptic connectivity and/or in the numbers, sizes, or development of synapses. Such changes could lead to alterations in the total tissue mass of the brain. These changes are regulated by the organism's "arousal" level (see, e.g. Bexton *et al.*, 1954; Hebb, 1955; Jasper *et al.*, 1958; Welch, 1965; Glassman and Wilson, 1972; DeFeudis, 1974b; Walsh and Cummins, 1975) and limbic system response (see, e.g. Sheer, 1961; Koikegami, 1964; Delgado, 1967; Kamback, 1973; Izquierdo, 1975a; Ely *et al.*, 1976), and are reflected as changes in learning and memory that alter behavior. The link between the limbic (e.g. hippocampal) response and the "arousal" level is well expressed in the recent model of Kamback (1973), which suggests that the hippocampus is involved in motivational and memory processes as a "forebrain integrative mechanism" that modifies, or biases, the organism's response to changes in the *external* environment as a function of the state of the internal homeostatic milieu which varies according to the organism's level of "arousal". Hence, a reciprocal inhibitory relationship appears to exist between hippocampal and reticular-activating (RA) mechanisms (see refs in Kamback, 1973), as is the case between the neocortex and RA (e.g. Dell, 1963). This notion is further strengthened by results which have indicated that both hippocampal lesions and homeostatic shifts (e.g. changes in glucose level) lead to improved performances on simple tasks and to impaired performances on complex tasks. Furthermore, it should be noted that though human learning appears to be severely disturbed by hippocampal lesions, the hippocampus may not be involved in the *storage* of memory since old memories remain intact after such lesions (see P. M. Milner, 1970). In any case, the anatomical circuitry of the hippocampus, with its receiving areas for projections from diverse receptor systems (see, e.g. Cragg, 1961), certainly qualifies it for a significant role in learning and memory (see also Chapter 2 and Chapter 8, Part E).

Recent electrophysiological studies have shed further light on the possible role of the hippocampus in learning and memory. A feature of the hippocampus which predisposes it to a role in learning and memory is that its major excitatory inputs possess a remarkable degree of physiological and morphological *plasticity*. Long-lasting facilitation (potentiation), a physiological measure of plasticity, has been shown both *in vitro* following tetanization of slices of guinea-pig hippocampus (see Schwartzkroin and Wester, 1975) and *in vivo* by electrical stimulation of the perforant path-granule cell synapses of rabbit or rat hippocampus (see Bliss and Gardner-Medwin, 1973; Douglas and Goddard, 1975). Also, in the dentate gyrus the synaptic loss caused by removal of one afferent was compensated for by formation of additional synapses by intact afferents (see refs in Nadler *et al.*, 1976). Other *in vitro* data have indicated that glutamate and aspartate may be transmitters in hippocampal excitatory afferents (Nadler *et al.*, 1976). Such characteristics further substantiate a role for the hippocampus in information processing. Such synaptic plasticity indicates that this hippocampal pathway could be associated with memory storage, rather than merely with the transmission of learned information (see Bliss *et al.*, 1973). Further studies have revealed that electrical stimulation of this same monosynaptic pathway (rat entorhinal cortex-dentate gyrus projection; "perforant path"; see Bliss *et al.*, 1973 and Schubert *et al.*, 1976 for diagrams of these anatomical structures) elicited a release of ^{3}H-adenosine derivatives from central axon terminals to their postsynaptic neurones, after stereotaxic injection of ^{3}H-adenosine into the dorsal aspect of the entorhinal cortex (Schubert *et al.*, 1976). Hence, as suggested by Schubert *et al.*, (1976), the unusual physiological plasticity of these synapses; i.e. their potentiation for hours or even days after repeated stimulation (see above refs), could be related to adenosine-induced metabolic changes in granule cells.*

(7) A change occurs in the behavior emitted by the organism which further modifies the environment (see also Fig. 1).

An hypothetical scheme concerning the temporal relations of changes in synaptic connectivity that occur during plastic synaptic changes has been provided by Roberts and Mattysse (1970); see Fig. 81. From this scheme, it might be considered that in the scheme of Fig. 80 the steps 1 and 2 represent the "physiological" level (milliseconds), step 3 represents the "activational" level (seconds), steps 4 and 5 represent the "growth" level (minutes to hours), and steps 6 and 7 represent the level of "genetic expression" (minutes to years). Since the components have overlapping time scales in both schemes, it is suggested, in accord with Roberts and Matthysse (1970), that their sum, memory consolidation, is best represented as a continuum (see also Chapter 8, Part E).

It should be noted that the natures of behavioral responses, as well as the amounts and types of stresses that produce these responses, are governed by the characteristics of the new stimulus and are modified by previous experience (i.e. by the remembering and forgetting of previous events). Studies by the Berkeley group (see, e.g. Bennett *et al.*, 1964, 1973; Rosenzweig *et al.*, 1969) which have revealed that the effects of an "enriched" environment on cerebral anatomy and chemistry can be shown in blind, dark-reared, and anosmic animals have provided support for a role of the generalized "arousal" reaction (see

---

* The hippocampus-amygdala region of mammals contains high concentrations of heavy metals (e.g. lead, zinc), which may be of physiological significance (see Danscher, 1976).

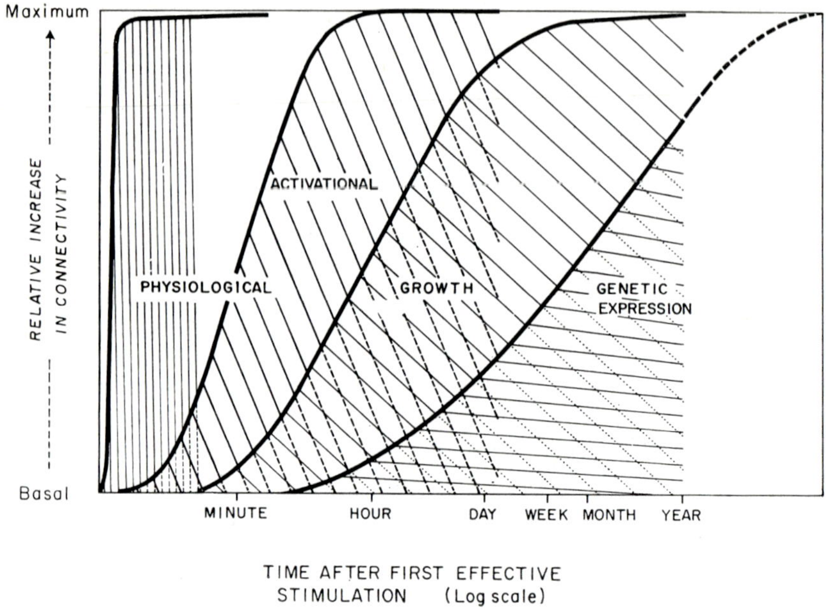

FIG. 81. Hypothetical time-relations of increases in synaptic connectivity from the basal to maximal levels during plastic synaptic changes. The dark curves indicate the increases in connectivities presumed to occur when a synapse is activated by a stimulus of uniform intensity occurring at a constant frequency; the lighter lines extending downward from points on the curves are intended to show the decreases in connectivities from points attained on the curves as a function of time after cessation of all stimulation. (Reproduced with permission from Roberts and Matthysse, 1970.)

Walsh and Cummins, 1975). Thus, this element of the *Behavioral Code* is largely independent of the sensory modality stimulated.

Much evidence supports an involvement of the limbic system in behavioral responses. For example, this region exhibited increased macromolecular synthesis during training and learning experiments (e.g. Nasello and Izquierdo, 1969; Bowman and Strobel, 1969; Glassman and Wilson, 1972; Hydén, 1973a; Izquierdo, 1975a, b; see Chapter 4). However, many of the neurochemical changes that govern behavior are not expected to be regionally specific since the whole nervous system, as well as many peripheral tissues, are likely to be activated by environmental changes. As a dramatic example of this, one might consider the global effects of electrically induced seizures in man. Such environmental insults cause an increased metabolic demand in the brain, and the compensatory mechanisms include increases in systemic blood pressure and cerebral blood flow, decreases in cerebral venous and cortical oxygen tension and a depletion of cerebral energy stores (e.g. Posner *et al.*, 1969). Some of these changes (e.g. hypertension) may well be correlated with the behavioral dysfunctions which accompany certain psychotic states in man (see also Chapter 8, Part C). Also, the hypotheses of Lashley (1950, 1958) should be recalled in relation to this lack of regional specificity (see Chapter 2, Part A).

The amount of new protein (both structural and enzymatic) synthesized for synaptic structures appears to be the major event which occurs during memory and which governs behavioral changes (see also Chapters 3, 4, and 8). This chemo-morphological change, which is triggered by nervous impulses and which affects membrane events and, subsequently, the regulation of gene expression, may persist in localized neural pathways by a form of specificity that could depend mainly upon the nature of the environmental stimulus. As suggested by Rosenzweig *et al.* (1972c), there may be several types of structural changes underlying learning and memory storage: (a) an increase in the number of synapses; (b) a decrease in the number of synapses; (c) an increase in the sizes of synaptic contact areas; (d) a decrease in the sizes of contact areas. Also, different types of changes could occur in different sets of synapses in the same brain depending on the nature of the environmental insult received by the animal.

Evidence does indeed exist to support the idea that all such changes may occur during various types of experience (e.g. Ramón y Cajal, 1899; Thesleff, 1962; Cragg, 1967, 1968, 1969; Fifkova, 1970; Møllgaard *et al.*, 1971; Ramón y Cajal cited by Jacobson, 1970b; Dawkins, 1971; see also Chapter 4).

"Enriched" and "impoverished" environments are expected to elicit positive and negative effects, respectively, on energy metabolism (e.g. incorporation of glucose carbon atoms into brain; see Chapter 5) and on protein and RNA synthesis (see, e.g. Chapters 4 and 5). DNA synthesis also represents an important element of the *Behavioral Code*, since postnatal neurogenesis and gliogenesis occur in brain (see, e.g. Altman, 1969). Biochemical, physiological, and morphological differences that exist between the brains of "isolated" and "grouped" animals, and the difference in their behavior, can be fitted to this scheme. The *Behavioral Code* can be studied by performing a detailed analysis of any of the events shown in Fig. 80, but in no case can it be explained by any one of these events.

8. General Conclusions and Related Hypotheses; Relevancy to Man

A. Environmental Considerations

Although much further study is required for complete elucidation of the *Behavioral Code*, its two main elements, *environment* and *cerebral subcellular structure*, have been long known. Further consideration of this relationship appears to be required for solving human problems associated with social dysfunction and population dynamics.

Alterations of social environment cause quantitative (and perhaps, qualitative) chemical and morphological changes in synaptic structures of animals that have been correlated with alterations in their perception and behavior and with modifications of their responses to centrally active drugs. Such behavioral changes thus become elements of the social environment (see Figs 1 and 80). In general, it may be assumed that an optimal level of environmental stimulation exists for an organism (Welch, 1965), and that this is reflected in its optimal level of "arousal" and associated performance (Hebb, 1955). Therefore, in many of the studies discussed in the preceding chapters of this monograph, both the decreases and increases in level of environmental stimulation of the experimental animals caused stresses upon the development of their brains and behaviors.

At the biochemical level, changes in the cerebral incorporation of carbohydrate atoms produced by "long-term" differential housing in mice, were reversible in 1·5 h (DeFeudis, 1973). Also, increases in cerebral cortical weights of rats that were induced by 15–60 days of exposure to an "enriched" environment exhibited some spontaneous reversal. Such changes were most pronounced at 30 days and declined thereafter (Rosenzweig, 1967). Other studies have revealed that though the weights of the cerebral cortices of rats were decreased or increased by 80-day exposures to "impoverished" or "enriched" environments, respectively, these effects could be readily overcome by subjecting the animals to the opposite rearing condition for 50 days (Rosenzweig, 1967; see also Chapter 2, Part B). The transient natures of some environmentally induced effects were also demonstrated by Gyllensten *et al.* (1965), who showed in light- *vs* dark-reared mice, that the differences in cortical thickness (in favor of light-reared animals) decreased from about 5·5% at two months to about 4% at four months, and that the difference in the diameters of cell nuclei decreased from about 8% at 20 to 30 days of age, to about 3% at two or three months and to about 1% at four months. Hence, environmentally induced brain growth appeared to decline with increased visual experience. Also in this connection, Levitan *et al.* (1972b) showed that hippocampal protein synthesis in rats was increased after 4 days, but not after 30 days, of training (see also Chapter 4, Part C2).

Most of the above mentioned findings provide examples of temporary or transient environmentally induced changes. However, other types (permanent changes; permanent changes having temporary components) could occur (see Rosenzweig, 1967), especially when environmental changes occur during periods of rapid cerebral development. Changes that occur in the total protein and in the numbers of ligand "binding" sites present in synaptic particles (see, e.g. Figs 22 and 23) that might be involved in the transduction of informational input into "long-term" memory, are perhaps not so readily reversible since these involve cellular structure. At the behavioral level, Leigh and Hofer (1975) have shown that altered early social relationships, caused by reducing rat litter size to a single pup at the twelfth postnatal day (pre-weaning isolation), can produce enduring effects on physiological responses and behavior which can be measured when these animals achieve adulthood (see also Leigh and Hofer, 1973 and Fig. 82). Also, other studies have revealed that children reared in emotionally deprived environments can exhibit dwarfism, presumably due to abnormalities in their sleep-wakefulness patterns which can affect the release of growth hormone and other pituitary hormones (see Gardner, 1973, and Chapter 4, Part E2).

The possible reversibility of such environmentally induced changes, which most likely involve structural proteins of synapses, (see Chapter 4), provides a basic rationale for devising methods for reversing behavioral disturbances in man. Many experientially induced neurochemical correlates of behavioral modification may be reversible. This contention is supported by the present widespread use of various psychopharmaca (e.g. neuroleptics) for treating behavioral disorders in man, but perhaps therapies based on nutritional and/or socio-environmental modification would better serve this purpose. Some examples of nutritional therapy have been discussed in Chapter 1 (Part F) and some examples of environmental therapy have been outlined above (see also refs in Chapter 1, Part H; Chapter 2; Suomi *et al.*, 1974). Other studies have revealed that behavioral deficits produced by hormonal deficiencies (e.g. experimental cretinism in rats) can also be reversed by environmental stimulation (see, e.g. Davenport *et al.*, 1976).

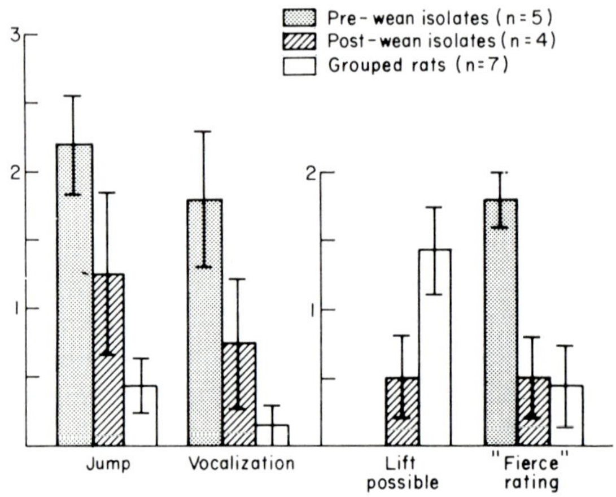

FIG. 82. Reactions of pre-wean and post-wean isolated rats and grouped rats to a "gloved hand" attempting to pick up the experimental animal. Means ± S.E.M.; numbers of animals are indicated. (Reproduced with permission from Leigh and Hofer, 1975.)

Reductions in the numbers of GABA and glycine "binding" sites and in the protein content of cerebral nerve-ending fractions and increases in the binding capacities of these fractions for psycho-active drugs (see Chapter 3, Part D; Part F, below; Figs 22 and 24) that were caused by subjecting mice to environmental impoverishment may have been produced by a decrease in biological maturation (or ageing) of the brain (DeFeudis, 1975b; Part G, below) which has features in common with the developmental lag caused by undernutrition (e.g. Smart, 1975). This decrease in "binding" sites for inhibitory transmitter suspects has been corroborated to some extent by recent findings which have indicated that an increase in the GABA content of the olfactory bulbs of "killer" rats (produced by direct injections of either GABA, or of inhibitors of GABA catabolism) prevented their killing responses (P. Mandel and colleagues, 1975, personal communication), and that bulbectomy-induced increases in muricidal behavior of rats appeared to be related to the destruction of tissue that might exert a tonic inhibitory influence on the rest of the limbic system (Hull and Homan, 1975). Such an isolation-induced decrease in "binding" sites for GABA and glycine could reflect a decrease in the development of cerebral inhibitory processes which could explain the altered "arousal" level of "isolated" mice (e.g. their hyper-irritability and increased aggressiveness when presented with novel stimuli or with con-specifics). The results of P. Mandel and colleagues (1975; personal communication), together with other findings (see Chapter 7), further support the proposed role of limbic areas (e.g. hippocampus) in emotional control, learned responses, memory and behavior (see, e.g. Meissner, 1966; Douglas, 1967; P. M. Milner, 1970; B. Milner, 1970; Glassman and Wilson, 1972, 1973; Hydén, 1973a; Izquierdo, 1975a, b; Ely *et al.*, 1976). However, this does not mean that central responses to environmental modification are specific to the hippocampus, but rather that the response of the hippocampus to a given amount of environmental stress appears to be greater than that of other cerebral structures, especially with regard to the initial events of information processing.

At the human level, and in line with some earlier words of Lashley (1951), Miller (1972) has revealed the relationship that exists between learned factors and acquired mental illness:

> The effect of learning on psychoses is demonstrated quite clearly by the contents of delusions. People no longer are likely to be deluded into thinking that they are Napoleon Bonaparte or Franklin Delano Roosevelt, but their delusions have other, more modern contents.

Such learned factors are obviously derived from the environment.

B. Implications of Structural Plasticity; Ageing

From the foregoing discussion (see especially Chapters 2 and 3), it seems evident that considerable potential for structural plasticity exists in the mammalian CNS, thus providing the morphological basis of the *Behavioral Code*. Recent studies have revealed further that group-living can promote increases in brain weight well into old age, at least in the rat (see Cummins *et al.*, 1973; Welch *et al.*, 1974a). Thus, adaptable chemical and morphological elements of the CNS may exist as long as an animal possesses the abilities to become aroused, to learn, to memorize, or to display adaptive behavior (see Ordy and Schjeide, 1973).

Some reports have indicated that neurones and synaptic connections may be lost from the human brain throughout adult life (e.g. Brody, 1955, 1970). However, this idea is not universally accepted (e.g. Comfort, 1971: Tomasch, 1971; Hanley, 1974; Meier-Ruge, 1976), especially with regard to lower animals. In any case, if such a neuronal loss does occur during senescence, it could be correlated with an increase in cerebral DNA concentration (Perez and Moore, 1970), with decreases in cerebral metabolic rate, O_2 consumption, blood flow (e.g. Frehan *et al.*, 1951; Scheinberg *et al.*, 1953; Kety, 1955; Quastel and Quastel, 1961; Himwich, 1973; Fig. 83; see below), and cerebral protein synthesis (e.g. Meier-Ruge, 1976; see below), and possibly with an abnormal ganglioside pattern in cerebral gray matter (e.g. Bowen

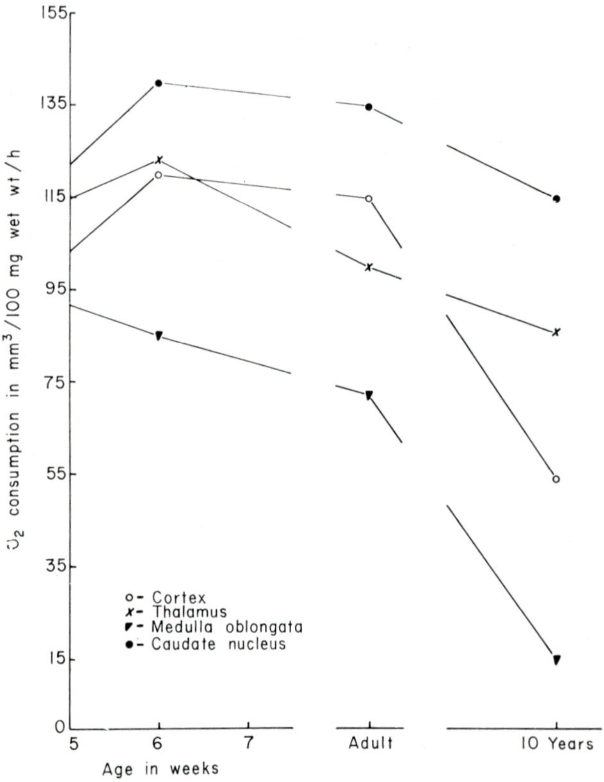

FIG. 83. Oxygen consumption of various regions of dog brain during development. (Reproduced with permission from Himwich and Himwich, 1959.)

et al., 1973; Op Den Velde and Hooghwinkel, 1975). Such a neuronal loss would most likely be "non-specific" because most, or all, neurones could be similarly affected by the above mentioned changes. If these changes are found to be associated with the normal mechanism of *stored* ("long-term") memories, or their loss, support will be gained for the concepts of Lashley (see Chapter 2, Part B) since there would be no specific localization of the engram. This idea holds even when it is considered that electrical stimulation of discrete brain regions can elicit recall of specific memories in humans, since such memories persisted after extirpation of the regions involved (Penfield and Jasper, 1954).* *Short-term memory* processes and their loss, which appear to be linked to a general activation of the brain (see also Part E, below), could be even more readily associated with some of the above mentioned "non-specific", stress-related processes.

In relation to this, it has been shown that a gross proportionality exists between functional deficit and decreased cerebral blood flow (CBF) in humans afflicted with severe dementia, coma, stupor or major focal abnormalities (Plum and Posner, 1966; Brodersen and Jørgensen, 1974; Ingvar and Gadea-Ciria, 1975). Both CBF and cerebral metabolism are reduced in cases of dementia (e.g. Ingvar and Gustafson, 1970; Raichle *et al.*, 1975). In some patients, presenile Alzheimer's dementia has been correlated with a low mean CBF in gray matter which was most pronounced in pre- and post-central regions of the cerebral cortex ("association cortex"). During mental effort these regions appeared to be less activated than in normal persons, or even de-activated, causing a further decrease in CBF; similar changes occurred in human cases of low pressure hydrocephalus which were also characterized by symptoms of dementia (see Ingvar and Schwartz, 1975). Note that such regional changes in CBF and in associated carbohydrate metabolism, which appear to occur in relation to memory disorders in humans, argue against the "law of equipotentiality" of Lashley (1929) (see also Chapter 2, Part A).

---

* In the course of neurosurgical operations it was found that electrical stimulation of the temporal cortex produced *psychical experiences* in patients which could be memories, dreams or hallucinations (Penfield and Jasper, 1954). These experiences were auditory or visual, or both, and unfolded steadily as long as the stimulation was maintained. A single stimulation activated a single memory, and re-stimulation could produce an identical memory or a different memory, but two different psychical responses were never produced at the same time. In a few cases a specific experience (e.g. hearing a particular tune) seemed to be brought back as it had originally occurred, and the patients reported it as an experience rather than as a memory of one. One patient reported, "The song is not as real to me now, as I recall it to memory, as it was in the operating room. There, in the operating room, I actually heard it." (Penfield and Jasper, 1954, p. 132.) This statement was made after removal of the area from which the experience had been elicited by electrical stimulation. Penfield and Jasper (1954) also described cases in which removal of those areas from which stimulation produced specific, vivid recollections did not alter the retention of these same memories. Penfield and Jasper (1954) suggested further that such findings may be peculiar to persons having temporal lobe epilepsy; i.e. these memories are more easily elicited in cases in which some alteration occurs in the stimulatability of this cerebral area, since temporal lobe stimulation of patients with epileptic foci in different cortical areas did not elicit particular memories. However, it should be noted that in these latter patients such stimulation was quite limited and was carried out only on the surface of the temporal lobe.

In rats, an age-related decrease in dendritic spines of cerebral cortical cells has been reported (Feldman, 1974). Also, Scheibel *et al.* (1975) have shown that a progressive loss of horizontal elements of the dendritic system of cerebral cortical pyramidal cells, accompanied by, or preceded by, a "swelling" of soma and major dendrites, occurred during ageing in humans (Figs 84 and 85; see also Gibson *et al.*, 1976 and Chapter 3, Part B). This sequence of events appeared to retrace, in opposite direction, the original development of these dendrites (e.g. Ramón y Cajal, 1929; Conel, 1939–63; Scheibel and Scheibel, 1964). Some of these age-related dendritic changes were similar to those that characterize certain neuropathological conditions (see, e.g., refs in Scheibel *et al.*, 1975).

An involvement of cerebral protein with memory losses which occur during ageing and in certain neuropathological disorders also appears evident. Decreases in the activities of cerebral AChE, Na^+, K^+-ATPase, histone methylase, succinate dehydrogenase, lactate dehydrogenase and pyrophosphatase all occur during ageing (see refs in Meier-Ruge, 1976). These findings support the "error catastrophy theory" (Orgel, 1963, 1970) which suggests that during ageing the error rate in the genetic code for protein synthesis increases.

Other evidence interrelating ageing, memory and cerebral protein has been obtained using newly devised techniques for the separation of neurones and glia from human brain tissue (Iqbal and Tellez-Nagel, 1972; see also Norton and Poduslo, 1970), followed by preparation and purification of tubulin (Shelanski *et al.*, 1973) and neurofilament protein (Shelanski *et al.*, 1971) and then by polyacrylamide gel electrophoresis (e.g. Shapiro *et al.*, 1967; Maizel, 1971). It has been shown that one of the major changes that occurs during cerebral ageing; i.e. the presence of many intraneuronal argentophilic fibrillary tangles composed of bundles of fibrillar elements (termed "twisted tubules" (TT)), found to date only in human brain (Terry, 1963; Terry and Wisniewski, 1972), might be due to an alteration in protein metabolism. Twisted tubules are present in greatest extent in hippocampus (see, e.g. Iqbal *et al.*, 1974a).

Further studies with neurones isolated from the hippocampal cortices of normal patients and of patients afflicted with Alzheimer's presenile-senile dementia and Guam-Parkinsonism dementia revealed that a new protein was present in TT-enriched fractions (Iqbal *et al.*, 1974b, 1975; Fig. 86). Interestingly, the molecular weight of this protein (about 50 000 daltons) is about the same as those of the 14-3-2 protein (see, e.g. Moore, 1972) and protein II of Ueda *et al.* (1973). This "enriched fraction band" (EFB) protein, though abundant in TT-enriched fractions, was absent from control hippocampi.

Earlier studies which had indicated that TT might represent a pair

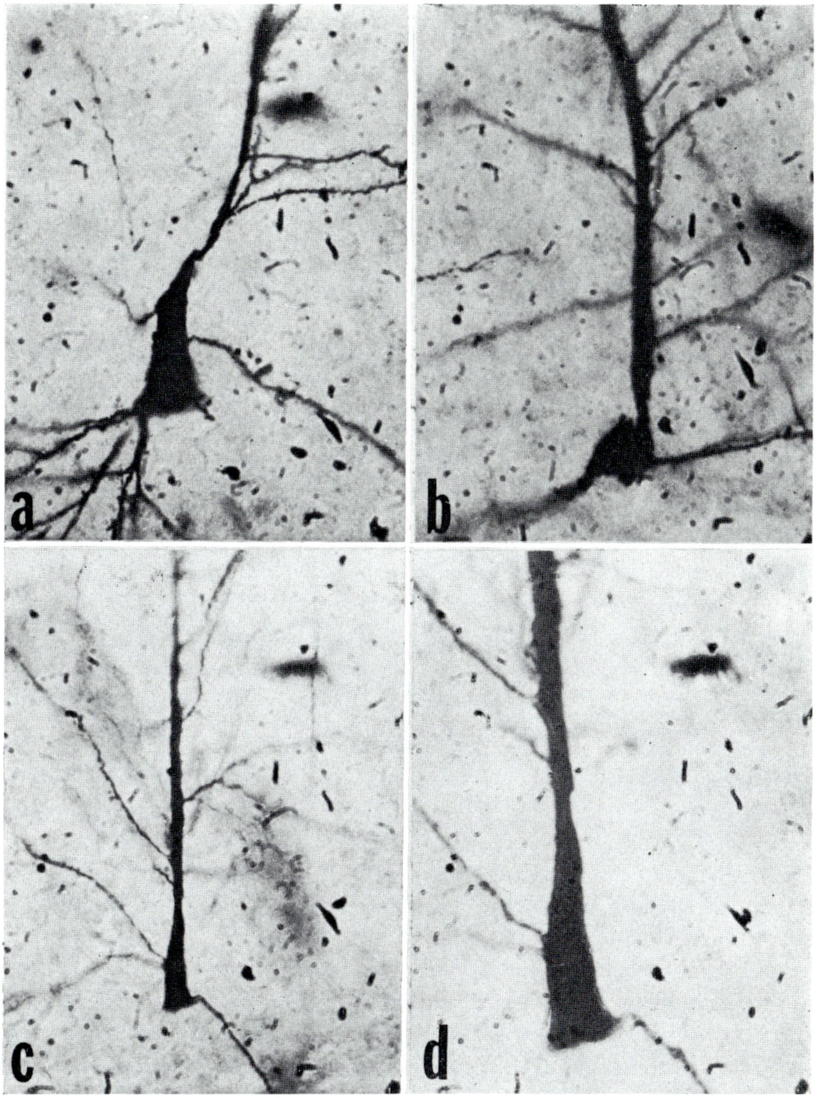

FIG. 84. Golgi-impregnated neurones showing progressive senile changes in pyramids. (a) Fifth layer pyramid in premotor cortex of 83-year-old man showing swelling of cell body and loss of basilar dendrites on one side; (b) upper fifth layer pyramid of motor cortex showing irregular swelling of cell body and apical shaft and partial loss of basilar dendrites in 83-year-old man; (c) and (d) fifth layer pyramid in prefrontal cortex showing characteristic distortion of cell-dendrite outlines and loss of basilar branches, 83-year-old man. Original magnifications; (a), (b) and (d), ×440; (c), ×300. (Reproduced with permission from Scheibel *et al.*, 1975.)

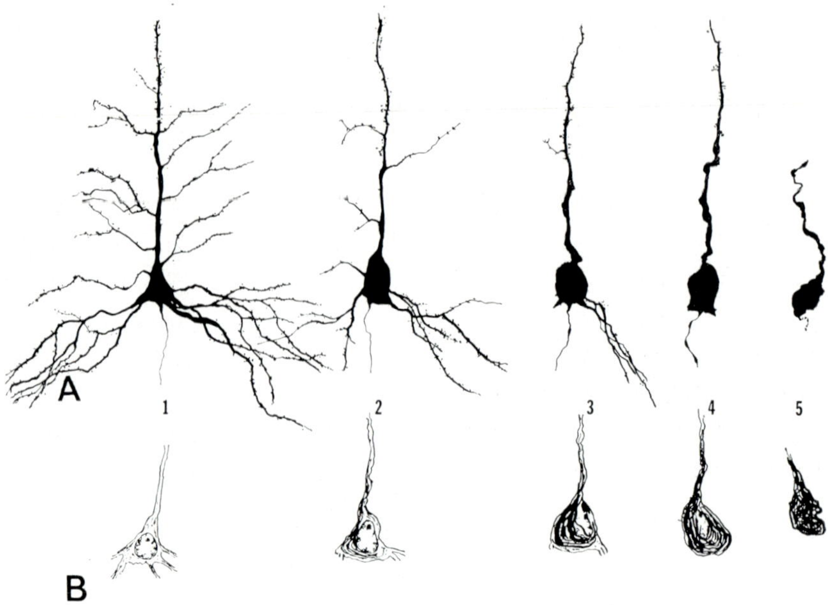

FIG. 85. Progression of senile changes in cortical pyramids as seen in Golgi impregnations (A) and in Bielschowsky stains (B). The authors suggested that the increasing swelling and distortion of the soma-dendrite silhouette and progressive loss of dendrites, first horizontal, then vertical, are closely related to the increased choking of the cytoplasmic space with abnormal tubular material. (Reproduced with permission from Scheibel *et al.*, 1975.)

of helically-wound filaments (Kidd, 1963) have been recently confirmed (Wisniewski *et al.*, 1976). Unlike normal neurotubules, TT resembles neurofilaments since it is insoluble in buffer and is structurally stable in autopsied tissue (see Iqbal *et al.*, 1974b). In this relation, it seems interesting that colchicine and vinblastine (inhibitors of fast axoplasmic transport) and aluminum (which is present in concentrations about 3- to 4-fold above normal in cerebral tissues of patients with Alzheimer's disease; Crapper *et al.*, 1973, 1976) all cause neurofilaments to proliferate in experimental animals (Shelanski and Taylor, 1970; Wisniewski and Terry, 1970; Wisniewski *et al.*, 1970). The possible role of aluminum in Alzheimer's disease provides an example of how a common cytotoxic factor of the *external* environment might be related to disturbances in memory.

Dendrites also appear to be involved in Alzheimer's disease (Wisniewski and Terry, 1973). The densities of apical dendrites on pyramidal cells of layers III and V of the cingulate gyrus and Ammon's horn were lower in cases of senile dementia and Alzheimer's disease than in control subjects at all distances from the cell body measured (see Mehraein *et al.*, 1975). Such a "de-afferentation" of cortical cells could be the primary morphological change underlying dementia. Since the density

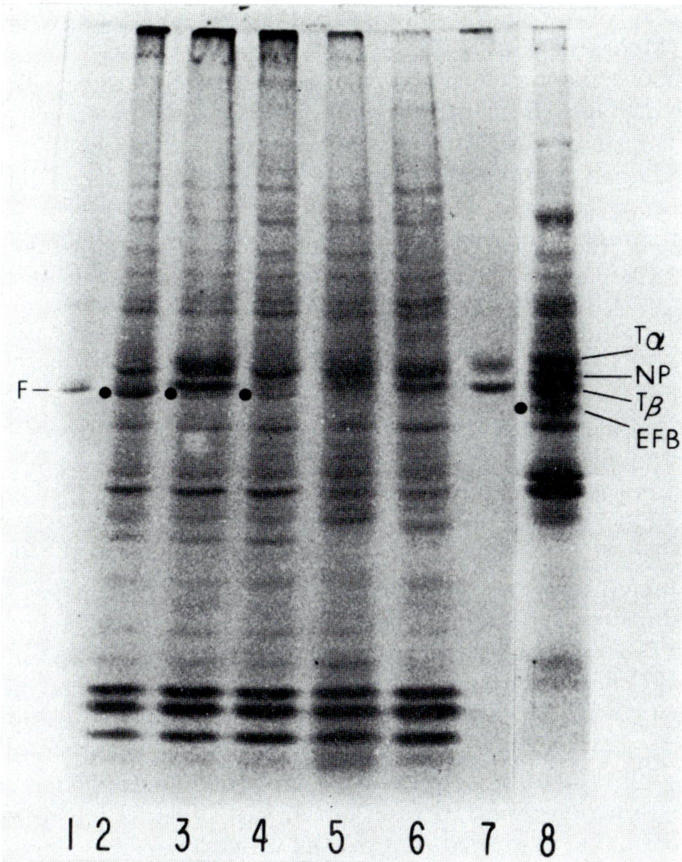

FIG. 86. Sodium dodecyl sulfate (SDS)-polyacrylamide slab gels showing protein patterns of: 1, purified bovine neurofilament protein (F). Molecular weight of major band is 51 000 daltons; 2, total neuronal proteins from a case of senile dementia mixed with the filament protein; 3, total neuronal proteins as in 2 but mixed with purified human brain tubulin; 4, 5, and 6, total neuronal proteins, respectively, from cases of senile dementia, normal adult control, and normal young (5-year-old) control; 7, purified human brain tubulin, α-monomer, mol. wt. 56 000 daltons, β-monomer, mol. wt. 53 000 daltons; 8, a separate but similar gel with TT-enriched fraction from another senile dementia patient. F is the major neurofilament protein subunit; T_α and T_β are the α and β-monomers of tubulin, respectively; NP is a neuronal protein which migrates between the two tubulin monomers; and EFB is the new band corresponding to the enrichment of twisted tubules. Dots on the left side of samples 2, 3, 4 and 8 indicate the position of the band (EFB). Other differences seen in the electrophoretic patterns between diseased and control brain fractions were not consistent. (Reproduced with permission from Iqbal *et al.*, 1974b.)

and extent of arborization of Purkinje cell dendrites were also reduced, cells throughout the nervous system might be affected in these diseases (Mehraein *et al.*, 1975). Also, since Schubert *et al.* (1972) showed that inhibition of dendritic transport in nerve cells can produce similar changes, an effect on this process could be involved in Alzheimer's disease (Mehraein *et al.*, 1975).

The above mentioned findings gain added significance when it is considered that neurofilaments could be involved in the force-generating mechanism of axoplasmic transport (Weiss and Mayr, 1971; Ochs, 1971, 1972a, b). Twisted tubules (if representing pairs of helically-wound neurofilaments) could inhibit this process, and thereby produce a loss of neuronal function which could lead to memory impairment. This suggestion holds even though other postulated mechanisms for axoplasmic transport involve neurotubules (e.g. Kreutzberg, 1969; Gross and Beidler, 1975) or vesicles (Schmitt, 1968), since the new protein of TT could be a constituent of any of these organelles. Therefore, TT appears to provide a link between ageing, memory loss and the axoplasmic transport mechanism. Furthermore, since neurofibrillary tangles (characterized by TT) have also been found in the brains of normal aged humans (see Terry and Wisniewski, 1970), they could also be involved in the memory loss which occurs during normal ageing.

Huntington's chorea (HC), is characterized by dementia and by neuronal losses in the cerebral cortex and basal ganglia (Tellez-Nagel *et al.*, 1973). Iqbal *et al.* (1974a) have shown that ribosomal fractions of neurones obtained from unfixed autopsy specimens of the cerebral cortices of cases of HC possessed abnormally high levels of three proteins (mol. wt. $\simeq$ 10 000 to 16 000 daltons). These three proteins, which accounted for about 50% of the total protein content of the neuronal ribosomal fraction from HC cases but for only about 10% of this fraction in normal or abnormal controls and which are likely to be histones, (normally synthesized by ribosomes and transported to the nucleus where they are bound to DNA; e.g. Robbins and Borum, 1967; Pederson and Robbins, 1970), may be related to the memory loss associated with HC.

In line with the changes in cerebral protein that appear to be associated with ageing and memory loss, Meier-Ruge (1976) has suggested that a reduction in the enzyme activities associated with cerebral aerobic energy metabolism along with its effect on ATP synthesis may be the basic event that mediates the ageing process. Thus, the age-related decrease in cerebral electrical activity could result from a decrease in the coupling of neuronal metabolism and electrical activity caused by the reduction in aerobic energy metabolism (see also Quastel and Quastel, 1961, and Chapter 5). Various theories of ageing, as well as the difficulties involved in conducting this type of research, have been discussed by Meier-Ruge (1976). Many of the neurochemical changes which occur during middle and late life dementias in humans have been reviewed by Embree *et al.* (1972).

In conclusion, it should be noted that the roles played by the CNS in adaptation to the environment and in the integration of mechanisms of the entire body appear to control the rate of ageing (Ordy *et al.*, 1975). Loss of "short-term" memory, decreases in cerebral electrical activity, blood flow and O_2 consumption, and impairment of behavioral regulation appear to be salient features of ageing (see, e.g. Kety, 1955; Quastel and Quastel, 1961; Birren, 1965; Keevil-Rogers and Schnore, 1969; Brody and Ordy. 1975; Ordy and Brizzee, 1975; Gershon and Terry, 1976; Meier-Ruge, 1976).

C. Environment Produces a General Somatic Effect

By simply considering that changes in hormonal mechanisms and blood flow are caused by diverse *external* and *internal* environmental stresses, one can readily understand that the accompanying, or ensuing, metabolic changes are not specific to the nervous system. The chemistries and morphologies of peripheral tissues (e.g. liver, adrenal glands, heart, kidneys) can also be modified by impoverished rearing conditions and by other environmental stresses (e.g. Welch, 1965; Christian *et al.*, 1965; Geller *et al.*, 1965; Welch and Welch, 1971; Samorajski *et al.*, 1971; Maengwyn-Davies *et al.*, 1973; Corley *et al.*, 1973; DeFeudis, 1975a; Figs 87 and 88, and Table 37). The reduction of isolation-induced aggressiveness in mice caused by adrenalectomy (Yen *et al.*, 1962; Sigg *et al.*, 1966; Harding and Leshner, 1972) indicates further that environmentally induced cerebral effects can even be controlled by somatic lesions that affect stress-related mechanisms. A clear indication that "isolation-rearing" is stressful to weanling rats stems from the results of Geller *et al.* (1965) who showed that adrenal weight and hepatic transaminase activities were increased by 30 days of such treatment (see also Geller, 1971 and Fig. 88). Such changes indicated that an activation of the pituitary-adrenal system is essential for mediating the cerebral and peripheral changes that underly environmentally induced behavioral modification (but see Rosenzweig *et al.*, 1972b, and Chapter 3, Part B4). The role of stress as an integral element of the *Behavioral Code* has been discussed (see Chapter 4, Part D; Chapter 6, Part B; Chapter 7).

Increased incorporation of precursors into RNA also appears to occur in many *non-neural* tissues during various types of environmental stimulation (see, e.g. Zemp *et al.*, 1966; refs in Glassman, 1969; refs in

TABLE 37. Effects of EC-IC on organ weights and corticosteroids in rats.

Parameter	EC	IC	p
Initial body wt. (g)	36·2 ± 1·0	36·1 ± 1·0	Originally matched
Final body wt. (g)	174·3 ± 3·2	164·3 ± 5·5	0·05
Liver wt. (g)	8·92 ± 0·22	9·02 ± 0·35	0·01
Adrenal wt. (g)	33·6 ± 1·2	35·0 ± 1·1	0·05
Brain wt. (g)	1·61 ± 0·01	1·54 ± 0·015	0·01
Adrenal corticosteroids (μg/g)	16·27 ± 2·93	13·14 ± 2·23	NS
Serum corticosteroids (μg/100 ml)	10·05 ± 1·58	9·13 ± 1·16	NS

Means ± S.E.M.; p-values derived from analysis of variance with one covariate; NS = not significant. (Reproduced with permission from Geller *et al.*, 1965.)

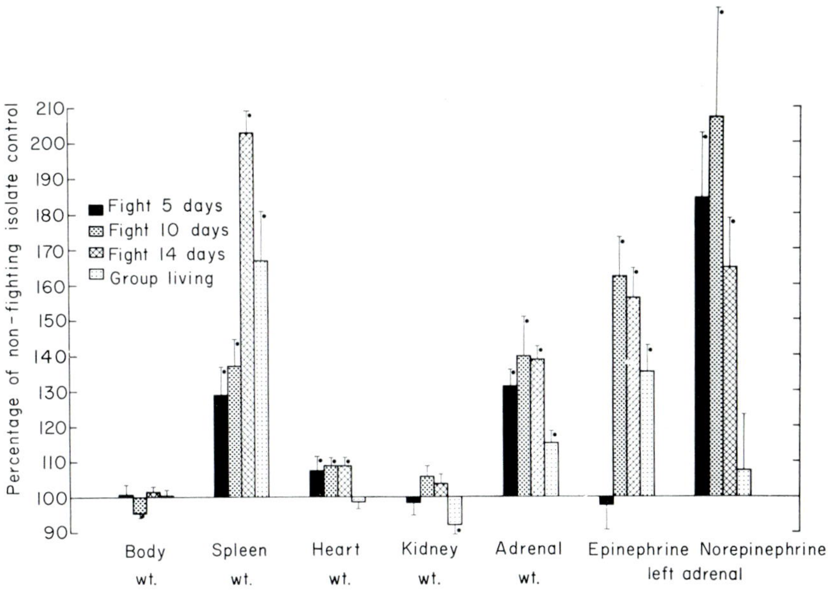

Fig. 87. Lasting effects of short daily fights on body, adrenal, heart, kidney and spleen weights, and upon adrenal catecholamines. (Reproduced with permission from Welch and Welch, 1969b.)

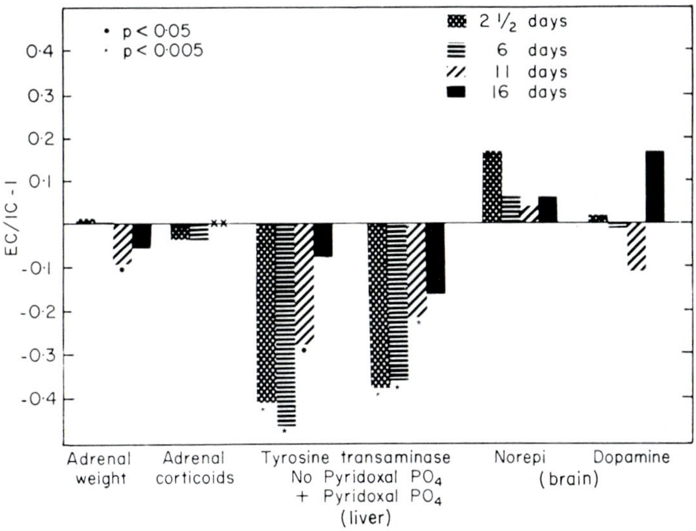

Fig. 88. Time course of differences in some stress-responsive parameters in rats reared under "enriched" (EC) or "impoverished" conditions (IC), expressed as (EC/IC)-1. Five pairs of animals were used at each time period. (Reproduced with permission from Geller, 1971.)

Nasello and Izquierdo, 1969). Also, differences in the incorporation of carbon atoms of injected D-glucose occurred in several extra-cerebral tissues of differentially housed mice (DeFeudis, 1971b; Table 29). From these findings, it has become evident that environmental modifications produce a general somatic effect that is reflected in behavior mainly because the nervous system (which control behavior) is among the organs affected. Thus, it becomes readily understood that the so-called "mental illnesses" of man may actually be reflections of various environmentally induced diseases or illnesses that affect various extra-cerebral tissues of the body (see also Szasz, 1973). Such environmentally induced changes that occur in peripheral tissues which are regenerative (e.g. liver) could be of some significance with respect to the diagnosis of behavioral dysfunctions since such tissues can be biopsied.

Other recent findings have indicated that the chemistries and morphologies of certain peripheral tissues exhibit adaptation. For example, the choline acetylase (ChAc) activity of mouse adrenal medulla and its preganglionic splanchnic innervation, which is largely cholinergic (Coupland, 1965), appear to adapt to environmental changes (Goldberg and Welch, 1972; Welch and Goldberg, 1973).

Regarding behavioral effects, facilitation or depression of conditioned or "pseudoconditioned" responses in rats have been achieved by pre-trial systemic administration of various agents that affect peripheral adrenergic or cholinergic mechanisms while not reaching the brain in significant amount (Orsingher and Fulginiti, 1973; Izquierdo, 1974; Izquierdo and Thaddéu, 1975). However, it should be realized that both conditioning and pseudoconditioning represent environmentally induced changes in behavior, and are thus elements of the *Behavioral Code*, as discussed herein. Furthermore, the involvement of a stress component in behavioral regulation (see Chapter 4, Part E; Chapters 5 and 7) indicates that any environmental change can cause behavioral modifications which may be reflected in the metabolism of various peripheral tissues. This becomes very clear when one considers that stress-related changes in emotionality necessarily involve changes in the sympathetic nervous system which is involved in the control of heart rate, blood vessel diameter, blood flow and many other autonomic mechanisms which can alter cerebral metabolism, function, and subsequent behavioral responses. In this context. it is noteworthy that many metabolic, physiologic, morphologic and behavioral changes can be caused by intracranial hypertension in man (Lundberg *et al.*, 1975).

As an example of a peripheral change produced by modification of the *internal* environment, one might consider the changes in hepatic steroidal enzymes that accompany the prolonged consumption of ethanol in experimental animals and humans (see, e.g. Rubin *et al.*, 1976 and refs therein). It should also be remembered that mental dysfunctions associated with some genetic defects can be explained by changes in peripheral tissues (e.g. the lack of hepatic phenylalanine hydroxylase in phenylketonuria).

The reader should realize that the notion that environment produces a general somatic effect is not new. Whole books have been written during the past 40 years about studies which have placed emphasis on the related concept that bodily changes are governed by emotional states (e.g. Dunbar, 1954; Young, 1961; Arnold, 1970; Levi, 1975). This concept has provided the basis for psychosomatic interrelationships and medicine. The more modern version of this "mind-brain, mind-body" issue seems to be "biofeedback" i.e. the training of one's own physiological processes (see, e.g. Shapiro *et al.*, 1973; Miller, 1975; Brown, 1975).

D. Parkinsonism as a Model for Learning, Memory and Behavior

It is easy to conceive that the macromolecular changes produced by learning and memory can represent alterations of neural structure. Therefore, the obvious morphological bases for certain extrapyramidal movement disorders (e.g. degeneration of the nigro-striatal pathway in Parkinsonism; see, e.g. Hornykiewicz, 1966, 1973; Lloyd *et al.*, 1975a; Poirier *et al.*, 1975; the additional loss of neurones of the putamen, caudate nucleus and cerebral cortex in Huntington's Chorea; e.g. Faurbye, 1970) might serve as models of the *Behavioral Code*, although the principal brain regions affected in these *somatic dysfunctions* (e.g. basal ganglia) could be more restricted than those involved in normal mechanisms of learning and memory. However, recently the association between the basal ganglia and "higher" neural functions in man and other mammals has become more strongly emphasized (e.g. Jung and Hassler, 1960; Buchwald *et al.*, 1961; Crosby *et al.*, 1962; Laursen and Mosfeldt, 1963; Campbell, 1965; Munkvad *et al.*, 1968; Sandler, 1972), adding further support to the suggested analogy.

The theory that destruction of neurones in the substantia nigra is the morphological substrate for Parkinsonism was advanced long ago (Blocq and Marinesco, 1893; Brissaud, 1895). Recent studies have revealed that this loss of nigral cells is accompanied by a considerable decrease in striatal dopamine (e.g. Ehringer and Hornykiewicz, 1960; see also Calne *et al.*, 1975; Lloyd *et al.*, 1975a). Therefore, Parkinsonian symptoms could be caused by an inadequate activation of dopamine (DA) receptors in the caudate nucleus (Hornykiewicz, 1973). Also, neuronal susceptibility to DA decreases, whereas that to acetylcholine (ACh) increases, in feline caudate nucleus depleted of biogenic amines by nigro-striatal lesions, indicating a change in the number of, or sensitivity of receptor sites for these agents (Spehlmann, 1975). As suggested recently by Cools and Van Rossum (1976), "there seems to exist an unmistakable correlation between central DA-activity and psychomotor diseases such as Parkinson's disease, Huntington's chorea, tardive dyskinesias and some psychotic disorders." However, other possible transmitters certainly appear to be involved in Parkinsonism (e.g. γ-aminobutyric acid (GABA); 5-hydroxytryptamine; see, e.g. DeFeudis, 1974b; Spehlmann, 1975; Poirier *et al.*, 1975; Cools and Van Rossum, 1976), as would be expected for all other complex behavioral phenomena.

Support for an involvement of GABA in Parkinsonism derives from findings that high levels of GABA and glutamate-α-decarboxylase (GAD) are found in the substantia nigra (SN) of several mammalian species (Muller and Langemann, 1962; Fahn and Coté, 1968; Okada *et al.*, 1971), and that certain brain lesions caused significant decreases in GABA or GAD of the SN (Kim *et al.*, 1971; McGeer *et al.*, 1971, 1973). "GABA-ergic" cell bodies appear to be present in the caudate nucleus (Yoshida and Precht, 1971), whereas their terminals may be localized to the SN (Hattori *et al.*, 1973). Furthermore, a high dose of haloperidol can decrease the GABA content of both the striatum and SN; such "GABA-ergic" strio-nigral systems could influence dopaminergic nigro-striatal systems (Kim and Hassler, 1975).

Other studies have supported the concept that a modulation of dopaminergic neurones by GABA might be involved in Huntington's chorea (HC). Perry *et al.*, (1975) have shown that low levels of GABA were present in the caudate nucleus, putamen and SN of patients who died from HC. Therefore, a hypoactive GABA system in the basal ganglia could be involved in this disease. Also, choreiform activity induced in rats by direct application of *d*-tubocurarine to the striatum was blocked by intra-striatal infusions of GABA, results which indicated that the *d*-tubocurarine-induced activity may have been due to blockade of GABA receptors in the striatum or to an increased DA release produced by blockade of GABA receptors (McKenzie and Viik, 1975). Other recent studies have indicated that the cerebral content of another candidate inhibitory transmitter, taurine, was markedly reduced in an unusual inherited neuro-psychiatric disorder characterized by mental depression and by Parkinsonian-like symptoms.

These recent findings support the contention that mechanisms underlying extrapyramidal diseases might be quite similar to those involved in normal learning, memory and behaviour. Further supporting evidence derives from recent studies which have revealed that passive avoidance in rats (see footnote on p. 99) is disrupted by unilateral electrical stimulation of the dopaminergic afferent to the caudate nucleus; i.e, the SN pars compacta (Routtenberg and Holzman, 1973). Also.

blockade of GABA-mediated caudato-nigral synaptic inhibition by intra-nigral picrotoxin injection produced a significant distruption in this same learning paradigm (Kim *et al.*, 1975). These post-trial intra-nigral injections of picrotoxin may have interfered with memory storage processes of the dopaminergic nigro-caudatal pathway that continue after learning (Kim *et al.*, 1975).

It should be remembered that changes in the metabolism of glucose, nucleic acids and proteins (including transmitter-associated enzymes) that accompany the effects of defective transmitter systems perhaps represent the more fundamental mechanisms involved in CNS dysfunction and in behavior. As examples, changes in glial and neuronal RNA that have been shown in biopsies from cases of Parkinsonism (Gomirato and Hydén, 1963; Hydén, 1964b, 1966, 1973c) resemble some changes that occur during learning and memory (see Chapter 4), and both choline acetylase (ChAc) and GAD activities appear to be reduced by about 15–25% in the SN of Parkinsonian brain (Lloyd *et al.*, 1975b). McGeer and McGeer (1976) have recently completed an extensive investigation on neurotransmitter enzymes in post-mortem brain regions of patients afflicted with Parkinsonism and Huntington's chorea. Parkinsonian cases showed decreased tyrosine hydroxylase (TH) activity in the SN, caudate nucleus and putamen and decreased GAD in the globus pallidus and SN. Choreics showed a substantial decrease in GAD in all extrapyramidal structures and some loss of ChAc in neostriatum and locus coeruleus. The decrease in striatal TH activity in Parkinsonian patients was also shown by Lloyd *et al.*, (1975a), and, in addition, their studies revealed a marked decrease of DOPA-decarboxylase and no change in monoamine oxidase and catechol-O-methyl transferase in striatal areas. With further regard to this suggested analogy, it should be noted that extrapyramidal symptoms characterize the central actions of both the "anti-psychotic" neuroleptic agents that block DA receptors (e.g. Janssen, 1967; Van Rossum, 1967) and the cholinomimetics (e.g. Connor *et al.*, 1966; Baker *et al.*, 1967; Lalley *et al.*, 1970; Costall *et al.*, 1972; see refs in DeFeudis, 1974b), and that prolonged neuroleptic therapy of humans can cause persistent extrapyramidal symptoms that are associated with organic brain lesions (e.g. Faurbye, 1970).

From the above discussion, it becomes evident that the cerebral structures and morphological changes known to be associated with these movement disorders and those that comprise the anatomical elements of the *Behavioral Code* show considerable overlap. A systematic analysis of chemical agents that either produce or counteract Parkinsonian-like tremors, together with the mental symptoms and cerebral morphological changes produced by these agents, should yield valuable information about the *Behavioral Code*, at both the cerebral and general somatic levels. Brain lesions that characterize extrapyramidal disorders and which involve neuronal losses are not expected to be more than exaggerated examples of those morphological changes that occur during normal or abnormal behavioral responses, though the latter changes are likely to be more subtle and generally reversible in nature.

Many other models for the *Behavioral Code*, based on modifications of the *external* environment of the organism (e.g. by changes in atmospheric gas composition; by temperature changes) could be constructed. As examples, Nicholson *et al.* (1970) have demonstrated that profound hypoxia caused histological changes in cells of the cerebral cortices and basal ganglia of monkeys which were correlated with disturbances of visual discrimination and spatial alternation behavior (Fig. 89), and Ford (1967) has shown that intravenously administered ^{3}H-lysine was incorporated to a lesser extent into anterior horn motoneurones and spinal gray matter of rats subjected to hyperbaric oxygenation (OHP; 3 atmospheres) than into these same regions of control rats. Subcellular changes in dendrites and axons, especially in the spinal gray matter of the anterior horn of experimental animals, have also been demonstrated after their exposure to OHP (e.g. Balentine, 1975). Also, the increased incorporation of ^{3}H-leucine into central neurones produced by physical exercise should be recalled (Altman, 1964; see also Chapter 4, Part C).

Other model systems, based on modifications of the *internal environment* (e.g. by drugs that produce rather selective lesions; by intracerebral stimulation) could also be used to study the morphological and biochemical elements of the *Behavioral Code*. As examples, chronic electrical stimulation of the cerebellar surface in monkeys produced dramatic changes in its morphology and chemistry (Tennyson *et al.*, 1975; Gilman *et al.*, 1975), administration of 6-hydroxydopamine destroyed central catecholaminergic nerve endings (e.g. Ungerstedt, 1968, 1974; Bloom *et al.*, 1969; Hökfelt and Ungerstedt, 1973), and 6-hydroxytryptamine can be used to damage or to destroy central indoleamine neurones (e.g. Nobin *et al.*, 1973; see also Chapter 1, Part G). Animals possessing genetically linked increases in sensitivity to visually or acoustically evoked epileptiform activity and other neurological mutants should also be of value in this regard (see, e.g. Rakic, 1975; Sotelo, 1975a, b; Schlesinger *et al.*, 1965, 1970; Suzuki and Zagoren, 1976). However, models based on changes in socio-environmental conditions appear to be most suitable since in these cases the animals adaptive responses mimic more closely those that naturally constitute their behaviors. Some studies have already revealed that monkeys reared in "enriched" environments performed better at learning tasks when tested in their home environments than when tested in an adjoining room and that these monkeys performed better than "partially-isolated" controls on certain tasks (Gluck *et al.*, 1973; but see Harlow *et al.*, 1969).

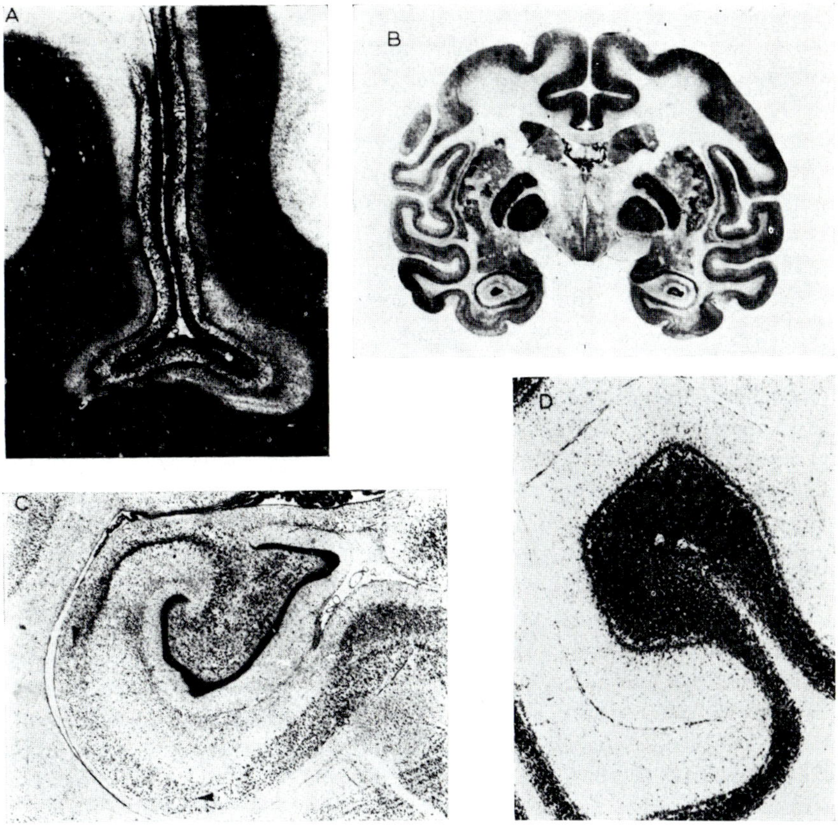

FIG. 89. Neuropathological changes produced by exposure of monkeys to profound hypoxia. A, left parieto-occipital sulcus showing laminar necrosis in the third layer; Phosphotungstic acid hematoxylin. B, coronal section showing patchy necrosis in each caudate nucleus and putamen and total necrosis in each globus pallidus; Cresyl Fast Violet (×1). C, left hippocampus showing loss of neurones with microglial proliferation in h.1 (between arrows); Cresyl Fast Violet (×10). D, cerebellar folium showing almost total loss of Purkinje cells with proliferation of the Bergmann glia; Cresyl Fast Violet (×28). (Reproduced with permission from Nicholson *et al.*, 1970.)

E. Memory Processes

At least two types of storage mechanism, "short-" and "long-term" memory (e.g. John, 1967; McGaugh and Dawson, 1971; Squire and Barondes, 1972; Glassman and Wilson, 1972, 1973; Kesner, 1973; Agranoff, 1974; Little and Shaw, 1975) are elements of the *Behavioral Code*. "Short-term" memory, which appears to be related to learning, itself, may be associated with changes in oxidative metabolism and with conformational changes in membrane macromolecules (step 3 of Fig.

80; Chapter 7). "Long-term" memory, the initial stages of which infer consolidation of information (McGaugh and Dawson, 1971), appears to involve changes in protein synthesis and in synaptic structure (step 5 of Fig. 80). "Short-" and "long-term" memories can be represented as steps of a continuum (Fig. 80) with events at step 3 (of Fig. 80) being reversible by accident, ECS or anesthesia since they involve oxidative metabolism, and events at later steps being more stable since macromolecular structures become involved.

The idea of a continuum was emphasized in a recently proposed statistical theory of "short-" and "long-term" memory (Little and Shaw, 1975; see also Roberts and Matthysse, 1970). "Short-term" memory was considered to endure from a fraction of a second to seconds, "intermediate-memory" to last up to hours (during which time imprinting into "long-term" memory can be affected by drugs or ECS; see McGaugh, 1966; McGaugh and Dawson, 1971), and "long-term" memory was considered to be permanent and to involve plastic changes (e.g. in synaptic strength and/or in the growth of new synapses). This model, based on Hebb's (1949) postulate that learning involves synaptic changes and correlated pre- and post-synaptic facilitation, suggested further that correlated pre- and post-synaptic facilitation on three time scales are crucial: (1) "active", or "short-term" memory corresponds to an *active* persistent firing pattern for the order of a second; (2) "intermediate-memory" occurs in a period (up to hours) during which facilitation of synaptic parameters (after (1), above) remains sizeable so that the pattern of (1) is easily re-excited in the network and persists as before; (3) "long-term" memory involves enduring plastic changes which enable patterns of (1) to be re-excited readily as in (2) (above). The important concepts are (a) that "short-term" memory is represented by persistent states or patterns that endure for the order of a second after excitation by a strong stimulus that had induced considerable facilitation and that was followed by synaptic depression (see Katz, 1966). (This concept could be based on the period of after-hyperpolarization of neurones (a function of neuronal membrane resistance) which changes in duration with increased synaptic firing, which may involve Ca^{2+}, and which is related to the Na^+, K^+-pump and to stimulation of O_2 uptake; see also Chapter 7); (b) that "intermediate" and "long-term" memories are associated with the connectivity of the network as a whole, and it is the values of the synaptic parameters which make these processes relatively insensitive to local damage and give them a *large storage capacity of non-local character*; (c) the model exhibits a single mechanism for "short", "intermediate" and "long-term" memories based on facilitation of the neuronal parameters, associative recall of memory, a robustness of the network against local damage, and a statisticial reliability. However, to date, no definitive evidence supports the idea that "short-" and "long-term" memory mechanisms form part of a continuum (see, e.g. Glassman and Wilson, 1972; Squire and Barondes, 1972; Dunn and Bondy, 1974). But, Deutsch (1973) has proposed that the idea of a quantitatively, rather than qualitatively, changing continuum is supported by the effects of ECS and other agents on memory. Bohdanecky *et al.* (1968) compared the amnestic effects of ECS and fluorothyl (hexafluorodiethyl ether) and found only a quantitative difference. Deutsch (1973) concluded that "ECS seems to be capable of interfering with memory up to an ill-defined point on a merely quantitatively changing continuum." In addition, recovery of memory has been reported after ECS treatments (Zinkin and Miller, 1967; Peeke and Herz, 1967; Kohlenberg and Trabasso, 1968). It has been shown further than the effects of ECS on memory are not immediate. Geller and Jarvick (1968) found that memory remained a few hours after ECS treatment but had disappeared by 24 hours, the time usually used for testing amnestic effects. These conclusions pose difficulties for the reverberatory (or electrical) hypothesis of consolidation since under such a mechanism the memory has been disrupted (Deutsch, 1973) (see also Chapter 7). "Long-" and "short-term" memories may coexist (Albert, 1966; McGaugh, 1966), and they could either be separate or could depend on one another in a variety of ways (Squire and Barondes, 1972).

"Short-term" memories may keep a record of events as they occur, all events perhaps leaving a trace for a short time, but only those of significance to the organism being recorded as "long-term" memories (see Young, 1965). Retrograde amnesia, which can be produced by ECS or by puromycin (given shortly after training), may depend upon the destruction of "short-term" memory, and is most pronounced for most recent and least fully recorded events (Russell, 1959). "Short-term" memory does not appear to depend exclusively upon "reverberatory circuit" activity since it is not impaired by hypothermia and ischaemia which cause flattening of the EEG (Gerard, 1955; Baldwin and Soltysik, 1969). Andjus *et al.* (1955) cooled rats to 0°C for periods up to an hour, thus abolishing all electrical activity of the CNS, but upon recovery these animals retained their memories for events that occurred prior to cooling as well as normal rats. Since seizures or EEG flattening (e.g. Gerard, 1955; McGaugh, 1966) or antibotics (Agranoff, 1974) do not abolish most stored memories, such memories resemble the so-called inherited or instinctive behaviors (see Dunn, 1975). From these findings it becomes evident that Lashley's (1958) hypothesis about destruction of memory by "reduction in the tonic level of the system" perhaps referred to "short-term" memory or early stages of memory consolidation (see also Chapter 2, Part A). Such an hypothesis necessarily requires a large-scale cerebral organization that can be disrupted by seizures or other environmental insults which affect cerebral O_2 consumption and carbohydrate (energy) metabolism. Stored memories, on the other hand, which are not affected by such insults, appear to require structural changes in synaptic networks which are relatively more anatomically restricted, but which also exhibit considerable overlap and redundancy. Furthermore, since retrograde amnesia can be produced by antibiotics which do not produce a loss of consciousness, as do ECS and traumatic injury, it seems unlikely that early memory can be disrupted merely by destruction of "reverberatory circuits" (see Agranoff, 1974).

Dunn (1975) has suggested that "short-term" changes in neuronal connectivity might be accomplished by modifications of pre-existing membrane molecules; e.g. by addition or removal of phosphate (see also Chapter 4, Part C3), carbohydrate, or alkyl groups, or by simple oxidation or reduction of alcohols, ketones, acids or sulfhydryl groups. In contrast, early stages of "long-term" memory, appear to depend upon gene-activation and macromolecular synthesis since the mechanisms involved can be prevented by inhibitors of protein or RNA synthesis given before or immediately after training (e.g. Barondes, 1970; Squire and Barondes, 1972; Glassman and Wilson, 1972; see below). Hence, both types of memory apparently involve changes in oxidative (energy-yielding) metabolism and in synaptic membrane structure, and, in accord with the theory of Lashley (1929), these processes may not be specifically localized in various regions of the brain.

To explain the relatively short periods required for learning and memory it should be considered that prolonged "reverberatory circuit" activity might occur in the neuronal network so that a single environmental event can activate each synaptic link spatio-temporally thousands of times within a few seconds (Hebb, 1949; Gerard, 1949; Konorski, 1961). After experiments of the later 1940s and early 1950s revealed that ECS could disrupt impulses and that low temperature could reduce drastically the number of nervous impulses while not affecting memories formed prior to the treatments, the "reverberatory circuit" hypothesis for "long-term" memory was abandoned and the residual change hypothesis (i.e. that physical or chemical changes in structure occurred) was revived. "Reverberatory circuit" activity was still useful for explaining the processing of information for brief intervals (seconds or minutes) after some initial conditioning stimulus, though it could not account for memories or conditioned behaviors that survived either a virtual suppression of all activity in the cerebral cortex (e.g. deep anesthesia, coma, extreme cold, concussion, or even deep sleep), or the converse; i.e. convulsive seizures of the whole cortex (Eccles, 1964). Also, it remains possible that "a reverberatory trace might cooperate with the structural change, and carry the memory until the growth change is made" (Hebb, 1949).

With regard to synaptic efficacy, Eccles (1964) has proposed that a given sensory input causes a uniquely patterned activation of central neurones, such that a subsequent re-presentation of this input tends to be channelled along the same pathways because of the increased efficacy of the synaptic actions exerted by all neurones initially activated. Further strengthening of the synapses responsible for the unique pattern of activation and response would occur which would cause a more effective channelling, cumulatively for each successive application of that sensory input. Such changes in synaptic efficacy must necessarily be of very long duration (days or weeks) since no mechanism exists whereby relatively brief durations of synaptic change for each synapse of a serial arrangement can summate to give a more prolonged change. Eccles (1964) has suggested that synaptic efficacy involves the notion that *disuse* can cause synapses to acquire special properties. This hypothesis was derived from studies concerning the regression of synaptic function that was produced by the disuse which resulted from severing afferent pathways (see Eccles and McIntyre, 1953; Eccles *et al.*, 1958).

Perhaps the most unsatisfactory feature of the attempt to explain learning and conditioning by changes in synaptic efficacy is that long periods of excess use or disuse are required in order to produce *detectable* synaptic changes (Eccles, 1964). However, the theories of Jaffé and Sharpless (1968) and Collier (1969) do involve the concept of "disuse" as a basis for their contention that adaptive mechanisms play a major role in drug addiction (see also Goldstein, 1975; Way *et al.*, 1975; Part F, below). Such mechanisms of adaptation in neurotransmitter systems could be involved in learning and memory (see also Mandell, 1975a; Part H, below). Also, since changes in neurotransmitter systems can be elicited rather rapidly by certain drugs and by metabolic insults (e.g. anoxia), it seems possible that such changes could occur during the time period required for memory processes.

Our problem is to characterize the synaptic growth that is associated with increases in synaptic efficacy that might occur during learning, memory and behavior. The search continues to find, when (in relation to learning) memory storage occurs, if and where memory processes may be localized in the CNS, and what type of molecular events might subserve these processes.

F. Hypothesis for Drug Addiction

1. *Introduction*

The three principal components of drug addiction (or "drug dependence") are tolerance, physical or psychological dependence, and compulsive craving for the drug. Tolerance implies that the animal needs larger doses of a drug to produce the effects formerly evoked by much smaller doses; i.e. that the animal has a decreased sensitivity to the drug. Physical and/or psychological dependence on the effects of a drug infers that it is essential for maintaining homeostasis; i.e. that an abstinence (withdrawal) syndrome occurs when the drug is withdrawn. Compulsive craving represents the overwhelming need to continue taking the drug and to continue obtaining a supply of the drug. "Habituation" also refers to emotional or psychological dependence on the drug. The motives for excessive drug use which lead to addiction in man likely include "social isolation", depression and mild despair. Thus, the person who might feel powerless in controlling the causative factors in his *external* environment may seek means of modifying his perception by experimenting with his *internal* environment; e.g. by using drugs (see, e.g. Freedman, 1968, 1970a, b). It is the writers' aim to provide here a biological explanation for such behavior and to indicate the relationships which exist between environmental impoverishment and some properties of certain centrally active ("mood-changing") drugs.

2. *The Link Between Environment and Drug Addiction*

Studies on the metabolic dispositions of centrally active drugs have revealed that central stimulants (e.g. *d*-amphetamine) are metabolized less rapidly (DeFeudis and Marks, 1973a, b; Fig. 73), and that central depressants (e.g. hexobarbital) are metabolized more rapidly in "isolated" mice (Baumel *et al.*, 1970; Lal *et al.*, 1972; Table 31) than in their grouped counterparts. Also, the duration of hexobarbital-induced narcosis was much shorter in mice subjected to "isolation" than in their grouped counterparts (Baumel *et al.*, 1969, 1970; Lal *et al.*, 1972), a finding which was correlated with an enhanced hepatic metabolism of the drug (Table 38). The finding that individually housed mice were more resistant to barbiturate-induced narcosis indicated a reduced sensitivity to these depressant drugs and should be contrasted with the results of Yanai and Ginsburg (1976) which indicated that individually housed mice are more susceptible to chronic ethanol treatment. Also, Welch and Welch (1973) showed that "grouped" mice were more tolerant to the stimulant, *d*-amphetamine (see below).

Since environmental impoverishment of mice increased the "binding" of potentially addictive psycho-active agents to cerebral nerve ending fractions, a change in the number of macromolecular binding sites (or in the number of sites relative to total protein of synaptic membranes) for these ligands appears to have occurred (see Chapter 3, Part D). In accord with the concept of "denervation supersensitivity" (Cannon and Rosenblueth, 1949), the decreased sensory input received

Table 38. Effects of differential housing[a] on the pharmacological actions of hexobarbital, pentobarbital, chloral hydrate, barbital and chlorpromazine.

Drug	Dose (mg/kg, i.p.)	"Grouped" mice	"Isolated" mice	p-value
		Narcosis duration (min)[b]		
Hexobarbital	130	98 ± 3·5 (9)	54 ± 1·4 (10)	$<0·005$
Pentobarbital	50	60 ± 4·7 (15)	39 ± 3·3 (7)	$<0·005$
Chloral Hydrate	350	76 ± 4·2 (18)	57 ± 3·9 (15)	$<0·005$
Barbital	300	107 ± 5·7 (17)	87 ± 8·8 (18)	$<0·005$
		Conditioned avoidance (%)[c]		
Chlorpromazine	0	90 (2)	90 (20)	NS
Chlorpromazine	2	50 (10)	60 (10)	$>0·05$
Chlorpromazine	5	0 (10)	40 (10)	$<0·05$

[a] Randomly-bred, Swiss albino mice were housed either individually or in groups of ten (procedure of Baumel *et al.*, 1969). [b] Narcosis was measured as loss of righting reflex. [c] One-trial test, after 5 training sessions. Results are expressed as means ± standard errors or as means; numbers of mice in parentheses. (Reproduced with permission from Lal *et al.*, 1972.)

by the "isolated" animals could have altered drug receptors, thereby increasing drug sensitivity. This hypothesis applies to both stimulant (e.g. *d*-amphetamine) and depressant (e.g. ethanol, barbiturates) centrally-active agents.

Recent experiments have revealed that both synaptosomal and crude synaptic membrane fractions of the brains of "isolated" male mice bind more ^{3}H-dihydromorphine (^{3}H-DHM) than identically prepared fractions from "aggregated" mice (DeFeudis *et al.*, 1976b). Although the total number of opiate sites did not differ between fractions of the brains of "isolated" and "aggregated" mice, a greater amount of DHM was bound per mg protein to brain preparations of the "isolated" mice (Table 10). This difference is explained by the decrease in protein content of nerve ending fractions caused by environmental impoverishment and appears to be most pronounced in the cerebral cortex (Table 10). Since these studies were carried out in the presence of a balanced bicarbonated-buffered medium containing about 150 mEq/liter Na^+, it is expected that even greater differences will be shown in the absence of added Na^+ (see, e.g. Pasternak and Snyder, 1974, 1975; Pert and Snyder, 1974).

Thirty minutes after subscutaneous injection of ^{3}H-DHM more label was found in the pons-medullae and spinal cords of "aggregated" mice than in these same regions of "isolated" mice (DeFeudis *et al.*, 1976b). However, behavioral studies have indicated that "isolated" mice are more susceptible to acutely induced morphine analgesia (a single dose of morphine) than their "aggregated" counterparts (DeFeudis *et al.*, 1976d; Fig. 90). With respect to this latter finding, it is of interest that recent studies with a conditioned aversion technique have revealed that a single low dose of morphine can induce manifestations of dependence in rats (Pilcher, 1976). This effect may be related to the altered "arousal" response observed in "isolated" mice (e.g. Welch, 1965), which could be a major factor involved in drug addiction. Other studies which have indicated that iontophoretically applied morphine exerts blocking actions on putative neurotransmitters in cells of the spinal cord and brain stem (e.g. Dostrovsky and Pomeranz, 1973) and that a depressant action of morphine on nociceptive spinal impulse transmission participates in its analgesic action (e.g. Jurna and Grossman, 1976) further support this liaison with the "arousal" mechanism. Most recently, Simantov *et al.* (1976a) have shown that, like the opiate receptor (Hiller *et al.*, 1973; Kuhar *et al.*, 1973), the endogenous morphine-like substance (enkephalin) seems to be associated primarily with extrapyramidal and limbic areas of the monkey CNS (see also Hughes, 1975a), adding further support to the recent findings with DHM binding in differentially housed mice and to their explanation based on "arousal" and emotional mechanisms (see above).

Using the above studies as a basis, one might assume that morphine and ethanol, which exerted greater effects on "isolated" mice than on "aggregated" mice (DeFeudis *et al.*, 1976b; Yanai and Ginsburg, 1976), acted as stimulants. This result for morphine brings to mind the hypotheses which have attempted to explain the mechanism of opiate tolerance and dependence as being due to the development of supersensitivity to

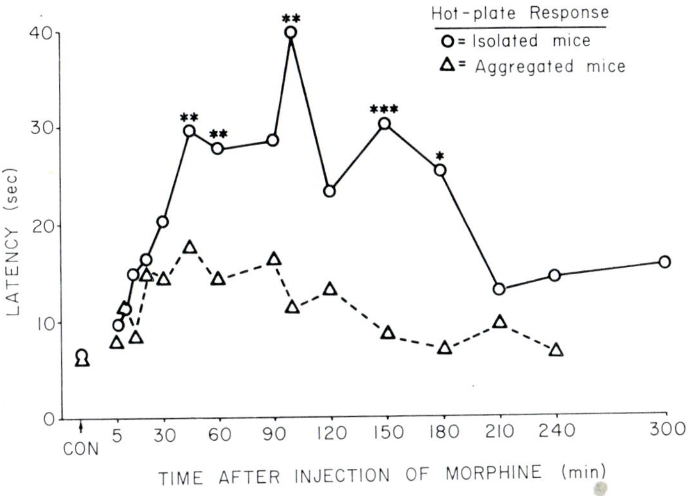

FIG. 90. Latencies of responses of differentially housed mice to a hot plate maintained at 57°C. Mean values; 5 mice in all cases; *, ** and *** indicate $p \simeq 0{\cdot}05$, $p < 0{\cdot}05$ and $p < 0{\cdot}01$ by comparison of these values with corresponding values for "aggregated" mice; Student's *t*-test; two-tailed. (Reproduced with permission from DeFeudis *et al.*, 1976d.)

neurotransmitters (see Jaffé and Sharpless, 1968; Collier, 1969). In the "disuse" hypothesis of Jaffé and Sharpless (1968) the cause of the latent hyper-excitability in morphine abstinence reflects the disuse or depression of neural pathways caused by repeated morphine administration, and as a compensatory response to this blockade, effector sites become more sensitive to their excitatory neurotransmitter. Withdrawal from the drug presumably releases the depression of the pathway allowing the released transmitter to act on the sensitized sites to produce the abstinence syndrome. This theory gains added impetus when one considers recent results which have indicated that a supersensitivity of cortical neurones of the rat to iontophoretic application of ACh and glutamate (two possible excitatory transmitters which appear to be involved in cortical "arousal", (see, e.g. Krnjević, 1967; DeFeudis, 1974b) developed in morphine tolerant/dependent rats (Satoh *et al.*, 1976).

Results of the studies with Li^+ and *d*-ampetamine, taken together with their *selective*, but *opposite*, effects on cerebral carbohydrate metabolism in "isolated" mice (DeFeudis and Paolino, 1972; DeFeudis *et al.*, 1972; P. A. DeFeudis *et al.*, 1975), could be explained by the difference in "arousal" level that exists between differentially housed mice (e.g. Welch, 1965), and could be used to explain the different susceptibilities of isolated animals to potentially addictive drugs (e.g. DeFeudis, 1975a).

3. *Studies with Amphetamines*

The ascending reticular-activating system is undoubtedly affected by sensory deprivation so that effective control over "arousal" and over inhibition of unwanted environmental signals is diminished. In support of this contention, Rosenzweig and Bennett (1972) have shown that rats treated with methamphetamine exhibited changes in cerebral morphology and chemistry that were similar to those caused by environmental "enrichment" (see also Chapter 2; Table 3, and Bennett *et al.*, 1973), and Welch and Welch (1973) found that group-housed mice were more resistant to the lethal effects of *d*-amphetamine and possessed greater tolerance to this drug (see also Welch and Welch, 1971). More recently, Sahakian *et al.* (1975) have shown that administration of *d*-amphetamine elicited a more pronounced stereotypy in socially deprived rats than in their controls (Fig. 91). Dopaminergic mechanisms in the corpus striatum appear to be involved in such stereotyped behavior (e.g. Fog *et al.*, 1967, 1970; Randrup and Munkvad, 1968; Fog and Pakkenberg, 1971).

Amphetamines are known to cause increases in "arousal" level and in perception of the environment (e.g. Shimizu and Himwich, 1968; Ibuka, 1971). In this connection, it seems noteworthy that chronic amphetamine intoxication produces mental and behavioral symptoms similar to those of paranoid schizophrenia in man and experimental "psychoses" in monkeys and other animals (see Chapter 3, Part B; Tatetsu *et al.*, 1956; Machiyama *et al.*, 1970, 1974; Angrist and Gershon, 1970; Angrist *et al.*, 1971; Utena, 1974). Furthermore, administration of amphetamines and other stimulants can enhance recovery from antibiotic induced amnesia (Barondes and Cohen, 1968; Serota *et al.*, 1972).

These findings, along with those which have indicated that amphetamines can alter the performance of ongoing behavior (e.g. Dews, 1955; Kelleher and Morse, 1968; Ibuka, 1971) and can produce both short-lasting (Verhave, 1958; Weiss and Laties, 1962; Wolthuis, 1972) and long-lasting (Kulig and Calhoun, 1972) effects on the learning of complex tasks, have revealed the value of these agents for studying the *Behavioral Code*. However, as with most drugs, amphetamines have several molecular mechanisms of action in brain. These agents appear to inhibit NE uptake, to deplete NE content (e.g. Glowinski and Axelrod, 1965; Glowinski *et al.*, 1966; Ross and Renyi, 1967) and to interfere with the deamination of NE by MAO (Cook and Schanberg, 1970). It should be noted further that in "short-term" (acute) studies, exposure of mice to "aggregation" (grouping) potentiaties both the behavioral

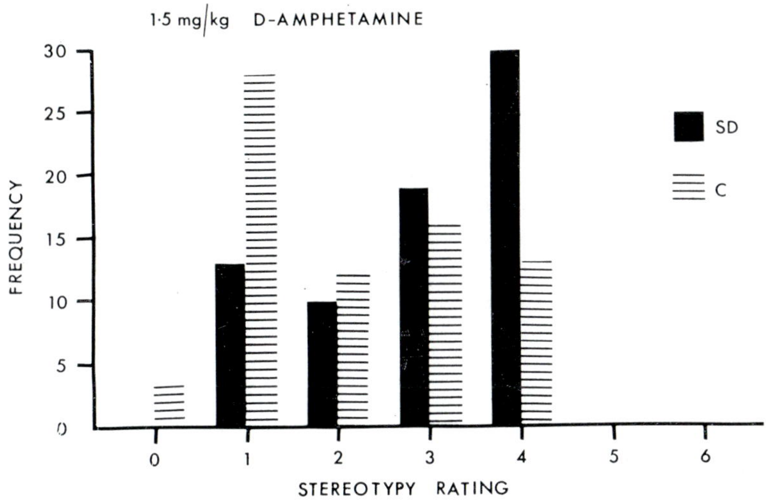

FIG. 91. Stereotypy after 1·5 mg/kg *d*-amphetamine in socially deprived (SD) rats and control (C) rats. The frequency of each rating is summed over the first hour of testing. (Reproduced with permission from Sahakian *et al.*, 1975.)

excitation and lethality due to injected amphetamines (Gunn and Gurd, 1940; Chance, 1947; Weaver and Kerley, 1962; Moore *et al.*, 1965; Bewsher *et al.*, 1966), whereas in "long-term" studies amphetamines are more toxic to "isolated" than to "aggregated" mice (Consolo *et al.*, 1965; Welch, 1965; Welch and Welch, 1966, 1971; Lal *et al.*, 1972; see review by DeFeudis, 1975a).

4. "*Enkephalin*"

Recent studies have revealed that a peptide(s) with analgesic properties ("enkephalin", or "endorphin") may exist in the mammalian CNS (Kosterlitz and Hughes, 1975; Hughes, 1975a, b; Hughes *et al.*, 1975a; Terenius and Wahlstrom, 1975; Pasternak *et al.*, 1975; Teschemacher *et al.*, 1975; Cox *et al.*, 1975). Hughes *et al.* (1975b) have shown that "enkephalin" is composed of two pentapeptides, H-TYR-GLY-GLY-PHE-MET-OH (MET-enkephalin) and H-TYR-GLY-GLY-PHE-LEU-OH (LEU-enkephalin), present in a ratio of about 3 : 1 in porcine brain. These peptides, like morphine, can inhibit both the electrically evoked contraction of guinea pig ileum and mouse vas deferens and the stereospecific receptor binding of ^{3}H-naloxone in cerebral homogenates (Hughes *et al.*, 1975b). Further studies by Belluzzi *et al.* (1976) have revealed that *in vivo* intraventricular injections of both peptides can induce a naloxone-sensitive analgesia in rats.

Recent studies have also revealed that analgesia elicited by focal electrical stimulation of the periaqueductal gray matter of the brains of rats can be partially antagonized by the opiate receptor antagonist, naloxone (Akil *et al.*, 1976). Such an effect, along with the reported antagonism by naloxone of acupuncture-induced analgesia in man (Mayer, 1975), could be mediated by an action of naloxone on the receptor sites for an endogenous "morphine-like" ligand. As suggested by Akil *et al.* (1976), activation of this endogenous system which mediates analgesia could be brought about either by direct receptor stimulation (via opiate administration) or by the stimulation-induced release of the endogenous "morphine-like" substance. It has been convincingly demonstrated, in mapping studies carried out with morphine micro-injection methods, that the brain region most enriched in analgesic sites is the periventricular-periaqueductal gray matter in rabbit (Tsou and Jang, 1964), rat (Jacquet and Lajtha, 1973) and monkey (Pert and Yaksh, 1974, 1975). Binding studies with radioactive narcotic analgesics have also revealed that this area is rich in opiate receptor sites in man (Hiller *et al.*, 1973) and in monkey (Kuhar *et al.*, 1973). These studies have paved the way for further studies on the possible roles that these peptides may serve as neurotransmitters or neuromodulators at synaptic junctions of central pathways that mediate analgesia and emotion.

These findings can be fitted to our addiction hypothesis as follows: In "isolated" animals, either the synthesis of "enkephalin" is decreased, or the number of its membrane receptor binding sites (per unit of cerebral structure, measured as synaptic protein) is increased. Either change would render animals more susceptible to addiction by externally applied opiates, a notion which finds support in recent studies on ^{3}H-DHM binding (see above and Table 10). "Enkephalin" could be an inhibitory transmitter involved in the mediation of analgesia and/or emotion, both of whose thresholds appear to be decreased by environmental impoverishment. Studies concerning the regional localization of enkephalin have indicated that it is primarily located in limbic and extrapyramidal areas (Hughes, 1975b; Simantov *et al.*, 1976a) which are associated with emotion, and perhaps with analgesia. It is of interest that Ungar and Cohen (1966) claimed to have produced a "chemical transfer" of morphine tolerance by brain extracts which contained peptides. Perhaps these workers had found "enkephalin" in these studies, since administration of this substance could be expected to increase morphine tolerance by reducing the number of "free" opiate receptor sites in the brain.

A relationship between the hormones involved in stress reactions and morphine has been long known. Morphine exerts pronounced effects on the pituitary-adrenal system (e.g. Selye, 1936; Briggs and Munson, 1955; Zimmerman and Critchlow, 1973), and ACTH modifies the central analgesic actions of morphine (e.g. Paroli, 1967). More recently, it has been shown that $ACTH_{1-24}$ (but not other ACTH-like peptides lacking corticotropic activity) when given one hour prior to testing for reduced responses of rats to electric foot-shock, inhibited the morphine-induced reduction; dexamethasone (the potent synthetic glucocorticoid) given four hours before testing, also antagonized this action of morphine (Gispen *et al.*, 1975b). It was concluded that the adrenal gland was essential for the antagonism by $ACTH_{1-24}$ of morphine-induced antinociception. Other studies have indicated that a naturally occurring vasopressin analog, desglycinamide-lysine-8-vasopressin (DG-LVP), facilitates the development of resistance to morphine-induced analgesia (Krivoy *et al.*, 1974). This finding was amplified by other studies which showed that rats with hereditary dibetes insipidus (DI; which lack the ability to synthesize vasopressin) exhibited an impaired development of tolerance to morphine analgesia (DeWied and Gispen, 1976). Since administration of endocrinologically inert DG-LVP (DeWied *et al.*, 1972) reversed this condition in DI rats, it was concluded that this action of vasopressin is related mainly to its influence on memory consolidation (see also Chapter 4, Part D2). These findings indicated that the development of tolerance (drug addiction) is a form of learning and memory (see also Cohen *et al.*, 1965; see below).

Since "enkephalin" appears to be an endogenous agonist of the cerebral opiate receptor and since vasopressin prevents against morphine action, all humans would qualify as potential drug addicts, as has been suggested previously in the absence of supporting experimental data. Therefore, a method for increasing the cerebral synthesis or concentration of "enkephalin" and/or vasopressin could be a cure for opiate-type addictions. Another more general substance which could be fitted to this scheme is phosphatidyl serine, a component of cerebral membranes which binds morphine in a stereospecific fashion (Abood and Hoss, 1975). The phosphatidyl serine concentration of brain could change in relation to environmentally induced alterations in cerebral metabolism in a manner that would favor a greater potential for addiction during exposure of animals to environmental impoverishment.

5. *Relation of Addictions to Learning and Memory*

Hebb (1949), in discussing "Hunger as an Addiction", in light of the previous works of Carlson (1916), Scott *et al.* (1938), Spragg (1940) and Morgan (1943), provided support for the notion that addictions require learning (and memory) by stating:

> We know, first, that there is always a period of "learning" or apprenticeship, in addiction; and we know also that an addiction lasts longer than the overt withdrawal symptoms of physiological disturbance. It may be proposed, therefore, that the learning is a genuine learning. . . . It follows that an addiction must be considered to be more than a metabolic reorganization: it includes a neural reorganization also, and one that tends to persist, since (as any tobacco addict or alcoholic knows) a "cure" is apt to be unreliable and the addiction can be reinstated much more promptly than acquired for the first time. (Hebb, 1949; p. 203)

These words also imply that a memory storage mechanism, "neural reorganization", is in play during the process of addiction. Hebb then went on to state:

> It is this very learning in addiction that we must know more about, to understand hunger. The learning can be described as an association between a sensory event and the subsequent euphoria: being at peace with the world follows a full belly, or an injection of morphine. The reference to "association", however, is misleading if it suggests that we understand the mechanism involved. We do not. (Hebb, 1949; p. 203)

Today, 27 years after these words of Hebb, we still do not understand completely this "association between a sensory event and the subsequent euphoria", but this problem has been partially solved since it is now known that changes in social environment (i.e. impoverishment) contribute to biochemical and morphological changes in brain structure (reflected as behavioral changes) that could be at the basis of drug addiction. This effect of an impoverished environment can lead to physiological adaptation of the organism's cerebral elements to behavioral patterns of low survival value (e.g. morphine addiction), even though increased learning and memory are involved. However, it should be understood that prior to the addiction, during the period of social isolation, a decrease in learning and memory presumably occurs. Somehow, this prior isolation primes the brain for adaptation to the "abnormal" environment. Some of the experimental evidence supporting these contentions has been discussed above; more follows below.

In accord with the studies of Welch and Welch (1973) on amphetamines, it has been shown that morphine tolerance in rats appears to be a learned response (Siegel, 1975). In support of the early view of Pavlov (1927) that narcotic tolerance results from the learning of an association between a drug's systemic effects and the environmental cues that reliably precede these effects, Siegal (1975) has shown that morphine tolerance can be extinguished by repeated presentation with a placebo of those environmental cues previously associated with morphine. It was shown further that experience with morphine in one environment did not facilitate the acquisition of morphine tolerance in another environment. Other studies with rats and monkeys have shown that drug self-administration can be affected by such environmental factors as dose, its schedule of administration, and conditioning of drug administration to secondary cues (see Weeks, 1975).

These findings are also in accord with our hypothesis that the environment exerts a major controlling influence on drug addiction. It is easy to conceive that the acquisition of other addictions (e.g. those to ethanol or barbiturates) can also be stimulated by environmental impoverishment. With specific regard to ethanol, it has already been shown that both C57BL/10 and DBA/1 mice subjected to individual housing exhibited a greater sensitivity to chronic ethanol treatment than their grouped counterparts (Yanai and Ginsburg, 1976). This environmentally induced difference in ethanol susceptibility could be mediated, at least in part, by a change in the level of thyrotropin-releasing hormone (TRH) which has been shown by Breese *et al.* (1974b) to antagonize ethanol-induced narcosis; i.e. a decrease in TRH release and/or synthesis could occur in "isolated" mice.

Such interactions between environment and an animal's physiological processes form the basis of "synthetic physiology", a discipline studied mainly by Eastern European and Soviet physiologists (see Bykov, 1960), and reveal further that drug addiction represents a behavioral adaptive response of the organism to changes in its *external* environment. Studies which have indicated that both morphine and ethanol dependence may be adaptive responses involving (cellular) neurotransmitter regulation have recently been reviewed (Way *et al.*, 1975; Goldstein, 1975), but it should be remembered that this concept is not new (see, e.g. Himmelsbach, 1943). Such adaptive responses could involve enzymes, receptors for neurotransmitters and/or drugs, or alternative neural pathways (see Goldstein, 1975), and could operate either by increasing tolerance or by enhancing susceptibility in the mechanism(s) of drug addiction. Clinical evidence has also indicated an adaptation of neuronal regulatory mechanisms to psycho-active drugs (e.g. Goodwin *et al.*, 1975).

6. *Conclusions*

In a recent review of the literature on opiate tolerance and dependence, Clouet and Iwatsubo (1975) concluded that "no known neurotransmitter is uniquely involved in tolerance and dependence, although the expression of pharmacological effects of opiates, including tolerance and dependence must be mediated via neurotransmitter systems". Such a conclusion further substantiates the view that such addictions are drives which may depend upon learning and memory—requiring extensive regions of cerebral organization which necessarily involve many (or all) neurotransmitter systems and being sensitive to (or triggered by) environmental influences. Other studies have indicated that learning is involved in the food drive (e.g. Hebb. 1949; Ghent, 1951; Miller and Kessen, 1952; see above), in sexual behavior (e.g. Beach, 1939) and in responses to pain (e.g. Melzack, 1954; see Hebb, 1955). Alterations in the effective numbers of opiate receptor sites caused by environmental restriction could be at the basis of solving the problems of opiate tolerance and dependence; such changes in receptors would be of a relatively "long-term" nature (and would require relatively long periods of environmental stress for their initiation).

G. "Chronological" versus "Biological" Time

Another important element of the *Behavioral Code* is *time*; i.e. *chronological* versus *biological* time (or maturation). Modification of time by manipulating the amount of experience (or nutrition) received by the organism represents an interaction of environment (or nutrition) and cerebral maturation. Not only the perception of time, but also the actual growth and development of the CNS can be modified by environmental manipulation, thus providing a means for increasing or decreasing the "biological age" of the CNS. Kohn (1971) has suggested that biological ageing could involve changes in all of the organs, tissues, cells, and molecules of the body. As emphasized above, the positive growth of cerebral synaptic structures appears to increase as the amount of information received from the environment is increased, regardless of the sensory modalities that are stimulated. This information appears to be stored by a system involving protein synthesis, since its structural basis can be shown by measuring the protein contents of synaptic elements. Conversely, brain growth and development appear to be decreased by a lack of this informationally synthesized protein (see Fig. 23).

In testing for whether the behavioral effects of social isolation in rats are due to a change in brain development, Morgan and Einon (1976) concluded, after treating the mothers of these animals with n-propylthiouracil (PTU) until litters were weaned, that the behavioral effects of social isolation were not due to a slowing of brain development. PTU cause a profound and enduring decrease in brain and body weights, but did not affect open-field behavior, as tested, thus differing from the effects caused by social isolation. This result is expected since one can hardly believe that the slowing of cerebral biological maturation caused by restricted experience can be due to decreased function of a single hormone, and furthermore, Rosenzweig *et al.* (1972b) have suggested that changes caused by differences in environmental complexity are not mediated by changes in thyroid output. These findings are of interest since many of the cerebral changes which characterize neonatal thyroidectomy; i.e. retardation of dendritic arborization (Eayrs, 1955; Legrand, 1967), reduction of cortical axonal density (Eayrs, 1955), general reduction in cerebral cell size (e.g. Pasquini *et al.*, 1967; Balázs *et al.*, 1968), altered cerebral enzyme activities (e.g. Valcana and Timiras, 1969) and reduced myelination (e.g. Balázs *et al.*, 1969; Walvarens and Chase, 1969), all seem to indicate that a reduction of synaptic contacts occurs. This may also be produced by social isolation. It is also noteworthy that thyroid-deficiency in rats leads to a decreased maturation of those cerebral metabolic compartments characterized by a high rate of glucose utilization and a predominance of glucose as oxidative substrate, which appear to be associated with dendrites and nerve terminals (see Cocks *et al.*, 1970). Social isolation, which also causes a decrease in cerebral glucose utilization (see, e.g. DeFeudis, 1975a), may be associated with similar metabolic changes.

Since the *Behavioral Code* appears to involve synaptic proteins that are associated with the control of "biological ageing", it can be studied by measuring changes in neuromorphology (DeFeudis, 1975b; see, e.g. Chapters 1–3). In accord with this contention, a genetic analysis of brain weights in mice revealed a significant directional dominance effect toward larger brains in F_1 crosses among animals reared under "enriched" environments, as compared with those kept in standard colonies (Henderson, 1973). The role of nutrition in this connection cannot be overstated (see also Chapter 1, Part F). Even at the human level, the biochemical growth of the brains of children who died between birth and two years of age was slower (in terms of protein, RNA and DNA contents of the cerebrum, cerebellum and brain stem) in those suffering from severe malnutrition ("marasmus") than in normal controls; the effect on cell number (DNA concentration) was much greater in cerebrum and cerebellum than in brain stem (Winick *et al.*, 1970). Also, the studies of Marin-Padilla (1972) and Purpura (1975) which indicated that "maturational arrest" might be involved in the production of dendritic spine abnormalities that characterize mental retardation in children should be recalled (see Chapter 3, Part B5). Environmental impoverishment could lead to similar morphological changes in the brain and such changes could well underly the resultant behavioral dysfunctions.

An excellent review of life-span changes in neurochemistry of the primate brain is available (Ordy *et al.*, 1975). Since such changes are reflected as behavioral modifications, they constitute part of the *Behavioral Code.* Furthermore, evolutionary studies on the human brain have indicated that changes in its organization (morphology) are more significant than changes in its size (Holloway, 1968, 1969; Passingham, 1975). Interestingly, a recent analysis based on the distribution of mass within the brain and encephalization ratios for various divisions of the brains of 63 species (data of Stephan *et al.*, 1970), revealed that the human brain resembles that of the spider monkey more closely than those of the chimpanzee or gorilla (Douglas and Marcellus, 1975). An extensive statistical study about the cranial morphology of the lesser apes has recently appeared (Creel and Preuschoft, 1976).

H. Directions of Future Research

More refined techniques are now available for the separation of neurones, nerve endings and glia (e.g. Rose, 1969b; Blomstrand and Hamberger, 1969, 1970; Sellstrom *et al.*, 1975), for subcellular fractionation (e.g. Morgan *et al.*, 1973; Whittaker, 1973), for the detection of biologically active substances (such as mass fragmentography (e.g. Hanin, 1969; Campbell *et al.*, 1970) and affinity chromatography (e.g. Susz *et al.*, 1973; Zanetta *et al.*, 1975; see also Lowe and Dean, 1974)), for determining cell populations and neuronal connectivity (e.g. Schubert and Kreutzberg, 1975b; Hökfelt and Ljungdahl, 1975; Laemle, 1975; Heimer and Lohman, 1975), for quantitative electron microscopy (e.g. Karlsson, 1966; Conradi, 1969; Guillery, 1969; Raisman, 1969; Westman, 1971; Pappas and Purpura, 1972; Ralston and Chow, 1973; Davis and Nunnemacher, 1974; Vrensen and DeGroot, 1973, 1974b, 1975; Hendrickson, 1975; Nakamura, 1975; Figs 92 and 93) and for scanning electron microscopy (see, e.g. Hayat, 1975a, b). These techniques, when used in conjunction with on-line perfusions of various cerebral structures of awake, behaving animals (e.g. Delgado, 1967; DeFeudis *et al.*, 1970) and computer simulation of cerebral metabolism (e.g. Garfinkel, 1973; Garfinkel *et al.*, 1975) should enable further exploration into the changes in neurochemistry and behavior that are elicited by environmental, nutritional, and genetic modifications and that occur during ontogeny. The central role of glucose metabolism and the role of nutrition, in general, should perhaps be emphasized in such studies. For example, the use of different radioactive metabolites for labelling the large and small metabolic compartments of the brain (e.g. glucose or lactate for the large pool; acetate or leucine for the small pool; see Berl *et al.*, 1961, 1962; Berl and Clarke, 1969), should provide further insight into the metabolic and subcellular mechanisms that underly various environmental insults.

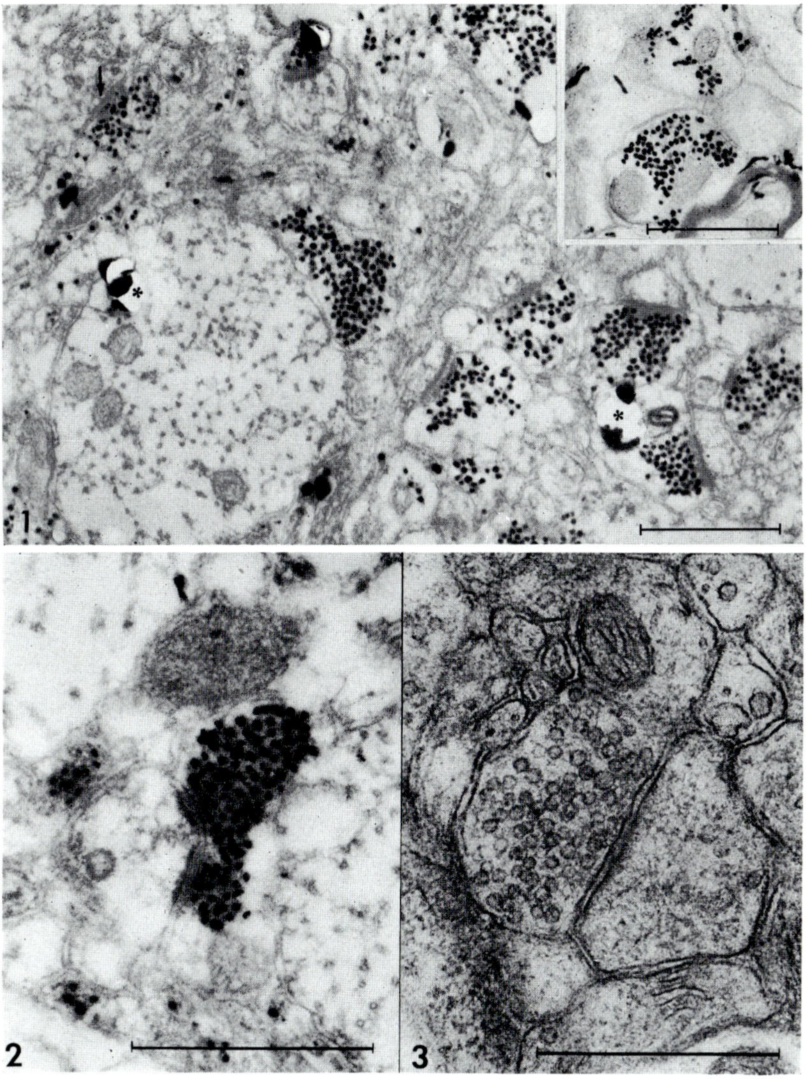

FIG. 92. Quantitative study of central synapses. All micrographs were taken from the visual cortices of rabbits treated with osmium zinc iodide (OZI) and aqueous-phosphotungstic acid (PTA) or with OsO_4-lead citrate. Bars indicate 1 μm. 1, Low power micrograph illustrating the clear-cut visualization of axo-dendritic and axo-somatic (↑) synapses with their heavily stained vesicles and aqueous-PTA contrasted contact zone. *: unspecific dense grains giving rise to cutting artifacts. *Insert*: OZI-stained tissue not counterstained with aqueous-PTA. Note the absence of clear-cut synaptic contact zones in the terminals. 2 and 3 illustrate the improvement of recognition of synaptic vesicles in the OZI picture (2) as compared with the OsO_4 picture (3). The aqueous-PTA counterstaining clearly outlined the synaptic contact zones. (Reproduced with permission from Vrensen and DeGroot, 1974b.)

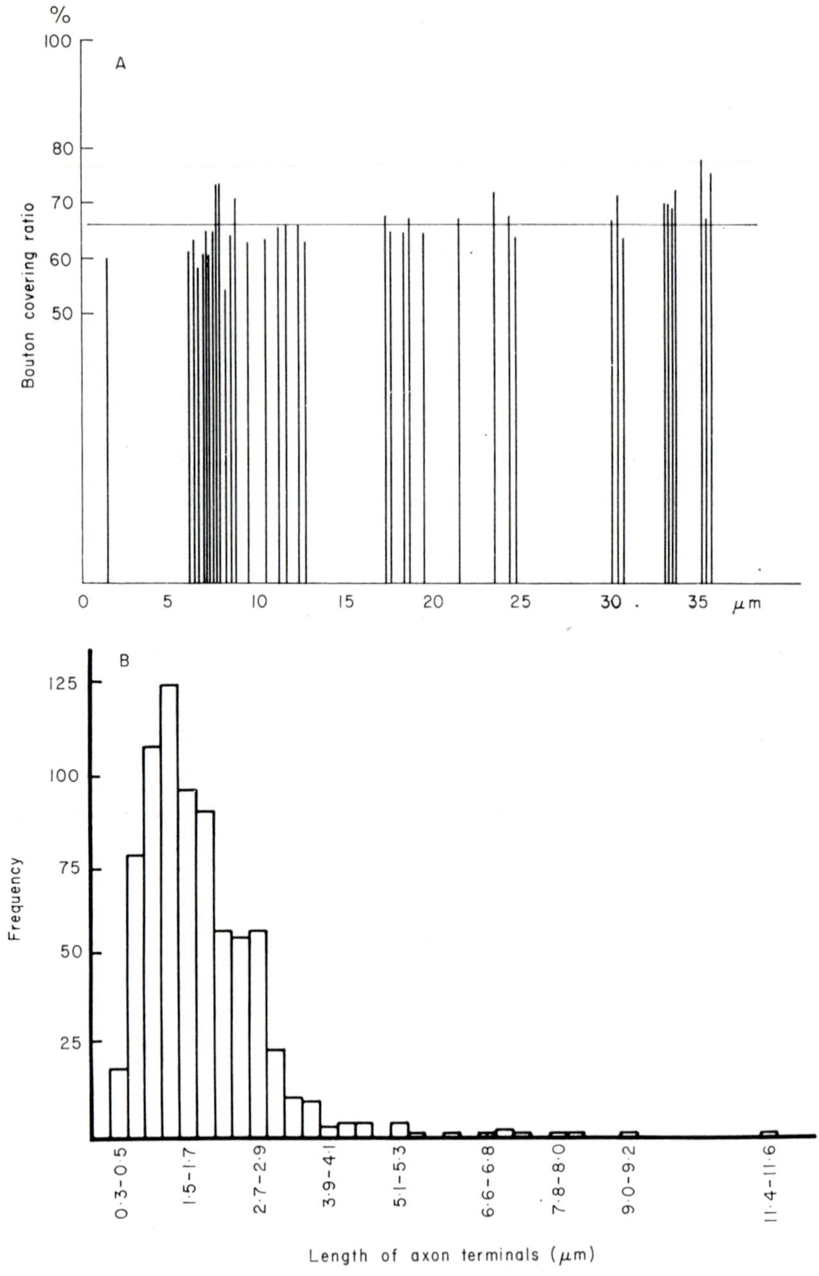

Fig. 93. Methods of quantitative electron microscopy. A, the bouton covering ratio (ordinate) of a magnocellular neurone of the red nucleus of a cat measured in a series of 38 semi-serial sections. The numbers on the abscissa are estimated distances of each section from the initial level of the serial section. Mean value is indicated by the dotted line: 65·7%. B, a histogram showing the frequency of the length of contacting area of axosomatic terminals upon a magnocellular neurone; the same neurone described in A (above). The total number of measured axon terminals is 757. (Reproduced with permission from Nakamura, 1975.)

The information about the *Behavioral Code* presented herein should be useful for devising remedies for those social and medical problems of man that are caused by environmental, genetic, or nutritional influences. Since nutritional deficiencies cause cerebral morphological changes which have been correlated with behavioral dysfunctions, dietary modifications could be the best method for reversing some behavioral dysfunctions in man. This suggestion gains added impetus when one considers that dietary changes, like the psychopharmaca, can modify an animal's perception of its *external* environment. Hence, the environment, itself, need not be changed.

There remains little doubt that there exists a cerebral morphological basis for behavioral changes that involves proteins and nucleic acids and that this may apply to humans, as well as to other vertebrates. It seems likely that early stages of information processing in the nervous system are not regionally specific and that these involve changes in energy metabolism, whereas information storage ("long-term" memory) may be more anatomically specific since it appears to involve changes in structural protein of nerve ending membranes. Significant findings have indicated that the total protein content of the brain does not appear to change whereas the protein associated with nerve endings does change with respect to the level of environmental stimulation received by the animal (see Chapter 4, Part C and Fig. 23). Furthermore, "binding" studies have indicated that the decrease in synaptic protein caused by environmental impoverishment is reflected as a decrease in the number of "binding' sites for inhibitory neurotransmitter suspects and as increases in the amounts of potentially addictive drugs bound per unit protein (see Chapter 3, Part D; Chapter 6, Part B3; Part F, above). Such changes in synaptic protein could also represent a decrease in the synthesis and subsequent axoplasmic flow of transmitter-synthesizing enzymes, caused by the environmental impoverishment (see Fig. 4; Mandell, 1975b).

Further use of animal models, especially primate models, for studying the bases of psychiatric and neurologic disorders should be encouraged. It is already well known that social isolation of primates can induce depressive behavior and other psychopathological disturbances which mimic closely certain behavioral dysfunctions in humans (e.g. Spitz, 1946; Bowlby, 1960; Suomi and Harlow, 1972; McKinney *et al.*, 1972; Harlow *et al.*, 1973; McKinney, 1974), but to date no systematic correlative neurochemical or neuromorphological studies have been performed at this level. However, this should not be taken to mean that valuable basic information about the *Behavioral Code* will not be obtained from studies on simple invertebrate preparations. Such preparations have already been useful for showing that experience modifies some plastic properties of identified neurones (see, e.g., Horridge, 1962,

1965; Galeano, 1972; Oliver, 1973; Kerkut *et al.*, 1973; Hoyle, 1975; Murphey and Matsumoto, 1976).

Perhaps the major concept that has emerged from the present analysis is that socio-environmental stress, especially during early development of the animal, is a major determinant of behavioral adaptation, and convincing experimental evidence now exists to support this concept. Selye (1956) has emphasized the important role of adaptation to the environment as a means of preventing illness and unhappiness:

> The secret of health and happiness lies in successful adjustment to the ever-changing conditions on this globe; the penalties for failure in this great process of adaptation are disease and unhappiness.

In line with Selye's words, it is now being realized that various environmental stresses (e.g. nutritional modifications, drugs, radiation) are likely to be the causes of cancer (see Fraumeni, 1975). It is easy to conceive that environmental modifications (social or physical) could be important causes of all health problems of man. Therefore, this should be a major direction for further research.

References

Abood, L. G. and Hoss, W. (1975). Stereospecific morphine adsorption to phosphatidyl serine and other membranous components of brain. *Eur. J. Pharmacol.* **32,** 66–75.

Adair, L. B., Wilson, J. E. and Glassman, E. (1968). Brain function and macromolecules. IV. Uridine incorporation into polysomes of mouse brain during different behavioral experiences. *Proc. Nat. Acad. Sci., U.S.A.* **61,** 917–922.

Adams, R. D., Collins, G. H. and Victor, M. (1962). Troubles de la mémoire et de l'apprentissage chez l'homme; leurs relations avec des lésions des lobes temporaux et du diencéphale. *In* "Physiologie de l'Hippocampe" (P. Passouant, Ed.) pp. 273–291. Centre Nationale de Recherche Scientifique, Paris.

Ader, R. and DeWied, D. (1972). Effects of lysine vasopressin on passive avoidance learning. *Psychonom. Sci.* **29,** 46–48.

Adlard, B. P. F. and Dobbing, J. (1972). Vulnerability of developing brain. VIII. Regional acetylcholinesterase activity in the brains of adult rats undernourished in early life. *Brit. J. Nutr.* **29,** 139–143.

Adlard, B. P. F., Dobbing, J. and Smart, J. L. (1970). Undernutrition and the development of certain enzymes in rat brain. *Biochem. J.* (Lond.) **119,** 46P.

Agranoff, B. W. (1967). Actinomycin D blocks formation of memory of shock-avoidance in goldfish. *Science* **158,** 1600–1601.

Agranoff, B. W. (1970). Recent studies on the stages of memory formation in the goldfish. *In* "Molecular Approaches to Learning and Memory" (W. L. Byrne, Ed.), pp. 35–39. Academic Press, New York and London.

Agranoff, B. W. (1973). Biochemical approaches to learning and memory. *In* "Macromolecules and Behaviour" (G. B. Ansell and P. B. Bradley, Eds.), pp. 143–149. Macmillan, London.

Agranoff, B. W. (1974). Biochemical concomitants of the storage of behavioral information. *In* "Biochemistry of Sensory Functions" (L. Jaenicke, Ed.), pp. 597–623. Springer-Verlag, Berlin.

Agranoff, B. W. (1975). Neurotransmitters and synaptic transmission. *Fed. Proc.* **34,** 1911–1914.

Agranoff, B. W., Davis, R. E. and Brink, J. J. (1965). Memory fixation in the goldfish. *Science* **146,** 952–953.

Agranoff, B. W., Davis, R. E. and Brink, J. J. (1966). Chemical studies on memory fixation in goldfish. *Brain Res.* **1,** 303–309.

Agranoff, B. W., Davis, R. E., Casola, L. and Lim, R. (1968). Actinomycin D blocks formation of memory of shock-avoidance in goldfish. *Science* **158,** 1600–1601.

Agrawal, H. C., Davis, J. M. and Himwich, W. A. (1966/67). Postnatal changes in free amino acid pool of rabbit brain. *Brain Res.* **3,** 374–380.

Agrawal, H. C., Fox, M. W. and Himwich, W. A. (1967). Neurochemical and behavioral effects of isolation-rearing in the dog. *Life Sci.* **6,** 71–78.

Agrawal, H. C., Davis, J. M. and Himwich, W. A. (1968). Water content of dog brain parts in relation to maturation of the brain. *Amer. J. Physiol.* **215,** 846–848.

Aillon, G. A. (1971). Biochemistry of affective disorders. *Psychosomatics* **12,** 260–272.

Akil, H., Mayer, D. J. and Liebeskind, J. C. (1976). Antagonism of stimulation-produced analgesia by naloxone, a narcotic antagonist. *Science* **191,** 961–962.

Al-Ani, A. T., Tunnicliff, G., Rick, J. T. and Kerkut, G. A. (1970). GABA production, acetylcholinesterase activity and biogenic amine levels in brain for mouse strains differing in spontaneous activity and reactivity. *Life Sci.* **9** (Part I), 21–27.

Albert, D. J. (1966). The effect of polarizing currents on the consolidation of learning. *Neuropsychologia* **4,** 65–78.

Alevizos, B., Spengos, M., Vassilopoulos, D. and Costas, S. (1976). γ-Glutamyl transpeptidase. *J. Neurol. Sci.* **28,** 225–231.

Allikmets, L. H. (1974). Cholinergic mechanisms in aggressive behaviour. *Med. Biol.* **52,** 19–30.

Altman, J. (1963). Differences in the utilization of tritiated leucine by single neurones in normal and exercised rats: An autoradiographic investigation with microdensitometry. *Nature* (*Lond.*) **199,** 777–780.

Altman, J. (1964). The use of fine-resolution autoradiography in neurological and psychobiological research. *In* "Response of the Nervous System to Ionizing Radiation" (T. J. Haley and R. S. Snider, Eds), 2nd Edn, p. 336. Little, Brown and Co., Boston.

Altman, J. (1967). Postnatal growth and differentiation of the mammalian brain, with implications for a morphological theory of memory. *In* "The Neurosciences" (G. C. Quarton, T. Melnechuk and F. O. Schmitt, Eds), pp. 723–743. Rockefeller University Press, New York.

Altman, J. (1969). DNA metabolism and cell proliferation. *In* "Handbook of Neurochemistry" (A. Lajtha, Ed.), Vol. 2, pp. 137–182. Plenum Press, New York.

Altman, J. (1972). Autoradiographic examination of behaviorally induced changes in the protein and nucleic acid metabolism of the brain. *In* "Macromolecules and Behavior" (J. Gaito, Ed.), 2nd Edn, pp. 305–334. Appleton-Century-Crofts, New York.

Altman, J. (1973). Experimental reorganization of the cerebellar cortex. IV. Parallel fiber reorientation following regeneration of the external germinal layer. *J. Comp. Neurol.* **149,** 181–191.

Altman, J. and Das, G. D. (1964). Autoradiographic examination of the effects of enriched environments on the rate of glial multiplication in the adult rat brain. *Nature* (*Lond.*) **204,** 1161–1163.

Altman, J. and Das, G. D. (1966). Behavioral manipulations and protein metabolism of the brain: Effects of motor exercise on the utilization of leucine-H^3. *Physiol. Behav.* **1,** 105–108.

Altman, J., Das, G. D. and Chang, J. (1966). Behavioral manipulations and protein metabolism of the brain: Effects of visual training on the utilization of leucine-H^3. *Physiol. Behav.* **1,** 111–115.

Altman, J., Wallace, R. B., Anderson, W. J. and Das, G. D. (1968). Behaviorally induced changes in length of cerebrum in rats. *Develop. Psychobiol.* **1,** 112–117.

Altman, J., Das, G. D. and Sudarshan, K. (1970). The influence of nutrition on neural and behavioral development. I. Critical review of some data on the growth of the body and the brain following dietary deprivation during gestation and lactation. *Develop. Psychobiol.* **3,** 281–301.

Altman, J., Das, G. D., Sudarshan, K. and Anderson, J. B. (1971a). The influence of nutrition on neural and behavioral development. II. Growth of body and brain in infant rats using different techniques of undernutrition. *Develop. Psychobiol.* **4,** 55–70.

Altman, J., Sudarshan, K., Das, G. D., McCormick, N. and Barnes, D. (1971b). The influence of nutrition on neural and behavioral development: III. Development of some motor, particularly locomotor patterns during infancy. *Develop. Psychobiol.* **4,** 97–114.

Altschule, M. D. (1970). Carbohydrate metabolism in mental disease: Associated changes in phosphate metabolism. *In* "Biochemistry, Schizophrenias and Affective Illnesses" (H. E. Himwich, Ed.) pp. 333–360. Williams and Wilkins Co., Baltimore.

Amano, M., Messier, B. and Leblond, C. P. (1959). Specificity of labeled thymidine as a deoxyribonucleic acid precursor in radioautography. *J. Histochem. Cytochem.* **7,** 153–155.

Ambron, R. T., Goldman, J. E. and Schwartz, J. H. (1975). Effect of inhibiting protein synthesis on axonal transport of membrane glycoproteins in an identified neuron of *Aplysia. Brain Res.* **94,** 307–323.

Andén, N.-E., Dahlström, A., Fuxe, K., Larsson, K., Olson, L. and Ungerstedt, U. (1966). Ascending monoamine neurons to the telencephalon and diencephalon. *Acta Physiol. Scand.* **67,** 313–326.

Andén, N.-E., Carlson, A. and Haggendahl, J. (1969). Adrenergic mechanisms. *Ann. Rev. Pharmacol.* **9,** 119–134.

Andjus, R. K., Knopfelmacher, F., Russell, R. W. and Smith, A. U. (1955). Effects of hypothermia on behaviour. *Nature* (*Lond.*) **176,** 1015–1016.

Angrist, B. M. and Gershon, S. (1970). The phenomenology of experimentally induced amphetamine psychosis—Preliminary observations. *Biol. Psychiat.* **2,** 95–107.

Angrist, B. M., Shopsin, B. and Gershon, S. (1971). Comparative psychotomimetic effects of stereoisomers of amphetamine. *Nature* (*Lond.*) **234,** 152–153.

Ansell, G. B. and Bradley, P. B. (Eds) (1973), "Macromolecules and Behaviour." University Park Press, London, 300 pp.

Appel, S. H. (1972). Macromolecular synthesis in synapses. *In* "Macromolecules and Behavior" (J. Gaito, Ed.), 2nd Edn, pp. 185–203. Appleton-Century-Crofts, New York.

Appel, S. H. and Locher, C. (1973). Changes in synapse membrane protein phosphorylation following electroconvulsive shock. *Neurology* **23,** 410.

Appel, S. H., Davis, W. and Scott, S. (1967). Brain polysomes: Response to environmental stimulation. *Science* **157,** 836–838.

Appel, S. H., Autilio, L., Festoff, B. W. and Escueta, A. V. (1969). Biochemical studies of synapses *in vitro*. III. Ionic activation of protein synthesis. *J. Biol. Chem.* **244,** 3166–3172.

Arese, P., Gremo, F. and Marchisio, P. C. (1975). Energy metabolism and fast axoplasmic transport in growing nervous fibers. Presented at First European Neurosciences Meeting, Munich, 9–13 September, Abstr. No. 14.

Arnold, M. B. (Ed.) (1970). "Feelings and Emotions". Academic Press, New York and London, 339 pp.

Austin, J. H. (1972). Disorders of glycogen and related macromolecules in the nervous system. *In* "Handbook of Neurochemistry" (A. Lajtha, Ed.), Vol. 7, pp. 1–15. Plenum Press, New York.

Austin, L. and Morgan, I. G. (1967). Incorporation of ^{14}C-labelled leucine into synaptosomes from rat cerebral cortex *in vitro*. *J. Neurochem.* **14,** 377–388.

Autilio, L. A., Appel, S. H., Pettis, P. and Gambetti, P. L. (1968). Biochemical studies of synapses *in vitro*. I. Protein synthesis. *Biochemistry* **7,** 2615–2622.

Autilio-Gambetti, L., Gambetti, P. and Shafer, B. (1973). RNA and axonal flow. Biochemical and autoradiographic study in the rabbit optic system. *Brain Res.* **53,** 387–398.

Avis, H. H. and Carlton, P. L. (1968). Retrograde amnesia produced by hippocampal spreading depression. *Science* **161,** 73–75.

Axelrod, J. (1972). Biogenic amines and their impact in psychiatry. *Seminars in Psychiatry* **4,** 199–210.

Ayala, G. F., Johnston, D., Lin, S. and Dichter, H. (1976). The mechanism of action of Diphenylhydantoin on invertebrate neurones: II. Effects on synaptic mechanisms. Submitted for publication; cited by Raabe and Ayala (1976).

Baenninger, L. P. (1967). Comparison of behavioral development in socially isolated and grouped rats. *Anim. Behav.* **15,** 312–323.

Bailey, P. and Bremer, F. (1938). A sensory cortical representation of the vagus nerve. *J. Neurophysiol.* **1,** 405–412.

Baker, W. W., Connor, J. D., Rossi, G. V. and Lalley, P. M. (1967). Production of tremor by intracaudate cholinergic agents and its suppression by locally administered catecholamines. *Progr. Neuro-Genetics,* Vol. 1, pp. 390–403, *Int. Congr. Series No. 175,* Excerpta Medica, Amsterdam.

Balázs, R. (1970). Carbohydrate metabolism. *In* "Handbook of Neurochemistry" (A. Lajtha, Ed.), Vol. 3, pp. 1–36. Plenum Press, New York.

Balázs, R. and Cocks, W. A. (1967). RNA metabolism in subcellular fractions of brain tissue. *J. Neurochem.* **14,** 1035–1055.

Balázs, R., Kovács, S., Teichgräber, P., Cocks, W. A. and Eayrs, J. T. (1968). Biochemical effects of thyroid deficiency on the developing brain. *J. Neurochem.* **15,** 1335–1349.

Balázs, R., Brooksbank, B. W. L., Davison, A. N., Eayrs, J. T. and Wilson, D. A. (1969). The effect of neonatal thyroidectomy on myelination in the rat brain. *Brain Res.* **15,** 219–232.

Baldwin, B. A. and Soltysik, S. S. (1969). The effect of cerebral ischaemia or intracarotid injection of methohexitone on short term memory in goats. *Brain Res.* **16,** 105–120.

Baldwin, D. M., Haun, C. H. and Sawyer, C. H. (1974). Effect of intraventricular infusion of $ACTH^{1-24}$ and $ACTH^{4-10}$ on LH release, ovulation and behavior in the rabbit. *Brain Res.* **80,** 291–301.

Balentine, J. D. (1975). Dendritic degeneration following hyperbaric oxygen exposure. *In* "Advances in Neurology, Vol. 12, Physiology and Pathology of Dendrites" (G. W. Kreutzberg, Ed.), pp. 471–481. Raven Press, New York.

Banks, P., Mangnall, D. and Mayor, D. (1969). The redistribution of cytochrome oxidase, noradrenaline and adenosine triphosphate in adrenergic nerves constricted at two points. *J. Physiol.* (*Lond.*) **200,** 745–762.

Banks, P., Mayor, D., Mitchell, M. and Tomlinson, D. (1971). Studies on the translocation of noradrenaline containing vesicles in post ganglionic sympathetic neurones *in vitro.* Inhibition of movement by colchicine and vinblastine and evidence for the involvement of axonal microtubules. *J. Physiol.* (*Lond.*) **216,** 625–639.

Barker, J. L., Hoffman, P. N., Gainer, H. and Lasek, R. J. (1975). Rapid transport of proteins in the sonic motor system of the toadfish. *Brain Res.* **97,** 291–301.

Barley, J., Ginsburg, M., Greenstein, B. D., MacLusky, N. J. and Thomas, P. J. (1974). A receptor mediating sexual differentiation? *Nature* (*Lond.*) **252,** 259–260.

Barlow, H. B. (1975). Visual experience and cortical development. *Nature* (*Lond.*) **258,** 199–204.

Barondes, S. H. (1966). On the site of synthesis of the mitochondrial protein of nerve endings. *J. Neurochem.* **13,** 721–727.

Barondes, S. H. (1968a). Further studies of the transport of proteins to nerve endings. *J. Neurochem.* **15,** 343–350.

Barondes, S. H. (1968b). Incorporation of radioactive glucosamine into macromolecules at nerve endings. *J. Neurochem.* **15,** 699–706.

Barondes, S. H. (1969a). Axoplasmic transport. *In* "Handbook of Neurochemistry" (A. Lajtha, Ed.), Vol. 2, pp. 435–446. Plenum Press, New York.

Barondes, S. H. (1969b). Two sites of synthesis of macromolecules in neurons. *Symp. Int. Soc. Cell Biol.* **8,** 351–364.

Barondes, S. H. (1970). Cerebral protein synthesis inhibitors block long-term memory. *Int. Rev. Neurobiol.* **12,** 177–205.

Barondes, S. H. (1972). Memory transfer. *Science* **176,** 631–632.

Barondes, S. H. (1974). Synaptic macromolecules: Identification and metabolism. *Ann. Rev. Biochem.* **43,** 147–168.

Barondes, S. H. and Cohen, H. D. (1967). Delayed and sustained effect of acetoxylcycloheximide on memory in mice. *Proc. Nat. Acad. Sci., U.S.A.* **58,** 157–164.

Barondes, S. H. and Cohen, H. D. (1968). Arousal and the conversion of "short-term" to "long-term" memory. *Proc. Nat. Acad. Sci., U.S.A.* **61,** 923–929.

Barraco, R. A. and Irwin, L. N. (1976). Cerebral protein patterns from trained and naive pigeons. *J. Neurochem.* (In press).

Bateson, P. P. G. (1971). Imprinting. *In* "The Ontogeny of Vertebrate Behaviour" (M. Moltz, Ed.), pp. 369–387. Academic Press, London and New York.

Bateson, P. P. G., Horn, G. and Rose, S. P. R. (1969). Effect of an imprinting procedure on regional incorporation of tritiated lysine into protein of chick brain. *Nature* (*Lond.*) **223,** 534–535.

Bateson, P. P. G., Horn, G. and Rose, S. P. R. (1972). Effects of early experience on regional incorporation of precursors into RNA and protein in the chick brain. *Brain Res.* **39,** 449–465.

Battistin, L., Grynbaum, A. and Lajtha, A. (1969). Distribution and uptake of amino acids in various regions of the cat brain *in vitro*. *J. Neurochem.* **16,** 1459–1468.

Baumel, I., DeFeo, J. and Lal, H. (1969). Decreased potency of CNS depressants after prolonged social isolation in mice. *Psychopharmacologia* **15,** 153–158.

Baumel, I., DeFeo, J. J. and Lal, H. (1970). Alterations in brain sensitivity and barbiturate metabolism unrelated to aggression in socially deprived mice. *Psychopharmacologia* **18,** 320–324.

Baumgarten, H. G. and Lachenmayer, L. (1972). 5-7-Dihydroxytryptamine: Improvement in chemical lesioning of indoleamine neurons in the mammalian brain. *Z. Zellforsch.* **135,** 399–414.

Baxter, B. L. and Riesen, A. H. (1961). Electroretinogram of the visually deprived cat. *Science* **134,** 1626–1627.

Baxter, C. F. (1966). Regional changes in amino acid composition during anesthesia in the central nervous system of rats. *Proc. Western Pharmacol. Soc.* **9,** 52–53.

Baxter, C. F. (1970). The nature of γ-aminobutyric acid. *In* "Handbook of Neurochemistry" (A. Lajtha, Ed.), Vol. 3, pp. 289–353. Plenum Press, New York.

Baxter, C. F. (1976). Effect of GABA on protein metabolism in the nervous system. *In* "GABA in Nervous System Function" (E. Roberts, T. N. Chase and D. B. Tower, Eds), pp. 89–102. Raven Press, New York.

Beach, F. A. (1939). The neural basis of innate behavior. III. Comparison of learning ability and instinctive behavior in the rat. *J. Comp. Psychol.* **28,** 225–262.

Beach, F. A. and Jaynes, J. (1954). Effects of early experience upon the behavior of animals. *Psychol. Bull.* **51,** 239–263.

Beach, F. A., Hebb, D. O., Morgan, C. T. and Nissen, H. W. (Eds) (1960). "The Neuropsychology of Lashley (Selected Papers of K. S. Lashley)". McGraw-Hill Book Co., Inc., New York, 564 pp.

Beach, G., Emmens, M., Kimble, D. P. and Lickey, M. (1969). Autoradiographic demonstration of biochemical changes in the limbic system during avoidance training. *Proc. Nat. Acad. Sci., U.S.A.* **62,** 692–696.

Bebenick, G. and Monnier, M. (1972). Nuclear size variations in cells of locus coeruleus during sleep, arousal and stress. *Exptl. Neurol.* **35,** 1–12.

Beitel, R. E. and Porter, P. B. (1968). Deficit in short- and long-term retention and impairments in learning induced by severe hypothermia in mice. *J. Comp. Physiol. Psychol.* **66,** 53–59.

Bell, R. W., Miller, C. E., Ordy, J. M. and Rolsten, C. (1971). Effects of population density and living space upon neuroanatomy, neurochemistry and behavior in the C57BL/10 mouse. *J. Comp. Physiol. Psychol.* **75,** 258–263.

Belluzzi, J. D., Grant, N., Garsky, V., Sarantakis, D., Wise, C. D. and Stein, L. (1976). Analgesia induced *in vivo* by central administration of enkephalin in rat. *Nature* (*Lond.*) **260,** 625–626.

Benda, P., Lightbody, J., Sato, G., Levine, L. and Sweet, W. (1968). Differentiated rat glial cell strain in tissue culture. *Science* **161,** 370–371.

Benjamin, A. M. and Quastel, J. H. (1972). Locations of amino acids in brain slices from the rat; tetrodotoxin-sensitive release of amino acids. *Biochem. J.* (*Lond.*) **128,** 631–646.

Benjamin, A. M. and Quastel, J. H. (1975). Metabolism of amino acids and ammonia in rat brain cortex slices *in vitro*: A possible role of ammonia in brain function. *J. Neurochem.* **25,** 197–206.

Benjamins, J. A. and McKhann, G. M. (1972). Neurochemistry of development. *In* "Basic Neurochemistry" (R. W. Albers, G. J. Siegel, R. Katzman and B. W. Agranoff, Eds), pp. 269–298. Little, Brown and Co., Boston.

Bennett, E. L. (1973). Personal communication to Walsh and Cummins (1975).

Bennett, E. L. (1976). Cerebral effects of differential experience and training. *In* "Neural Mechanisms of Learning and Memory" (M. R. Rosenzweig and E. L. Bennett, Eds), pp. 279–287. The M.I.T. Press, Cambridge, Mass.

Bennett, E. L., Diamond, M. C., Krech, D. and Rosenzweig, M. R. (1964). Chemical and anatomical plasticity of the brain. *Science* **146,** 610–619.

Bennett, E. L. and Rosenzweig, M. R. (1971). Chemical alterations produced in brain by environment and training. *In* "Handbook of Neurochemistry" (A. Lajtha, Ed.), Vol. 6, pp. 173–210. Plenum Press, New York.

Bennett, E. L., Rosenzweig, M. R. and Wu, S. -Y. C. (1973). Excitant and depressant drugs modulate effects of environment on brain weight and cholinesterases. *Psychopharmacologia* **33,** 309–328.

Bennett, E. L., Rosenzweig, M. R., Diamond, M. C., Morimoto, H. and Hebert, M. (1974). Effects of successive environments on brain measures. *Physiol. Behav.* **12,** 621–631.

Berl, S. and Clarke, D. D. (1969). Compartmentation of amino acid metabolism. *In* "Handbook of Neurochemistry" (A. Lajtha, Ed.), Vol. 2, pp. 447–472. Plenum Press, New York.

Berl, S., Lajtha, A. and Waelsch, H. (1961). Amino acid and protein metabolism. VI. Cerebral compartments of glutamic acid metabolism. *J. Neurochem.* **7,** 186–197.

Berl, S., Takagaki, G., Clarke, D. D. and Waelsch, H. (1962). Metabolic compartmentation *in vivo*: Ammonia and glutamic acid metabolism in brain and liver. *J. Biol. Chem.* **237,** 2562–2569.

Berl, S., Clarke, D. D. and Schneider, D. (Eds) (1975)." Metabolic Compartmentation and Neurotransmission: Relation to Brain Structure and Function." Plenum Press, New York, 721 pp.

Berlucchi, G. and Buchtel, H. A. (1975). Some trends in the neurological study of learning. *In* "Handbook of Psychobiology" (M. S. Gazzaniga and C. Blakemore, Eds), pp. 481–498. Academic Press, New York and London.

Bernard, C. (1859). "Leçons sur les Propriétés physiologiques et les Altérations pathologiques des Liquides de l'organisme." Baillière, Paris.

Bernard, C. (1878). "Leçons sur les Phénomènes de la Vie communs aux Animaux et aux Végétaux." Vol. 1. Baillière, Paris.

Berry, R. M. and Jack, C. E. (1971). The effect of temperature upon shock-elicited aggression in rats. *Psychonom. Sci.* **23,** 341–343.

Berry, R. W. (1969). Ribonucleic acid metabolism of a single neuron: Correlation with electrical activity. *Science* **166,** 1021–1023.

Bertolini, A., Vergoni, W., Gessa, G. L. and Ferrari, W. (1969). Induction of sexual excitement by the action of adrenocorticotrophic hormone in brain. *Nature* (*Lond.*) **221,** 667–669.

Besser, G. M. (1974). Hypothalamus as an endocrine organ. I. *Brit. Med. J.* **3,** 560–564.

Bewsher, P. D., Hillman, C. C. and Ashmore, J. (1966). Studies on the hypoglycemic effect of *d*-amphetamine in aggregated mice. *Biochem. Pharmacol.* **15,** 2079–2085.

Bexton, W. H., Heron, W. and Scott, T. H. (1954). Effects of decreased variation in the sensory environment. *Can. J. Psychol.* **8,** 70–76.

Biederman, G. B. (1974). The search for the chemistry of memory: Recent trends and the logic of investigation in the role of cholinergic and adrenergic transmitters. *Progr. Neurobiol.* **2,** 289–307.

Bindra, D. (1959). "Motivation, a Systematic Reinterpretation." Ronald, New York.

Bindra, D. (1968). Neuropsychological interpretation of the effects of drive and incentive motivation on general activity and instrumental behavior. *Psychol. Rev.* **75,** 1–22.

Bingham, W. E. and Griffiths, W. J. (1952). The effect of different environments during infancy on adult behavior in the rat. *J. Comp. Physiol. Psychol.* **45,** 307–312.

Birren, J. E. (1965). Age changes in speed of behavior: Its central nature and physiological correlates. *In* "Behavior, Aging and the Nervous System" (A. T. Welford and J. E. Birren, Eds), pp. 191–216. C. C. Thomas, Springfield, Illinois.

Bjorklund, A. and Stenevi, U. (1971). Growth of central catecholamine neurones into smooth muscle grafts in the rat mesencephalon. *Brain Res.* **31,** 1–20.

Bjorklund, A., Katzman, R., Stenevi, U. and West, K. A. (1971). Development and growth of axonal sprouts from noradrenaline and 5-hydroxytryptamine neurones in the rat spinal cord. *Brain Res.* **31,** 21–33.

Bjorklund, A., Nobin, A. and Stenevi, U. (1973). Regeneration of central serotonin neurons after axonal degeneration induced by 5,6-dihydroxytryptamine. *Brain Res.* **50,** 214–220.

Bjorklund, A., Baumgarten, H.-G. and Nobin, A. (1974). Chemical lesioning of central monoamine axons by means of 5,6-dihydroxytryptamine and 5,7-dihydroxytryptamine. *In* "Advances in Biochemical Psychopharmacology," Vol. 10, pp. 13–33. Raven Press, New York.

Blakemore, C. (1974). Development of functional connexions in the mammalian visual system. *Brit. Med. Bull.* **30,** 152–157.

Blakemore, C. and Cooper, G. F. (1970). Development of the brain depends on the visual environment. *Nature* (*Lond.*) **228,** 477–478.

Blakemore, C. and Mitchell, D. E. (1973). Environmental modification of the visual cortex and the neural basis of learning and memory. *Nature* (*Lond.*) **241,** 467–468.

Blakemore, C. and Van Sluyters, R. C. (1974). Reversal of the physiological effects of monocular deprivation in kittens: Further evidence for a sensitive period. *J. Physiol.* (*Lond.*) **237,** 195–216.

Blakemore, C. and Van Sluyters, R. C. (1975). Innate and environmental factors in the development of the kitten's visual cortex. *J. Physiol.* (*Lond.*) **248,** 663–716.

Blaschko, H. and Muscholl, E. (1972). "Catecholamines: Handbook of Experimental Pharmacology", XXXIII. Springer-Verlag, Berlin, 1054 pp.

Blinzinger, K. and Kreutzberg, G. (1968). Displacement of synaptic terminals from regenerating motoneurons by microglial cells. *Z. Zellforsch. Mikrosk. Anat.* **85,** 145–157.

Bliss, E. L. and Ailion, J. (1969). Response of neurogenic amines to aggregation and strangers. *J. Pharmacol. Exptl. Therapeut.* **168,** 258–263.

Bliss, E. L., Ailion, J. and Zwanziger, J. (1968). Metabolism of norepinephrine, serotonin and dopamine in rat brain with stress. *J. Pharmacol. Exptl. Therapeut.* **164,** 122–134.

Bliss, T. V. P. and Gardner-Medwin, A. R. (1973). Long-lasting potentiation of synaptic transmission in the dentate area of the unanaesthetized rabbit following stimulation of the perforant path. *J. Physiol.* (*Lond.*) **232,** 357–374.

Bliss, T. V. P. and Lømo, T. (1973). Long-lasting potentiation of synaptic transmission in the dentate area of the anaesthetized rabbit following stimulation of the perforant path. *J. Physiol.* (*Lond.*) **232,** 331–356.

Bliss, T. V. P., Gardner-Medwin, A. R. and Lømo, T. (1973). Synaptic plasticity in the hippocampal formation. *In* "Macromolecules and Behaviour" (G. B. Ansell and P. B. Bradley, Eds), pp. 193–203. University Park Press, London.

Bloch, V. (1970). Facts and hypotheses concerning memory consolidation processes. *Brain Res.* **24,** 561–575.

Blocq, P. and Marinesco, G. (1893). Sur un cas de tremblement parkinsonien hémiplégique symptomatique d'une tumeur du pédoncule cérébrale. *C.R. Soc. Biol. Paris* **5,** 105–111.

Blomstrand, C. and Hamberger, A. (1969). Protein turnover in cell-enriched fractions from rabbit brain. *J. Neurochem.* **16,** 1401–1407.

Blomstrand, C. and Hamberger, A. (1970). Amino acid incorporation *in vitro* into protein of neuronal and glial cell-enriched fractions. *J. Neurochem.* **17,** 1187–1195.

Blondaux, C., Juge, A., Sordet, F., Chouvet, G., Jouvet, M. et Pujol, J. (1973). Modification du metabolisme de la serotonine (5-HT) cérébrale induit chez le rat par administration de 6-hydroxydopamine. *Brain Res.* **50,** 101–114.

Bloom, F. E., Algerie, S., Groppetti, A., Revuelta, A. and Costa, E. (1969). Lesions of central norepinephrine terminals with 6-OH-dopamine: Biochemistry and fine structure. *Science* **166,** 1284–1286.

Bloom, F. E., Iversen, L. L. and Schmitt, F. O. (Eds) (1970). Macromolecules in synaptic function. *Neurosci. Res. Progr. Bull.* **8,** 325–455.

Bodian, D. (1965). A suggestive relationship of nerve cell RNA with specific synaptic sites. *Proc. Nat. Acad. Sci., U.S.A.* **53,** 418–425.

Boegman, R. J. and Wood, P. L. (1976). Monoamines as possible mediators in the regulation of fast axoplasmic flow. *J. Neurochem.* **26,** 737–740.

Bogoch, S. (1965). Brain proteins in learning: Findings in the experiments in progress discussed at the NRP work session, June 27–29, 1965. *Neurosci. Res. Progr. Bull.* **3,** 38–41.

Bogoch, S. (1968). "The Biochemistry of Memory", Oxford University Press, New York and Oxford, 254 pp.

Bogoch, S. (1972). Brain glycoprotein 10B: Further evidence of the "sign-post" role of brain glycoproteins in cell recognition, its change in brain tumor, and the presence of a "distance factor". *In* "Functional and Structural Proteins of the Nervous System" (A. N. Davison, P. Mandel and I. G. Morgan, Eds), pp. 39–52. Plenum Press, New York.

Bogoch, S. (1973). Brain glycoproteins and learning: New studies supporting the "sign-post" theory. *In* "Current Biochemical Approaches to Learning and Memory" (W. B. Essman and S. Nakajima, Eds), pp. 147–157. Spectrum-Halstead Press, New York.

Bogoch, S. (1974). Glycoproteins and brain circuitry: The "sign-post" theory in normal memory function and in the regressive states of brain tumors and the psychoses. *Biol. Psychiat.* **9,** 73–88.

Bogoch, S., Dussik, K. T. and Conran, P. (1964). Nervous-system glycoproteins in mental disorders. *New Engl. J. Med.* **264,** 521–529.

Bogolepov, N. N. and Pushkin, A. S. (1975). Submicroscopic changes of cortex nerve cells in chronic mirror epileptic focus in rat. *Brain Res.* **94,** 173–184.

Bohdanecky, Z., Kopp, R. and Jarvik, M. E. (1968). Comparison of ECS and Flurothyl-induced retrograde amnesia in mice. *Psychopharmacologia* **12,** 91–95.

Bohus, B. and Korányi, L. (1969). Hormonal conditioning of adaptive behavioral processes. *In* "Recent Developments of Neurobiology in Hungary" (K. Lissak, Ed.), Vol. II, pp. 50–74. Akademiai Kiado, Budapest.

Bonaventure, N., Goswamy, S. and Karli, P. (1971). Electroretinogram and visual evoked response in rabbits reared in total darkness or continuous illumination. *Doc. Ophthal.* **30,** 339–347.

Bondy, S. C. (1971). Axonal transport of macromolecules. II. Nucleic acid migration in the central nervous system. *Exptl. Brain Res.* **13,** 135–139.

Bondy, S. C. and Madsen, C. J. (1973). Axoplasmic transport of RNA. *J. Neurobiol.* **4,** 535–542.

Bondy, S. C. and Madsen, C. J. (1974). The extent of axoplasmic transport during development, determined by migration of various radioactively-labelled materials. *J. Neurochem.* **23,** 905–910.

Bondy, S. C. and Marchisio, P. C. (1973). Development of axonal transport of RNA and its precursors in the optic pathway of the chick embryo. *Exptl. Neurol.* **41,** 29–37.

Bondy, S. C. and Morelos, B. S. (1971). Stimulus deprivation and cerebral blood flow. *Exptl. Neurol.* **31,** 200–206.

Bondy, S. C. and Purdy, J. L. (1975). Migration of ribosomes along the axons of the chick visual pathway. *Biochim. Biophys. Acta* **390,** 332–341.

Bondy, S. C., Lehman, R. A. and Purdy, J. L. (1974). Visual attention affects brain blood flow. *Nature (Lond.)* **248,** 440–441.

Booth, D. A. and Simson, P. C. (1973). Aversion to a cue acquired by its association with effects of an antibiotic in the rat. *J. Comp. Physiol. Psychol.* **84,** 319–323.

Borisy, G. C. and Taylor, E. W. (1967). The mechanism of action of colchicine. Binding of colchicine-^{3}H to cellular protein. *J. Cell Biol.* **34,** 525–533.

Bourgault, P. C., Karczmar, A. G. and Scudder, C. L. (1963). Contrasting behavioral, pharmacological, neurophysiological, and biochemical profiles of C57 Bl/6 and SC-I strains of mice. *Life Sci.* No. 8, 533–553.

Bowen, D. M., Smith, C. B. and Davison, A. N. (1973). Molecular changes in senile dementia. *Brain* **96,** 849–856.

Bowlby, J. (1960). Grief and mourning in early infancy and early childhood. *Psychoanalyt. Stud. Child.* **15,** 9–32.

Bowman, R. E. and Kottler, P. D. (1970). Regional brain RNA metabolism as a function of different experiences. *In* "Biochemistry of Brain and Behaviour" (R. E. Bowman and S. P. Datta, Eds), pp. 301–326. Plenum Press, New York.

Bowman, R. E. and Strobel, D. A. (1969). Brain RNA metabolism in the rat during learning. *J. Comp. Physiol. Psychol.* **67,** 448–456.

Bradley, P. B., Eayrs, J. T. and Schmalbach, K. (1960). The electroencephalogram of normal and hypothyroid rats. *EEG Clin. Neurophysiol.* **12,** 467–477.

Brady, J. V. (1960). Temporal and emotional effects related to intracranial electrical self-stimulation. *In* "Electrical Studies on the Unanesthetized Brain" (E. R. Ramey and D. S. O'Doherty, Eds), pp. 52–77. Hoeber, New York.

Brady, J. V., Boren, J. J., Conrad, D. and Sidman, M. (1957). The effect of food and water deprivation upon intracranial self-stimulation. *J. Comp. Physiol. Psychol.* **50,** 134–137.

Brady, R. O. (1973). Inborn errors of lipid metabolism. *Advances in Enzymology* **38,** 293–315.

Brain, P. F. and Nowell, N. W. (1971a). Isolation versus grouping effects on adrenal and gonadal function in Albino mice. I. The male. *Gen. Comp. Endocr.* **16,** 149–154.

Brain, P. F. and Nowell, N. W. (1971b). Isolation versus grouping effects on adrenal and gonadal function in Albino mice. II. The female. *Gen. Comp. Endocr.* **16,** 155–159.

Bray, J. J. and Austin, L. (1968). Flow of protein and ribonucleic acid in peripheral nerve. *J. Neurochem.* **15,** 731–740.

Brazier, M. A. B. (Ed.) (1975). "Growth and Development of the Brain: Nutritional, Genetic, and Environmental Factors", International Brain Research Organization Monograph Series, Vol. 1. Raven Press, New York, 412 pp.

Brdar, B., Rifkin, D. B. and Reich, E. (1973). Studies on Rous sarcoma virus: Effects of nucleoside analogues on virus synthesis. *J. Biol. Chem.* **248,** 2397–2408.

Breckenridge, W. C. and Morgan, I. G. (1972). Common glycoproteins of synaptic vesicles and the synaptosomal plasma membranes. *FEBS Lett.* **22,** 253–256.

Breese, G. R., Cooper, B. R., Prange, A. J., Jr., Cott, J. M. and Lipton, M. A. (1974a). Interactions of thyrotropin-releasing hormone (TRH) with centrally acting drugs. *In* "The Thyroid Axis, Drugs, and Behavior" (A. J. Prange, Jr., Ed.), pp. 115–127. Raven Press, New York.

Breese, G. R., Cott, J. M., Cooper, B. R., Prange, A. J., Jr. and Lipton, M. A. (1974b). Antagonism of ethanol narcosis by thyrotropin releasing hormone. *Life Sci.* **14,** 1053–1063.

Breese, G. R., Cott, J. M., Cooper, B. R., Prange, A. J., Jr., Lipton, M. A. and Plotnikoff, N. P. (1975). Effects of thyrotropin releasing hormone (TRH) on the actions of pentobarbital and other centrally acting drugs. *J. Pharmacol. Exptl. Therapeut.* **193,** 11–22.

Bresnahan, E. and Routtenberg, A. (1972). Memory disruption by unilateral low level, sub-seizure stimulation of the medial amygdaloid nucleus. *Physiol. Behav.* **9,** 513–525.

Breuer, J. and Freud, S. (1892). On the theory of hysterical attacks. "Complete Works of Sigmund Freud", Standard Edition, Vol. 1.

Brierley, J. B., Brown, A. W. and Meldrum, B. S. (1971a). The nature and time course of the neuronal alterations resulting from oligaemia and hypoglycaemia in the brain of *Macaca Mulatta. Brain Res.* **25,** 483–499.

Brierley, J. B., Brown A. W. and Meldrum, B. S. (1971b). The neuropathology of insulin-induced hypoglycaemia in a primate (*M. mulatta*); topography and cellular nature. *In* "Brain Hypoxia" (J. B. Brierly and B. S. Meldrum, Eds), pp. 225–230. S.I.M.P./William Heinemann Medical Books Ltd.

Briggs, F. N. and Munson, P. L. (1955). Studies on the mechanism of stimulation of ACTH secretion with the aid of morphine as a blocking agent. *Endocrinology* **57,** 205.

Brimijoin, S. (1976). Cycloheximide alters axonal transport and subcellular distribution of dopamine-β-hydroxylase activity. *J. Neurochem.* **26,** 35–40.

Brink, F. (1954). The role of calcium ions in neural processes. *Pharmacol. Rev.* **6,** 243–298.

Brink, J. J., Davis, R. E. and Agranoff, B. W. (1966). Effects of puromycin, acetoxycycloheximide and actinomycin D on protein synthesis in goldfish brain. *J. Neurochem.* **13,** 889–896.

Brissaud, E. (1895). "Leçons sur les Maladies Nerveuses". Masson and Cie, Paris.

Broadbent, D. E. (1971). "Decision and Stress". Academic Press, London and New York.

Brodersen, P. and Jørgensen, E. (1974). Cerebral blood flow and oxygen uptake and cerebrospinal fluid biochemistry in severe coma. *J. Neurol. Neurosurg. Psychiat.* **37,** 384–391.

Brody, H. (1955). Organization of the cerebral cortex. III. A study of aging in the human cerebral cortex. *J. Comp. Neurol.* **102,** 511–556.

Brody, H. (1970). Structural changes in the aging nervous system. *Interdisc. Top. Geront.* **7,** 9–21.

Brody, H. H. D. and Ordy, J. M. (Eds) (1975). "Clinical, Morphologic, and Neurochemical Aspects in the Aging Central Nervous System", Aging, Vol. 1. Raven Press, New York, 240 pp.

Bronson, F. H. and Chapman, V. M. (1968). Adrenal-oestrous relationships in grouped or isolated female mice. *Nature* (*Lond.*) **218,** 483–484.

Brown, B. B. (1975). "New Mind, New Body: Bio-Feedback: New Directions for the Mind". Hodder and Stoughton, London, 464 pp.

Brown, C. M. and Chambers, J. W. (1974). Neurotransmitter regulation of growth hormone and ACTH release. *J. Pharmacol.* Suppl. 2, p. 12.

Brown, D. J. and Stone, W. E. (1973). Glutamate and aspartate in cortical subarachnoid fluid in relation to convulsive activity. *Brain Res.* **54,** 143–148.

Brunngraber, E. G. (1969). Brief proposals: The possible role of glycoproteins in neural function. *Perspect. Biol. Med.* **12,** 467–470.

Bryans, W. A. (1959). Mitotic activity in the brain of the adult rat. *Anat. Rec.* **133,** 65–71.

Bubenik, G. A. and Brown, G. M. (1973). Morphologic sex differences in primate brain areas involved in regulation of reproductive activity. *Experientia* **29,** 619–621.

Buchwald, N., Wyers, C., Lauprecht, C. and Heuser, G. (1961). The "caudate-spindle". IV. A behavioral index of caudate-induced inhibition. *EEG Clin. Neurophysiol.* **13,** 531–537.

Bureš, J. and Burešová, O. (1960). The use of Leão's spreading depression in the study of interhemispheric transfer of memory traces. *J. Comp. Physiol. Psychol.* **53,** 558–565.

Bureš, J. and Burešová, O. (1963). Cortical spreading depression as a memory disturbing factor. *J. Comp. Physiol. Psychol.* **56,** 268–272.

Bureš, J., Burešová, O. and Krivanek, J. (1974). "The Mechanism and Application of Leão's Spreading Depression of Electroencephalographic Activity". Academic Press, New York and London.

Bureš, J., Burešová, O. and Zahorová, A. (1958). Conditioned reflexes and Leão's spreading cortical depression. *J. Comp. Physiol. Psychol.* **51,** 263–268.

Burešová, O. and Bureš, J. (1969). The effect of prolonged cortical spreading depression on learning and memory in rats. *J. Neurobiol.* **1,** 135–146.

Burgus, R. and Guillemin, R. (1970). Chemistry of thyrotropin releasing factor (TRF). *In* "Hypophysiotropic Hormones of the Hypothalamus: Assay and Chemistry" (J. Meites, Ed.), pp. 227–241. Williams and Wilkins, Baltimore.

Burns, N. M. and Kimura, D. (1963). Isolation and sensory deprivation. *In* "Unusual Environments and Human Behavior" (N. M. Burns, R. M. Chambers and E. Hendler, Eds), pp. 167–192. Free Press of Glencoe, Collier-Macmillan, London.

Burt, D. R. and Larrabee, M. G. (1973). Subcellular site of the phosphatidylinositol effect: Distribution on density gradients of labelled lipids from resting and active sympathetic ganglia of the rat. *J. Neurochem.* **21,** 255–272.

Buryakova, A. V. and Sytinsky, I. A. (1975). Amino acid composition of cerebrospinal fluid in acute neuroinfections in children. *Arch. Neurol.* **32,** 28–31.

Bykov, K. M. (Ed.) (1960). "Text-book of Physiology". Foreign Language Publishing House, Moscow.

Calissano, P. (1973). Specific properties of the brain-specific protein S-100. *In* "Proteins of the Nervous System" (D. J. Schneider, Ed.), pp. 13–26. Raven Press, New York.

Calissano, P. and Bangham, A. D. (1971). Effect of two brain specific proteins (S-100 and 14.3.2) on cation diffusion across artificial membranes. *Biochem. Biophys. Res. Commun.* **43,** 504–509.

Calne, D., Chase, T. N. and Barbeau, A. (Eds) (1975). "Advances in Neurology, Vol. 9, Dopaminergic Mechanisms." Raven Press, New York, 427 pp.

Cammermeyer, J. (1969). Species differences in acute retrograde neuronal reaction of the facial and hypoglossal nuclei. *J. Hirnforsch.* **11,** 13–29.

Cammermeyer, J. (1972). Nonspecific changes of the central nervous system in normal and experimental material. *In* "The Structure and Function of Nervous Tissue" (G. H. Bourne, Ed.), Vol. 6, pp. 131–251. Academic Press, New York and London.

Cammermeyer, J. (1975). The effect of cortisone treatment and reoperation on reactive changes in the facial nucleus after axotomy. *Brain Res.* **95,** 241–252.

Campbell, H. J. (1965). "Correlative Physiology of the Nervous System." Academic Press, London and New York, 313 pp.

Campbell, L. B., Hanin, I. and Jenden, D. J. (1970). Gas chromatographic evaluation of the effects of some muscarinic and anti-muscarinic drugs on acetylcholine levels in rat brain. *Biochem. Pharmacol.* **19,** 2053–2059.

Campbell, R. J., Bogoch, S., Scolaro, M. J. and Belval, P. C. (1967). Cerebrospinal fluid glycoproteins in schizophrenia. *Amer. J. Psychiat.* **123,** 952–962.

Cannon, W. B. (1932). "The Wisdom of the Body". Norton, New York.

Cannon, W. B. and Rosenblueth, A. (1949). "The Supersensitivity of Denervated Structures". Macmillan, New York, 245 pp.

Caputto, B. L., Maccioni, A. H. R. and Caputto, R. (1975). Light inhibits the labelling of gangliosides in chicken retina. *Nature* (*Lond.*) **257,** 492–493.

Cardo, B. (1965). Rôle de certains noyaux thalamiques dans l'éboration et la conservation de divers conditionnements. *Psychol. France,* **10,** 344–351.

Cardo, B. (1967). Effets de la stimulation du noyau parafasciculaire thalamique sur l'acquisition d'un conditionnement d'évitement chez le rat. *Physiol. Behav.* **2,** 245–248.

Carlson, A. J. (1902–1903). Changes in the Nissl's substance of the ganglion and the bipolar cells of the retina of the Brandt cormorant *Phalacrocorax pencillatus* during prolonged normal stimulation. *Amer. J. Anat.* **2,** 341–347.

Carlson, A. J. (1916). "The Control of Hunger in Health and Disease". University of Chicago Press, Chicago.

Casola, L., Lim, R., Davis, R. E. and Agranoff, B. W. (1968). Behavioral and biochemical effects of intracranial injection of cytosine arabinoside in goldfish. *Proc. Nat. Acad. Sci., U.S.A.* **60,** 1389–1395.

Castellano, C. and Oliverio, A. (1976). Early malnutrition and postnatal changes in brain and behavior in the mouse. *Brain Res.* **101,** 317–325.

Catravas, G. N., McFarland, W. L. and Teitelbaum, H. (1973). Neurochemical changes in specific regions of the rat brain during different phases of conditioning. *Physiol. Behav.* **11,** 359–363.

Ceccarelli, B., Hurlbut, W. P. and Mauro, P. (1972). Depletion of vesicles from frog neuromuscular junctions by prolonged tetanic stimulation. *J. Cell. Biol.* **54,** 30–38.

Ceccarelli, B., Hurlbut, W. P. and Mauro, A. (1973). Turnover of transmitter and synaptic vesicles at the frog neuromuscular junction. *J. Cell. Biol.* **57,** 499–524.

Cerda, M. and Israel, Y. (1969). Effect of ethanol on synthesis and active accumulation of gamma-aminobutyric acid in rat brain slices. *Arch. Biol. Med. Exp.* **6,** 62–64.

Chambers, J. W. and Brown, G. M. (1976). Neurotransmitter regulation of growth hormone and ACTH in the Rhesus monkey: Effects of biogenic amines. *Endocrinology* **98,** 420–428.

Chance, M. R. A. (1947). Factors influencing the toxicity of sympathomimetic amines in solitary mice. *J. Pharmacol. Exptl. Therapeut.* **89,** 289–296.

Chang, C. M. (1972). Effect of colchicine and cytochalsin B on axonal particle movement and outgrowth *in vitro*. *J. Cell. Biol.* **55,** 37a.

Chang, M. M., Leeman, S. E. and Niall, H. D. (1971). Amino-acid sequence of substance P. *Nature, New Biol.* **232,** 86–87.

Chase, H. P., Lindsley, W. F. B. and O'Brien, D. (1969). Undernutrition and cerebellar development. *Nature* (*Lond.*) **221,** 554–555.

Chen, G., Bohner, B. and Bratton, A. C., Jr. (1963). The influence of certain central depressants on fighting behavior of mice. *Arch. Int. Pharmacodyn.* **142,** 30–34.

Cheney, D. L., Gubler, C. J. and Jaussi, A. W. (1969). Production of acetylcholine in rat brain following thiamine deprivation and treatment with thiamine antagonists. *J. Neurochem.* **16,** 1283–1291.

Chitre, V. S., Chopra, S. P. and Talwar, G. P. (1964). Changes in the ribonucleic acid content of the brain during experimentally induced convulsions. *J. Neurochem.* **11,** 439–448.

Chiu, J. F. and Sung, S. C. (1973). Intracellular distribution of DNA polymerase activities in developing and adult rat brain. *J. Neurochem.* **20,** 617–620.

Chow, K. L. (1973). Neuronal changes in the visual system following visual deprivation. *In* "Handbook of Sensory Physiology" (R. Jung, Ed.), Vol. VII/3, Part A, pp. 599–627. Springer, Berlin.

Chow, K. L., Riesen, A. H. and Newell, F. W. (1957). Degeneration of retinal ganglion cells in infant chimpanzees reared in darkness. *J. Comp. Neurol.* **107,** 27–48.

Christensen, B. N. (1976). Morphological correlates of synaptic transmission in lamprey spinal cord. *J. Neurophysiol.* **39,** 197–212.

Christian, J. J., Lloyd, J. A. and Davis, D. E. (1965). The role of endocrines in self-regulation of mammalian populations. *Rec. Progr. Hormone Res.* **21,** 501–578.

Christomanos, A. and Scholtz, W. (1975). Klinische Beobachtungen und Pathologische-anatomische Befundum Zentralnervensystem bei mit Thiophen vergifteten Hunden. *Z. Neurol.* **144,** 1–20.

Chronister, R. B., Bernstein, J. J., Zornetzer, S. F. and White, L. E., Jr. (1973). Synaptic boutons in the hippocampus: Changes are produced by age and experience. *Experientia* **29,** 588–589.

Cicero, T. J., Cowan, W. M., Moore, B. W. and Suntzeff, V. (1970). The cellular localization of the two brain-specific proteins, S-100 and 14-3-2. *Brain Res.* **18,** 25–34.

Cliffe, E. E. and Waley, S. G. (1958). Acidic peptides of the lens. 4. The biosynthesis of ophthalmic acid. *Biochem. J.* (*Lond.*) **69,** 649–655.

Clouet, D. H. and Iwatsubo, K. (1975). Mechanisms of tolerance to and dependence on narcotic analgesic drugs. *Ann. Rev. Pharmacol.* **15,** 49–71.

Cocks, J. A., Balázs, R., Johnson, A. L. and Eayrs, J. T. (1970). Effect of thyroid hormone on the biochemical maturation of rat brain: Conversion of glucose-carbon into amino acids. *J. Neurochem.* **17,** 1275–1285.

Codish, S. D. (1971). Actinomycin D injected into the hippocampus of chicks: Effects upon imprinting. *Physiol. Behav.* **6,** 95–96.

Cohen, E. L. and Wurtman, R. J. (1976). Brain acetylcholine: Control by dietary choline. *Science* **191,** 561–562.

Cohen, H. D. and Barondes, S. H. (1967). Puromycin effects on memory may be due to occult seizure. *Science* **157,** 333–334.

Cohen, H. D. and Barondes, S. H. (1968). Cycloheximide impairs memory of an appetitive task. *Commun. Behav. Biol.* **1A,** 337–340.

Cohen, H. D., Ervin, F. and Barondes, S. H. (1966). Puromycin and cycloheximide: Different effects on hippocampal electrical activity. *Science* **154,** 1557–1558.

Cohen, M., Keats, A. S., Krivoy, W. A. and Ungar, G. (1965). Effect of actinomycin on morphine tolerance. *Proc. Soc. Exptl. Biol. Med.* **119,** 381–384.

Cole, M. and Maltzman, I. (Eds) (1969). "A Handbook of Contemporary Soviet Psychology". Basic Books, Inc., New York, 887 pp.

Coleman, P. D. (1965). Effects of rearing in the dark on dendritic fields of stellate cells in visual cortex. Presented at Meeting of Amer. Assn. Adv. Sci., Berkeley, California, Dec. 28.

Coleman, P. D. and Riesen, A. H. (1968). Environmental effects on cortical dendritic fields. I. Rearing in the dark. *J. Anat.* (*Lond.*) **102,** 363–374.

Collier, H. O. J. (1969). Humoral transmitters, supersensitivity, receptors and dependence. *In* "Scientific Basis of Drug Dependence" (H. Steinberg, Ed.), pp. 49–66. Churchill, London.

Colmenares, J. L., Wurtman, R. J. and Fernstrom, J. D. (1975). Effects of ingestion of a carbohydrate-fat meal on the levels and synthesis of 5-hydroxyindoles in various regions of the rat central nervous system. *J. Neurochem.* **25,** 825–829.

Colonnier, M. and Guillary, R. W. (1964). Synaptic organization in the lateral geniculate nucleus of the monkey. *Z. Zellforsch.* **62,** 333–355.

Comfort, A. (1971). Neuromythology? *Nature* (*Lond.*) **229,** 282.

Conde, F. P. and DeFeudis, F. V. (1977). Polypeptides of cerebral subcellular fractions of differentially-housed mice. *Experientia* (In press).

Conel, J. L. (1939–1963). "Postnatal Development of the Human Cerebral Cortex", Vols. 1–6. Harvard University Press, Cambridge, Mass.

Connor, J. D., Rossi, G. V. and Baker, W. W. (1966). Characteristics of tremor in cats following carbachol injections into the caudate nucleus. *Exptl. Neurol.* **14,** 371–382.

Conradi, S. (1969). Ultrastructure and distribution of neuronal and glial elements of the surface of the proximal part of motoneuron dendrite, as analyzed by serial sections. *Acta Physiol. Scand.* Suppl. **332,** 49–64.

Consolo, S. and Valzelli, L. (1970). Brain choline acetylase and monoamine oxidase activity in normal and aggressive mice. *Eur. J. Pharmacol.* **222,** 129–131.

Consolo, S., Garattini, S. and Valzelli, L. (1965). Amphetamine toxicity in aggressive mice. *J. Pharmacol. Exptl. Therapeut.* **17,** 53–54.

Cook, J. D. and Schanberg, S. M. (1970). The effects of methamphetamine on behavior and on the uptake, release and metabolism of norepinephrine. *Biochem. Pharmacol.* **19,** 1165–1179.

Cook, M. L. and Whitlock, D. G. (1975). Axoplasmic transport in the toad *Bufo Marinus*. *Brain Res.* **96,** 247–265.

Cools, A. R. and Van Rossum, J. M. (1976). Excitation-mediating and inhibition-mediating dopamine-receptors: A new concept towards a better understanding of electrophysiological, biochemical, pharmacological, functional and clinical data. *Psychopharmacologia* **45,** 243–254.

Corley, K. C., Shiel, F. O.'M., Path, M. R. C., Mauck, H. P. and Greenhoot, J. (1973). Electrocardiographic and cardiac morphological changes associated with environmental stress in squirrel monkeys. *Psychosom. Med.* **35,** 361–364.

Cornwell, A. C. and Sharpless, S. K. (1967). Electrophysiological retinal changes and visual deprivation. *Vision Res.* **8,** 389–401.

Corrodi, H., Fuxe, K. and Hökfelt, T. (1968). The effects of immobilization stress on the activity of central monoamine neurons. *Life Sci.* (Part I) **7,** 107–112.

Costa, E., Gessa, G. L. and Sandler, M. (Eds) (1974a). "Serotonin—New Vistas: Histochemistry and Pharmacology, Advances in Biochemical Psychopharmacology, Vol. 10". Raven Press, New York, 345 pp.

Costa, E., Gessa, G. L. and Sandler, M. (Eds) (1974b). "Serotonin—New Vistas:

Biochemistry and Behavioral and Clinical Studies, Advances in Biochemical Psychopharmacology, Vol. 11". Raven Press, New York, 446 pp.

Costall, B., Naylor, R. J. and Olley, J. E. (1972). Catalepsy and circling behaviour after intracerebral injections of neuroleptic, cholinergic and anticholinergic agents into the caudate-putamen, globus pallidus and substantia nigra of rat brain. *Neuropharmacology* **11,** 645–663.

Cotman, C. W. (1975). Molecular architecture of the CNS synapse. Fifth Int. Meeting, Int. Soc. for Neurochemistry, Barcelona, 2–6 September, p. 31.

Cotman, C. W., Banker, G., Zornetzer, S. F. and McGaugh, J. L. (1971). Electroshock effects on brain protein synthesis: Relation to brain seizures and retrograde amnesia. *Science* **173,** 454–456.

Coupland, R. E. (1965). Electron microscopic observations on the structure of the rat adrenal medulla. *J. Anat.* (*Lond.*) **99,** 231–272.

Cowan, W. M. (1970). Anterograde and retrograde transneuronal degeneration in the central and peripheral nervous system. *In* "Contemporary Research Methods in Neuroanatomy" (W. J. H. Nauta and S. O. E. Ebbeson, Eds), pp. 217–251. Springer-Verlag, New York.

Cowan, W. M. and Cuénod, M. (Eds) (1975). "The Use of Axonal Transport for Studies of Neuronal Connectivity". Elsevier, Amsterdam, 365 pp.

Cox, B. M., Opheim, K. E., Teschemacher, H. and Goldstein, A. (1975). A peptide-like substance from pituitary that acts like morphine. 2. Purification and properties. *Life Sci.* **16,** 1777–1782.

Cragg, B. G. (1961). Olfactory and other afferent connections of the hippocampus in the rabbit, rat and cat. *Exptl. Neurol.* **3,** 588–600.

Cragg, B. G. (1967). Changes in visual cortex on first exposure of rats to light. *Nature* (*Lond.*) **215,** 251–253.

Cragg, B. G. (1968). Are there structural alterations in synapses related to functioning? *Proc. Roy. Soc. B.* **171,** 319–323.

Cragg, B. G. (1969). The effects of vision and dark-rearing on the size and density of synapses in the lateral geniculate nucleus measured by electron microscopy. *Brain Res.* **13,** 53–67.

Cragg, B. G. (1970). What is the signal for chromatolysis? *Brain Res.* **23,** 1–21.

Cragg, B. G. (1972a). Plasticity of synapses. *In* "The Structure and Function of Nervous Tissue" (G. H. Bourne, Ed.), Vol. IV, pp. 1–60. Academic Press, New York and London.

Cragg, B. G. (1972b). The development of cortical synapses during starvation in the rat. *Brain* **95,** 143–150.

Crain, S. M. (1952). Development of electrical activity in the cerebral cortex of the albino rat. *Proc. Soc. Exptl. Biol. Med.* **81,** 49–51.

Crapper, D. R., Krishnan, S. S. and Dalton, A. J. (1973). Brain aluminum distribution in Alzheimer's Disease and experimental neurofibrillary degeneration. *Science* **180,** 511–513.

Crapper, D. R., Krishnan, S. S. and Quittkat, S. (1976). Aluminum, neurofibrillary degeneration and Alzheimer's disease. *Brain* **99,** 67–80.

Cravioto, R. O., Massieu, G. and Izquiero, J. J. (1951). Free amino-acids in rat brain during insulin shock. *Proc. Soc. Exptl. Biol. Med.* **78,** 856–858.

Creasey, W. A. and Malawista, S. E. (1971). Monosodium L-glutamate—Inhibition of glucose uptake in brain as a basis for toxicity. *Biochem. Pharmacol.* **20,** 2917–2920.

Creel, N. and Preuschoft, H. (1976). Cranial morphology of the lesser apes. *Gibbon and Siamang.* **4,** 219–303.

Cronly-Dillon, J. (1973). The effect of colchicine on memory. *J. Physiol.* (*Lond.*) **234,** 104P–105P.

Crosby, E. C., Humphrey, T. and Lauer, E. W. (1962). "Correlative Anatomy of the Nervous System". The Macmillan Co., New York.

Crow, T. J. (1972). Catecholamine-containing neurones and electrical self-stimulation. I. A review of some data. *Psychol. Med.* **2,** 414–421.

Crow, T. J. (1973). Catecholamine-containing neurones and electrical self-stimulation. II. A theoretical interpretation and some psychiatric implications. *Psychol. Med.* **3,** 66–73.

Cull, R. E. (1974). Role of nerve-muscle contact in maintaining synaptic connections. *Exptl. Brain Res.* **20,** 307–310.

Cull, R. E. (1975a). Role of axonal transport in maintaining central synaptic connections. *Exptl. Brain Res.* **24,** 97–101.

Cull, R. E. (1975b). Effect of sensory nerve division on the afferent synapses of axotomised motor neurones. *Exptl. Brain Res.* **22,** 421–425.

Culley, W. J. and Lineberger, R. O. (1968). Effect of undernutrition on the size and composition of the rat brain. *J. Nutr.* **96,** 375–381.

Culley, W. J., Saunders, R. N., Mertz, E. T. and Jolly, D. H. (1963). Effect of a tryptophan deficient diet on brain serotonin and plasma tryptophan level. *Proc. Soc. Exptl. Biol. Med.* **113,** 645–648.

Cummins, R. A., Berwick, M., Budtz-Olsen, O. E., Walsh, R. N. and Howard, M. J. (1975). Synaptic changes in differentially reared rats. Presented at 23rd Ann. Meeting, Australian Physiol. Pharmacol. Soc.

Cummins, R. A., Walsh, R. N., Budtz-Olsen, O. E., Konstantinos, T. and Horsfall, C. R. (1973). Environmentally-induced changes in the brains of elderly rats. *Nature* (*Lond.*) **243,** 516–518.

Curtis, B. A., Jacobson, S. and Marcus, E. M. (1972). "An Introduction to the Neurosciences". W. B. Saunders Co., Philadelphia, 878 pp.

Curtis, D. R. and Johnston, G. A. R. (1974). Amino acid transmitters in the mammalian central nervous system. *Ergebn. Physiol.* **69,** 97–188.

Dahl, D. R., Redburn, D. A. and Samson, F. E. (1970). Regional distribution of colchicine-binding (microtubular) protein in the rat brain. *J. Neurochem.* **17,** 1215–1219.

Dahlström, A. (1972). The axonal transport of monamine oxidases. *In* "Monoamine Oxidases—New Vistas" (E. Costa and M. Sandler, Eds), pp. 293–305. Raven Press, New York.

Dahlström, A. and Fuxe, K. (1964). A method for the demonstration of monoamine containing nerve fibers in the central nervous system. *Acta Physiol. Scand.* **60,** 293–295.

Dahlström, A. and Fuxe, K. (1965). Evidence for the existence of monoamine-containing neurons in the central nervous system. 1. Demonstration of monoamines in the cell bodies of brain stem neurons. *Acta Physiol. Scand.* **62** (Suppl. 232) 1–55.

Dahlström, A. and Haggendal, J. (1966). Studies on the transport and lifespan of amine storage granules in a peripheral adrenergic neuron system. *Acta Physiol. Scand.* **67,** 278–288.

Dahlström, A., Haggendal, J. and Atack, C. (1973). Localization and transport of serotonin. *In* "Serotonin and Behavior" (J. Barchas and E. Usdin, Eds), pp. 87–96, Academic Press, New York and London.

Daly, J. W. (1973). *In* "Frontiers in Catecholamine Research" (E. Usdin and S. H. Snyder, Eds), pp. 301–306. Pergamon Press, Oxford.

Damstra-Entingh, T., Entingh, D., Wilson, J. E. and Glassman, E. (1974). Environmental stimulation and fucose incorporation into brain and liver glycoproteins. *Pharmacol. Biochem. Behav.* **2,** 73–78.

Daneholt, B. and Brattgård, S. O. (1966). A comparison between RNA metabolism of nerve cells and glia in the hypoglossal nucleus of the rabbit. *J. Neurochem.* **13,** 913–921.

Danscher, G. (1976). Heavy Metals in the Hippocampal Region. Thesis, Institute of Anatomy, University of Aarhus, Denmark.

Daroqui, M. R. and Orsingher, O. A. (1973). Effect of α-methyl tyrosine pretreatment on drug-induced increase of hippocampal RNA. *Pharmacology* **7,** 366–370.

Daruna, J. H. and Kent, E. W. (1976). Comparison of regional serotonin levels and turnover in the brain of naturally high and low aggressive rats. *Brain Res.* **101,** 489–501.

Darwin, C. (1874). "The Descent of Man" (2nd Edn., p. 53). Rand McNally, Chicago.

Das, G. D. and Altman, J. (1966). Behavioral manipulations and protein metabolism of the brain: Effects of restricted and enriched environments on the utilization of leucine-H^3. *Physiol. Behav.* **1,** 109–110.

Davenport, J. W., Gonzalez, L. M., Carey, J. C., Bishop, S. B. and Hagquist, W. W. (1976). Environmental stimulation reduces learning deficits in experimental cretinism. *Science* **191,** 578–579.

Davidoff, R. A. (1972a). Diphenylhydantoin increases spinal presynaptic inhibition. *Trans. Amer. Neurol. Assn.* **97,** 193–196.

Davidoff, R. A. (1972b). Gamma-aminobutyric acid antagonism and presynaptic inhibition in the frog spinal cord. *Science* **174,** 331–333.

Davis, J. M., Himwich, W. A. and Pederson, V. C. (1970). Hypoglycemia and developmental changes in free amino acids of rat brain. *J. Appl. Physiol.* **29,** 219–222.

Davis, K. L., Berger, P. A. and Hollister, L. E. (1975). Choline for tardive dyskinesia. *New Engl. J. Med.* **293,** 152.

Davis, W. M. and Nunnemacher, R. F. (1974). Never fibers in the sympathetic trunk of *Rana pipiens*: A quantitative electron microscope study. *Anat. Rec.* **179,** 331–342.

Davison, A. N. and Dobbing, J. (1968). The developing brain. *In* "Applied Neurochemistry" (A. N. Davison and J. Dobbing, Eds), pp. 253–286. Blackwell, Oxford.

Davson, H. (1972). The blood-brain barrier. *In* "The Structure and Function of Nervous Tissue" (G. H. Bourne, Ed.), Vol. IV, pp. 321–445. Academic Press, New York and London.

Dawkins, R. (1971). Selective neuron death as a possible memory mechanism. *Nature* (*Lond.*) **229,** 118–119.

Dawson, R. G. and McGaugh, J. L. (1973). Drug facilitation of learning and memory. *In* "The Physiological Basis of Memory" (J. A. Deutsch, Ed.), pp. 77–111. Academic Press, New York and London.

Dawson, R. M. C. (1953). Cerebral amino acids in fluoroacetate-poisoned, anaesthetised and hypoglycaemic rats. *Biochem. Biophys. Acta* **11,** 548–552.

DeFeudis, F. V. (1969). Glutamate fluxes in peripheral nerves. *Second Int. Meeting Int. Soc. Neurochem.*, Milan (Abstract).

DeFeudis, F. V. (1971a). Effects of electrical stimulation on the efflux of L-glutamate from peripheral nerve *in vitro*. *Exptl. Neurol.* **30,** 291–296.

DeFeudis, F. V. (1971b). Effects of environmental changes on the incorporation of carbon atoms of D-glucose into mouse brain and other tissues. *Life Sci.* **10** (Part II) 1187–1194.

DeFeudis, F. V. (1972a). Effects of isolation and aggregation on the water content of mouse brain. *Biol. Psychiat.* **5,** 207–210.

DeFeudis, F. V. (1972b). Effects of isolation and aggregation on the incorporation of carbon atoms of D-mannose and D-glucose into mouse brain. *Biol. Psychiat.* **4,** 239–242.

DeFeudis, F. V. (1972c). A note on the subcellular distribution of brain protein in differentially housed mice. *Experientia* **28,** 1427–1428.

DeFeudis, F. V. (1972d). Binding of ^{3}H-acetylcholine and ^{14}C-γ-aminobutyric acid to subcellular fractions of the brains of differentially-housed mice. *Neuropharmacology* **11,** 879–888.

DeFeudis, F. V. (1972e). Effects of differential housing on the subcellular distribution of Li^{+} in the brains of mice. *Brain Res.* **43,** 686–689.

DeFeudis, F. V. (1973). Effects of *d*-amphetamine on the incorporation of carbon atoms of D-mannose into the brains and sera of differentially-housed mice; short-term reversibility of these effects. *Biol. Psychiat.* **7,** 3–9.

DeFeudis, F. V. (1974a). Differences in brain and body weights of mice caused by differential housing. *Experientia* **30,** 521–522.

DeFeudis, F. V. (1974b). "Central Cholinergic Systems and Behaviour". Academic Press, London and New York, 223 pp.

DeFeudis, F. V. (1974c). La biologie de la solitude. *La Recherche* (Paris) **6,** 344–356.

DeFeudis, F. V. (1975a). Cerebral biochemical and pharmacological changes in differentially-housed mice. *In* "Current Developments in Psychopharmacology" (W. B. Essman and L. Valzelli, Eds), Vol. 1, pp. 144–201. Spectrum Publ., Holliswood, New York.

DeFeudis, F. V. (1975b). An environmental-neurochemical model for aggressiveness and other psychiatric disturbances. *Proc. IV World Congress of Psychiatric Surgery*, Madrid, 7–12 September, (Abstr. No. 26).

DeFeudis, F. V. (1975c). Amino acids as central neurotransmitters. *Ann. Rev. Pharmacol.* **15,** 105–130.

DeFeudis, F. V. and Black, W. C. (1972). Effects of environment on the incorporation of radioactive carbon atoms of glucose into various structures of the brains of mice; evidence for diurnal rhythmicity. *Exptl. Neurol.* **36,** 41–45.

DeFeudis, F. V. and Marks, J. H. (1973a). Studies on the time course of entry and subcellular distribution of radioactivity of [^{3}H]*d*-amphetamine in the brains of differentially-housed mice. *Experientia* **29,** 1518–1520.

DeFeudis, F. V. and Marks, J. H. (1973b). Brain to serum distribution of radioactivity of injected [^{3}H]-*d*-amphetamine in differentially-housed mice. *Biol. Psychiat.* **6,** 85–88.

DeFeudis, F. V. and Paolino, R. M. (1972). Effects of environmental changes on the incorporation of carbon atoms of pyruvate into mouse brain and liver. *Experientia* **28,** 309–310.

DeFeudis, F. V., Delgado, J. M. R. and Roth, R. H. (1969). Content and release of amino acids and catecholamines in monkey brain. *Nature* (*Lond.*) **223,** 74–75.

DeFeudis, F. V., Delgado, J. M. R. and Roth, R. H. (1970). Content, synthesis and collectability of amino acids in various regions of the brains of Rhesus monkeys. *Brain Res.* **18,** 15–23.

DeFeudis, F. V., Paolino, R. M. and DeFeudis, P. A. (1972). Effects of injections of *d*-amphetamine and lithium chloride on the incorporation of carbon atoms of D-glucose and pyruvate into the brains of isolated and aggregated mice. *5th Int. Congr. Pharmacol., San Francisco*, p. 53.

DeFeudis, F. V., Ojeda, A., Madtes, P. and DeFeudis, P. A. (1975). Environmentally-sensitive nerve endings. *Naturwissenschaften* **62,** 585.

DeFeudis, F. V., Madtes, P., Ojeda, A. and DeFeudis, P. A. (1976a). "Binding" of GABA and glycine to synaptic particles of the brains of differentially-housed mice: Evidence for morphological changes. *Exptl. Neurol.* **52,** 285–294.

DeFeudis, F. V., Somoza, E., DeFeudis, P. A., Pugnaire, M. P., Orensanz-Muñoz, L. M., Gonzalez Portal, C. and Ibañez, A. E. (1976b). Environmental model for drug addiction; studies on the "binding" of dihydromorphine to synaptic particles of the brains of differentially-housed mice. *In* "International Symposium on Factors Affecting the Action of Narcotics, Mario Negri Monograph Series", Milan, 13–16 July (M. Adler, L. Manara and R. Samanin, Eds). Raven Press, New York.

DeFeudis, F. V., Madtes, P. and Gervas-Camacho, J. (1976c). Binding of glycine and γ-aminobutyric acid to synaptosomal fractions of the brains of differentially housed mice. *Exptl. Neurol.* **50,** 207–213.

DeFeudis, F. V., DeFeudis, P. A. and Somoza, E. (1976d). Altered analgesic responses to morphine in differentially-housed mice. *Psychopharmacologia* **49,** 117–118.

DeFeudis, P. A. (1968). The role of sensory factors in the organization of the instrumental response. M.A. Thesis, McGill University, Montreal.

DeFeudis, P. A., Paolino, R. M. and DeFeudis, F. V. (1975). Effects of *d*-amphetamine on the incorporation of carbon atoms of D-glucose into the brains of differentially-housed mice. *Psychopharmacologia* **45,** 167–169.

Delacour, J. (1970). Specific functions of a medial thalamic structure in avoidance conditioning in the rat. *In* 'Progress in Brain Research, Vol. 32; Pituitary, Adrenal, and the Brain" (D. DeWied and J. A. W. M. Weijnen, Eds), pp. 158–170. Elsevier, Amsterdam.

De la Paz, R. I., Dickman, S. R. and Grosser, B. I. (1975). Effects of stress on rat brain adenosine 3′, 5′-monophosphate *in vivo*. *Brain Res.* **85,** 171–175.

de la Torre, J. C. (1972). "Dynamics of Brain Monoamines". Plenum Press, New York, 210 pp.

Delgado, J. M. R. (1967). Limbic system and free behavior. *In* "Progress in Brain Research: Structure and Function of the Limbic System" (W. R. Adey and T. Tokizane, Eds), Vol. 27, pp. 48–68. Elsevier, Amsterdam.

Delgado, J. M. R. and DeFeudis, F. V. (Eds) (1976). "Behavioral Neurochemistry". Spectrum Publ. Inc., Holliswood, New York.

Delgado-Garcia, J. M., Grau, C., DeFeudis, P., del Pozo, F., Jiminez, J. M. and Delgado, J. M. R. (1976). Ultradian rhythms in the mobility and behavior of Rhesus monkeys. *Exptl. Brain Res.* **25,** 79–91.

Dell, P. (1963). Reticular homeostasis and critical reactivity. *In* "Progress in Brain Research" (G. Moruzzi, A. Fessard and H. H. Jasper, Eds), Vol. 1, pp. 82–114. Elsevier, Amsterdam.

DeLorenzo, R. J., Walton, K. G., Curran, P. F. and Greengard, P. (1973). Regulation of phosphorylation of a specific protein in toad-bladder membrane by antidiuretic hormone and cyclic AMP, and its possible relationship to membrane permeability changes. *Proc. Nat. Acad. Sci., U.S.A.* **70,** 880–884.

DeLorenzo, R. J. (1975). Possible correlation between protein D phosphorylation and membrane permeability changes to sodium ions in the toad bladder and its relationship to neuropharmacology. *Bull. Parent Drug Assn.* **29,** 250–260.

DeLorenzo, R. J. and Glaser, G. H. (1976). Effect of diphenylhydantoin on the endogenous phosphorylation of brain protein. *Brain Res.* **105,** 381–386.

DeLorenzo, R. J. and Greengard, P. (1973). Activation by adenosine 3′:5′-monophosphate of a membrane-bound phosphoprotein phosphatase from toad bladder. *Proc. Nat. Acad. Sci., U.S.A.* **70,** 1831–1835.

De Luca, B., Di Benedetta, C. and Cioffi, L. A. (1975). Brain glycoproteins during cortical spreading depression. *Presented at First European Neurosciences Meeting, Munich*, 9–13 September, Abstr. No. 96.

Den, H., Kaufman, B. and Roseman, S. (1970). Properties of some glycosyltransferases in embryonic chicken brain. *J. Biol. Chem.* **245,** 6607–6615.

Denenberg, V. H. and Morton, J. R. C. (1964). Infantile stimulation, prepubertal sexual-social interaction, and emotionality. *Animal Behav.* **12,** 11–13.

Dennis, M. J. and Gerschenfeld, H. M. (1969). Some physiological properties of identified mammalian neuroglial cells. *J. Physiol.* (*Lond.*) **203,** 211–222.

Denti, A., McGaugh, J. L., Landfield, P. W. and Shinkman, P. G. (1970). Effects of posttrial electrical stimulation of the mesencephalic reticular formation on avoidance learning in rats. *Physiol. Behav.* **5,** 659–662.

DeRobertis, E. (1964). "Histophysiology of Synapses and Neurosecretion". Pergamon Press, Oxford.

DeRobertis, E., Sellinger, O. Z., Rodriguez de Lores Arnaiz, G., Alberici, M. and Zieher, L. M. (1967). Nerve endings in methionine sulphoximine convulsant rats, a neurochemical and ultrastructural study. *J. Neurochem.* **14,** 81–89.

DeRobertis, E., Alberici, M. and Rodriguez de Lores Arnaiz, G. (1969). Astroglial swelling and phosphohydrolases in cerebral cortex of Metrazol convulsant rats. *Brain Res.* **12,** 461–466.

DeRobertis, E. D. P. and Carrea, A. (Eds) (1965). "Biology of Neuroglia, Progress in Brain Research", Vol. 15. Elsevier, Amsterdam.

Derry, D. M. and Wolfe, L. S. (1967). Gangliosides in isolated neurons and glial cells. *Science* **158,** 1450–1452.

Dessi-Fulgheri, F. and Lupi di Prisco, C. (1975). Effects of varying degrees of social deprivation on sex steroids in the rat. *Proc. XIVth Int. Ethological Conf., Parma* (Abstract).

Dessi-Fulgheri, F., Lupo di Prisco, C. and Verdarelli, P. (1975). Influence of long-term isolation on the production and metabolism of gonadal sex steroids in male and female rats. *Physiol. Behav.* **14,** 495–499.

Deutsch, J. A. (1971). The cholinergic synapse and the site of memory. *Science* **174,** 788–794.

Deutsch, J. A. (1972). Macromolecular change and the synapse. *In* "Macromolecules and Behavior" (J. Gaito, Ed.), 2nd Ed., pp. 279–303. Appleton-Century-Crofts, New York.

Deutsch, J. A. (1973). Electroconvulsive shock and memory. *In* "The Physiological Basis of Memory" (J. A. Deutsch, Ed.), pp. 113–124. Academic Press, New York and London.

Deutsch, J. A. (1975). Book review of "Dissension About Attention" (N. Moray). *Contemp. Psychol.* **20,** 397–398.

Deutsch, J. A., Hamburg, M. D. and Dahl, H. (1966). Anticholinesterase induced amnesia and its temporal aspects. *Science* **151,** 221–223.

DeWied, D. (1965). The influence of the posterior and intermediate lobe of the pituitary and pituitary peptides on the maintenance of a conditioned avoidance response in rats. *Int. J. Neuropharmacol.* **4,** 157–167.

DeWied, D. (1967). Opposite effects of ACTH and glucocorticosteroids on extinction of conditioned avoidance behavior. *Proc. 2nd Int. Congr. on Hormonal Steroids,* Milan, 1966, Excerpta Medica Fdn., **132,** 945–951.

DeWied, D. (1969). Effects of peptide hormones on behavior. *In* "Frontiers in Neuroendocrinology" (W. F. Ganong and L. Martini, Eds), pp. 97–140. Oxford University Press, Oxford.

DeWied, D. (1971). Long term effect of vasopressin on the maintenance of a conditioned avoidance response in rats. *Nature (Lond.)* **232,** 58–60.

DeWied, D. (1974). Pituitary-adrenal system hormones and behavior. *In* "The Neurosciences, Third Study Program" (F. O. Schmitt and F. G. Worden, Eds), pp. 653–666. M.I.T. Press, Cambridge, Mass.

DeWied, D. and Bohus, B. (1966). Long term and short term effects on retention of a conditioned avoidance response in rats by treatment with long acting Pitressin and α-MSH. *Nature (Lond.)* **212,** 1484–1486.

DeWied, D. and Gispen, W. H. (1976). Impaired development of tolerance to morphine analgesia in rats with hereditary diabetes insipidus. *Psychopharmacologia* **46,** 27–29.

DeWied, D., Greven, H. M., Lande, S. and Witter, A. (1972). Dissociation of the behavioral and endocrine effects of lysine vasopressin by tryptic digestion. *Brit. J. Pharmacol.* **45,** 118–122.

DeWied, D., Bohus, B. and Van Wimersma Greidanus, Tj. B. (1974). The hypothalamo-neurohypophysial system and the preservation of conditioned avoidance behavior in rats. *In* "Integrative Hypothalamic Activity, Progress in Brain Research" (D. F. Swaab and J. P. Schadé, Eds), Vol. 41, pp. 417–428. Elsevier, Amsterdam.

DeWied, D., Bohus, B. and Van Wimersma Greidanus, Tj. B. (1975a). Memory deficit in rats with hereditary diabetes insipidus. *Brain Res.* **85,** 152–156.

DeWied, D., Witter, A. and Greven, H. M. (1975b). Commentary: Behaviourally active ACTH analogues. *Biochem. Pharmacol.* **24,** 1463–1468.

Dews, P. B. (1955). Studies on behavior. II. The effects of pentobarbital, methamphetamine, and scopolamine on performances in pigeons involving discriminations. *J. Pharmacol. Exptl. Therapeut.* **115,** 380–389.

Deza, L. and Eidelberg, E. (1967). Development of cortical electrical activity in the rat. *Exptl. Neurol.* **16,** 425–438.

Diamond, J., Gray, E. G. and Yasargil, G. M. (1970). The function of the dendritic spines—an hypothesis. *In* "Excitatory Synaptic Mechanisms" (P. Anderson and J. K. S. Jansen, Eds), pp. 213–222. Universitetsforlaget, Oslo.

Diamond, M. C. (1967). Extensive cortical depth measurements and neuron size increases in the cortex of environmentally enriched rats. *J. Comp. Neurol.* **131,** 357–364.

Diamond, M. C., Krech, D. and Rosenzweig, M. R. (1964). The effects of an enriched environment on the histology of the rat cerebral cortex. *J. Comp. Neurol.* **123,** 111–120.

Diamond, M. C., Law, F., Rhodes, H., Lindner, B., Rosenzweig, M. R., Krech, D. and Bennett, E. L. (1966). Increases in cortical depth measurements and neuron size increases in the cortex of environmentally enriched rats. *J. Comp. Neurol.* **131,** 357–364.

Diamond, M. C., Johnson, R. E. and Ingham, C. (1971). Brain plasticity induced by environment and pregnancy. *Int. J. Neurosci.* **2,** 171–178.

Diamond, M. C., Rosenzweig, M. R., Bennett, E. L., Lindner, B. and Lyon, L. (1972). Effects of environmental enrichment and impoverishment on rat cerebral cortex. *J. Neurobiol.* **3,** 47–64.

Diamond, M. C., Johnson, R. E., Ingham, C., Rosenzweig, M. R. and Bennett, E. L. (1975). Effects of differential experience on neuronal nuclear and perikarya dimensions in the rat cerebral cortex. *Behav. Biol.* **15,** 107–111.

DiCara, L. V. (Ed.) (1974). "Limbic and Autonomic Nervous Systems Research". Plenum Press, New York, 428 pp.

DiCesare, J. L. and Dain, J. A. (1972). Localization, solubilization and properties of N-acetylgalactosaminyl and galactosyl ganglioside transferases in rat brain. *J. Neurochem.* **19,** 403–410.

DiChiara, G., Camba, R. and Spano, P. F. (1971). Evidence for inhibition by brain serotonin of mouse killing behaviour in rats. *Nature* (*Lond.*) **233,** 272–273.

Dingman, W. and Sporn, M. B. (1961). The incorporation of 8-azaguanine into rat brain RNA and its effect on maze learning by the rat. *J. Psychiat. Res.* **1,** 1–11.

Dinnendahl, V. (1975). Effects of stress on mouse brain cyclic nucleotide levels *in vivo*. *Brain Res.* **100,** 716–719.

Dismukes, K., Rogers, M. and Daly, J. W. (1976). Cyclic adenosine 3′, 5′-monophosphate formation in guinea-pig slices: Effect of H_1 and H_2-histaminergic agonists. *J. Neurochem.* **26,** 785–790.

Dobbing, J. (1968). Vulnerable periods in developing brain. *In* "Applied Neurochemistry" (A. N. Davison and J. Dobbing, Eds), pp. 287–316. Blackwell, Oxford.

Dobbing, J. (1976). Malnutrition et développement du cerveau. *La Recherche* (Paris) **7,** 139–145.

Dobbing, J. and Sands, J. (1973). The quantitative growth and development of the human brain. *Arch. Dis. Childh.* **48,** 757–767.

Dobbing, J. and Smart, J. L. (1973). Early undernutrition, brain development and behaviour. *In* "Ethology and Development. Clinics in Developmental Medicine No. 47" (S. A. Barnett, Ed.), pp. 16–36. Heinemann, London.

Dobbing, J. and Smart, J. L. (1974). Vulnerability of developing brain and behaviour. *Brit. Med. Bull.* **30,** 164–168.

Doemin, N. N. (1968). Effects of acetylcholine and adrenaline upon some sites of nucleic acid metabolism in single cells and subcellular fractions of nervous tissue. *In* "Macromolecules and the Function of the Neuron" (Z. Lodin and S. P. R. Rose, Eds), pp. 316–323. Excerpta Medica Fdn., Amsterdam.

Dominguez, M. and Longo, V. G. (1969). Taming effects of para-chlorophenylalanine on septal rats. *Physiol. Behav.* **4,** 1031–1033.

Dominic, J. A. (1973). Suppression of brain serotonin synthesis and metabolism by benzodiazepine minor tranquilizers. *In* "Serotonin and Behavior" (J. Barchas and E. Usdin, Eds), pp. 149–155. Academic Press, New York and London.

Domino, E. F. and Davis, J. M. (Eds) (1975). "Neurotransmitter Balances Regulating Behavior". Edwards Bros., Inc., Ann Arbor, 240 pp.

Donato, R. and Michetti, F. (1974). S-100 protein in cerebral cortex synaptosomes. *Experientia* **30,** 511–512.

Donato, R., Michetti, F. and Miani, N. (1975). Soluble and membrane-bound S-100 protein in cerebral cortex synaptosomes. Properties of the S-100 receptor. *Brain Res.* **98,** 561–573.

Dostrovsky, J. and Pomeranz, B. (1973). Morphine blockade of amino acid putative transmitters on cat spinal cord sensory interneurones. *Nature* (*Lond.*) **246,** 222–224.

Douglas, R. J. (1967). The hippocampus and behaviour. *Psychol. Bull.* **67,** 416–442.

Douglas, R. J. and Marcellus, D. (1975). The ascent of man: Deductions based on a multivariate analysis of the brain. *Brain Behav. Evolut.* **11,** 179–213.

Douglas, R. M. and Goddard, G. V. (1975). Long-term potentiation of the perforant path-granule cell synapse in the rat hippocampus. *Brain Res.* **86,** 205–215.

Dreyfus, P. M. and Geel, S. E. (1972). Vitamin and nutritional deficiencies. *In* "Basic Neurochemistry" (R. W. Albers, G. J. Siegel, R. Katzman and B. W. Agranoff, Eds), pp. 517–535. Little, Brown and Co., Boston.

Droz, B. (1973). Renewal of synaptic proteins. *Brain Res.* **62,** 383–394.

Droz, B. (1974). Conclusions générales. *Bull. de l'Académie Suisse des Sciences Médicales,* **30,** 124–137.

Droz, B. and Barondes, S. H. (1969). Nerve endings: Rapid appearance of labeled protein shown by electron microscope radioautography. *Science* **165,** 1131–1133.

Droz, B. and DiGiamberardino, L. (1974). Transport of macromolecules in nerve cells. *In* "Advances in Cytopharmacology" (B. Ceccarelli, F. Clementi and J. Meldolesi, Eds), Vol. 2, pp. 249–256. Raven Press, New York.

Droz, B. and LeBlond, C. P. (1963). Axonal migration of proteins in the central nervous system and peripheral nerves as shown by radioautography. *J. Comp. Neurol.* **121,** 325–337.

Droz, B., Koenig, H. L. and DiGiamberardino, L. (1973). Axonal migration of protein and glycoprotein to nerve endings. I. Radioautographic analysis of the renewal of protein in nerve endings of chicken ciliary ganglion after intracerebral injection of ^{3}H-lysine. *Brain Res.* **60,** 93–127.

Droz, B., Bruner, J. and Gerschenfeld, H. (1974). Transport axonal de protéines marquées dans le motoneurone géant de l'écrevisse. *Biochimie* **56,** 1613–1620.

Duffy, F. H., Snodgrass, S. R., Burchfiel, J. L. and Conway, J. L. (1976). Bicuculline reversal of deprivation amblyopia in the cat. *Nature* (*Lond.*) **260,** 256–257.

Duffy, M. J., Wong, J. and Powell, D. (1975). Stimulation of adenylate cyclase activity in different areas of human brain by Substance P. *Neuropharmacology* **14,** 615–618.

Dunbar, F. (1954). "Emotions and Bodily Changes". 4th Edn. Columbia University Press, New York, 1192 pp.

Dunn, A. (1971). Brain protein synthesis after electroshock. *Brain Res.* **35,** 254–259.

Dunn, A. (1973). The dependence of brain ATP content on cerebral electroshock current. *Brain Res.* **61,** 442–445.

Dunn, A. J. (1975). Biochemical correlates of training experiences: A discussion of the evidence. *In* "Neural Mechanisms of Learning and Memory" (M. R. Rosenzweig and E. L. Bennett, Eds), pp. 311–320. M.I.T. Press, Cambridge, Mass.

Dunn, A. J. and Bergert, B. J. (1976). Effects of electroconvulsive shock and cycloheximide on the incorporation of amino acids into proteins of mouse brain subcellular fractions. *J. Neurochem.* **26,** 369–375.

Dunn, A. J. and Bondy, S. C. (1974). "Functional Chemistry of the Brain". Spectrum Publ., Flushing, N.Y., 272 pp.

Dunn, A. J. and Hogan, E. L. (1975). Brain gangliosides: Increased incorporation of (1-^{3}H) glucosamine during training. *Pharmacol. Biochem. Behav.* **3,** 605–612.

Dunn, A., Giuditta, A. and Pagliuca, N. (1971). The effect of electroconvulsive shock on protein synthesis in mouse brain. *J. Neurochem.* **18,** 2093–2099.

Dunn, A. J., Entingh, D., Entingh, T., Gispen, W. H., Machlus, B., Perumal, R., Rees, H. D. and Brogan, L. (1974a). Biochemical correlates of brief behavioral experiences. *In* "The Neurosciences, Third Study Program" (F. O. Schmitt and F. G. Worden, Eds), pp. 679–684. M.I.T. Press, Cambridge, Mass.

Dunn, A. J., Giuditta, A., Wilson, J. E. and Glassman, E. (1974b). The effect of electroshock on brain RNA and protein synthesis and its possible relationship to behavioral effects. *In* "The Psychobiology of Convulsive Therapy" (M. Fink, J. McGaugh, S. Kety and T. A. Williams, Eds), pp. 185–197. Winston-Halsted-Wiley, Washington, D.C.

Eayrs, J. T. (1955). The cerebral cortex of normal and hypothyroid rats. *Acta Anat.* (Basel) **25,** 160–183.

Ebadi, M. S. and Costa, E. (Eds) (1972). "Role of Vitamin B_6 in Neurobiology, Advances in Biochemical Psychopharmacology", Vol. 4. Raven Press, New York, 248 pp.

Eccles, J. C. (1953). "The Neurophysiological Basis of Mind". Clarendon Press, Oxford, 314 pp.

Eccles, J. C. (1964). "The Physiology of Synapses". Academic Press, New York and London, 316 pp.

Eccles, J. C. and McIntyre, A. K. (1953). The effects of disuse and of activity on mammalian spinal reflexes. *J. Physiol.* (*Lond.*) **121,** 492–516.

Eccles, J. C., Krnjević, K. and Miledi, R. (1958). Delayed effects of peripheral severance of afferent nerve fibres on the efficacy of their central synapses. *J. Physiol.* (*Lond.*) **145,** 204–220.

Eckhert, C., Barnes, R. H. and Levitsky, D. A. (1976). The effect of protein-energy undernutrition induced during the period of suckling on cholinergic enzyme activity in the rat brain stem. *Brain Res.* **101,** 372–377.

Edström, J. E. (1957). Effects of increased motor activity on the dimensions and staining properties of the neuron soma. *J. Comp. Neurol.* **107,** 295–304.

Edström, A. and Mattsson, H. (1972). Rapid axonal transport *in vitro* in the sciatic system of the frog of fucose-, glucosamine- and sulphate-containing material. *J. Neurochem.* **19,** 1717–1729.

Ehringer, H. and Hornykiewicz, O. (1960). Verteilung von Noradrenaline und Dopamin (3-Hydroxytyramin) im Gehirn des Menschen und ihr Verhalten bei Erkrankungen des Extrapyramidalen Systems. *Klin. Wschr.* **38,** 1236–1239.

Ehrlich, Y. H. and Routtenberg, A. (1974). Cyclic AMP regulates phosphorylation of three protein components of rat cerebral cortex membranes for thirty minutes. *FEBS Lett.* 237–242.

Eichenbaum, H., Quenon, B. A., Heacock, A. and Agranoff, B. W. (1976). Differential behavioral and biochemical effects of regional injection of cycloheximide into mouse brain. *Brain Res.* **101,** 171–176.

Elam, J. S. (1975). Association of proteins undergoing slow axonal transport with goldfish visual system myelin. *Brain Res.* **97,** 303–315.

Elam, J. S. and Peterson, N. W. (1976). Axonal transport of sulfated glycoproteins and mucopolysaccharides in the garfish olfactory nerve. *J. Neurochem.* **26,** 845–850.

Eleftheriou, B. E. and Boehlke, K. W. (1967). Brain monoamine oxidase in mice after exposure to aggression and defeat. *Science* **155,** 1693–1694.

Eleftheriou, B. E. and Sprott, R. L. (Eds) (1975a). "Hormonal Correlates of Behavior", Vol. 1. Plenum Press, New York, 439 pp.

Eleftheriou, B. E. and Sprott, R. L. (Eds) (1975b). "Hormonal Correlates of Behavior", Vol. 2. Plenum Press, New York, pp. 441–806.

Ely, D. L., Greene, E. G. and Henry, J. P. (1976). Minicomputer monitored socia behavior of mice with hippocampus lesions. *Behav. Biol.* **16,** 1–29.

Embree, L. J., Bass, N. H. and Pope, A. (1972). Biochemistry of middle and late life dementias. *In* "Handbook of Neurochemistry" (A. Lajtha, Ed.), Vol. 7, pp. 329–369. Plenum Press, New York.

Emson, P. C., Björklund, A. and Stenevi, U. (1976). Possible regeneration of γ-aminobutyric acid-containing fibres into irides transplanted into the central nervous system. *Nature (Lond.)* **259,** 567–570.

Endo, M. (1970). Plasma growth hormone levels during insulin hypoglycemia in atypical psychosis. *Folia Endocr. Jap.* **45,** 1295–1296.

Entingh, D., Damstra-Entingh, T., Dunn, A., Wilson, J. E. and Glassman, E. (1974). Brain uridine monophosphate: Reduced incorporation of uridine during avoidance learning. *Brain Res.* **70,** 131–138.

Entingh, D., Dunn, A., Glassman, E., Wilson, J. E., Hogan, E. and Damstra, T. (1975). Biochemical approaches to the biological basis of memory. *In* "Handbook of Psychobiology" (M. S. Gazzaniga and C. Blakemore, Eds), pp. 201–238. Academic Press, New York and London.

Erickson, C. K. and Patel, J. B. (1969). Facilitation of avoidance learning by post-trial hippocampal electrical stimulation. *J. Comp. Physiol. Psychol.* **68,** 400–406.

Essman, W. B. (1968). Differences in locomotor activity and brain-serotonin metabolism in differentially-housed mice. *J. Comp. Physiol. Psychol.* **66,** 244–246.

Essman, W. B. (1969). Free and motivated behavior and amine metabolism in isolated mice. *In* "Aggressive Behaviour" (S. Garattini and E. B. Sigg, Eds), pp. 203–208. Wiley (Interscience), New York.

Essman, W. B. (1971a). Neurochemical changes associated with isolation and environmental stimulation. *Biol. Psychiat.* **3,** 141–147.

Essman, W. B. (1971b). Changes in cholinergic activity and avoidance behavior by nicotine in differentially housed mice. *Int. J. Neurosci.* **2,** 199–206.

Essman, W. B. (1971c). Drug effects and learning and memory processes. *Adv. Pharmacol. Chemother.* **9,** 241–330.

Essman, W. B. (1971d). Isolation-induced behavioral modification: Some neurochemical correlates. *In* "Brain Development and Behavior" (M. B. Sterman, D. J. McGinty and A. M. Adinolfi, Eds), pp. 265–276. Academic Press, New York and London.

Essman, W. B. (1972). Contribution of differential housing to brain development: Some implications for sleep behavior. *In* "Sleep and the Maturing Nervous System" (C. D. Clemente, D. P. Purpura and F. E. Mayer, Eds), pp. 99–107. Academic Press, New York and London.

Essman, W. B. (1973). "Neurochemistry of Cerebral Electroshock". Spectrum Publ., Flushing, N.Y., 181 pp.

Essman, W. B. (1974). *In* "Psychobiology of Convulsive Therapy" (M. Fink, J. McGaugh, S. Kety and T. A. Williams, Eds), pp. 237–249. Winston-Halsted-Wiley, New York.

Essman, W. B. and Smith, G. E. (1967). Behavior and neurochemical differences between differentially housed mice. *Amer. Zoologist* **7,** 793.

Eterović, V. A., Ferchmin, P. A. and Caputto, R. (1973). Influence of rat exploratory activity on brain oxygen uptake *in vitro*. *Int. J. Neurosci.* **5,** 133–137.

Ewen, L. M. and Griffiths, J. (1973). γ-Glutamyl transpeptidase—Elevated activities in certain neurologic diseases. *Amer. J. Clin. Pathol.* **59,** 2–9.

Fahn, S. and Coté, L. J. (1968). Regional distribution of gamma-aminobutyric acid (GABA) in brains of the Rhesus monkey. *J. Neurochem.* **15,** 209–213.

Fairén, A. and Valverde, F. (1973). Centros visuales en roedores: Estructura normal y efectos de la deprivacíon sensorial sobre la morfología dendritica. *Trabajos del Instituto Cajal de Investigaciones Biológicas,* **LXV,** 87–135.

Faurbye, A. (1970). The structural and biochemical basis of movement disorders in treatment with neuroleptic drugs and in extrapyramidal diseases. *Comprehensive Psychiat.* **11,** 205–225.

Feldman, M. L. (1974). Degenerative changes in aging dendrites. *Gerontologist* **14,** 36 (Abstr.)

Feldman, R. E. and Dement, W. (1968). Possible relationship between REM sleep and memory consolidation. *Psychophysiology* **5,** 243.

Ferchmin, P. A., Eterović, V. A. and Caputto, R. (1970). Studies of brain weight and RNA content after short periods of exposure to environmental complexity. *Brain Res.* **20,** 49–57.

Ferchmin, P. A., Bennett, E. L. and Rosenzweig, M. R. (1975). Direct contact with enriched environment is required to alter cerebral weights in rats. *J. Comp. Physiol. Psychol.* **88,** 360–367.

Ferguson, J., Henricksen, S., Cohen, H., Mitchell, G., Barchas, J. and Dement, W. (1970). "Hypersexuality" and behavioral changes in cats caused by administration of p-chlorophenylalanine. *Science* **168,** 499–501.

Fernandez, H. L., Burton, P. R. and Samson, F. E. (1971). Axoplasmic transport in the crayfish nerve cord. The role of fibrillar constituents of neurones. *J. Cell Biol.* **51,** 176–192.

Fernstrom, J. D. and Wurtman, R. J. (1971). Brain serotonin content: Increase following ingestion of carbohydrate diet. *Science* **172,** 1023–1025.

Ferrari, W., Gessa, G. L. and Vargiu, L. (1963). Behavioral effects induced by intracisternally injected ACTH and MSH. *Ann. N.Y. Acad. Sci.* **104,** 330–343.

Ferrendelli, J. A., Rubin, E. H. and Kinscherf, D. A. (1976). Influence of divalent cations on regulation of cyclic GMP and cyclic AMP levels in brain tissue. *J. Neurochem.* **26,** 741–748.

Festoff, B. W., Appel, S. H. and Day, E. (1970). Unpublished data cited by Appel (1972).

Festoff, B. W., Appel, S. H. and Day, E. (1971). Incorporation of [^{14}C]glucosamine into synaptosomes *in vitro*. *J. Neurochem.* **18,** 1871–1886.

Fifková, E. (1970). Changes of axosomatic synapses in the visual cortex of monocularly deprived rats. *J. Neurobiol.* **2,** 61–71.

Fifková, E. (1972). Effect of visual deprivation and light on synapses of the inner plexiform layer. *Exptl. Neurol.* **35,** 458–469.

Fillenz, M. (1972). Hypothesis for a neural mechanism involved in memory. *Nature* (*Lond.*) **238,** 41–43.

Fjerdingstad, E. J. (1973). Transfer of learning in rodents and fish. *In* "Current Biochemical Approaches to Learning and Memory" (W. B. Essman and S. Nakajima, Eds), pp. 73–98. Spectrum Publ., Flushing, New York.

Flament-Durand, J. and Dustin, P. (1972). Studies on the transport of secretory granules in the magnocellular hypothalamic neurons. I. Action of colchicine on axonal flow and neurotubules in the paraventricular nuclei. *Z. Zellforsch. Mikrosk. Anat.* **130,** 440–454.

Flexner, J. B. and Flexner, L. B. (1970). Adrenalectomy and the suppression of memory by puromycin. *Proc. Nat. Acad. Sci., U.S.A.* **66,** 48–52.

Flexner, J. B. and Flexner, L. B. (1971). Pituitary peptides and the suppression of memory by puromycin. *Proc. Nat. Acad. Sci., U.S.A.* **68,** 2519–2521.

Flexner, J. B., Flexner, L. B. and Stellar, E. (1963). Memory in mice as affected by intracerebral puromycin. *Science* **141,** 157–159.

Flexner, L. B., Flexner, J. B., De la Haba, G. and Roberts, R. B. (1965). Loss of memory as related to inhibition of cerebral protein synthesis. *J. Neurochem.* **12,** 535–541.

Flexner, L. B., Flexner, J. B. and Roberts, R. B. (1967). Memory in mice analyzed with antibiotics. *Science* **155,** 1377–1383.

Flood, J. F., Rosenzweig, M. R., Bennett, E. L. and Orme, A. E. (1973). The influence of duration of protein synthesis inhibition on memory. *Physiol. Behav.* **10,** 555–562.

Florey, E. (1960). Physiological evidence for naturally occurring inhibitory substances. *In* "Inhibition in the Nervous System and γ-Aminobutyric Acid" (E. Roberts, C. F. Baxter, A. Van Harreveld, C. A. G. Wiersma, W. R. Adey and K. F. Killam, Eds), pp. 72–84. Pergamon Press, London.

Fog, R. and Pakkenberg, H. (1971). Behavioral effects of dopamine and p-hydroxyamphetamine injected into corpus striatum of rats. *Exptl. Neurol.* **31,** 75–86.

Fog, R. L., Randrup, A. and Pakkenberg, H. (1967). Aminergic mechanisms in corpus striatum and amphetamine-induced stereotyped behavior. *Psychopharmacologia* **11,** 179–183.

Fog, R., Randrup, A. and Pakkenberg, H. (1970). Lesions in corpus striatum and cortex in rat brains and the effect on pharmacologically induced stereotyped, aggressive and cataleptic behavior. *Psychopharmacologia* **18,** 346–356.

Folbergrova, J., Lowry, O. H. and Passonneau, J. V. (1970). Changes in metabolites of the energy reserves in individual layers of mouse cerebral cortex and subjacent white matter during ischaemia and anaesthesia. *J. Neurochem.* **17,** 1155–1162.

Folbergrova, J., Norberg, K., Quistorff, B. and Siesjo, B. K. (1975). Carbohydrate and amino acid metabolism in rat cerebral cortex in moderate and extreme hypercapnia. *J. Neurochem.* **25,** 457–462.

Ford, D. H. (1967). Changes in brain accumulation of amino acids and adenine associated with changes in the physiologic state. *In* "Progress in Brain Research" (A. Lajtha and D. H. Ford, Eds), Vol. 29, pp. 401–413. Elsevier, Amsterdam.

Forgays, D. G. and Forgays, J. W. (1952). The nature of the effect of free-environmental experience in the rat. *J. Comp. Physiol. Psychol.* **45,** 322–328.

Forman, D. S. and Ledeen, R. W. (1972). Axonal transport of gangliosides in the goldfish optic nerve. *Science* **177,** 630–632.

Forstronen, R. F. (1963). The origin and morphogenetic significance of the external granular layer of the cerebellum, as determined experimentally in chick embryo. *Acta Neurol. Scand.* (Suppl.) **39,** 314–316.

Foster, M. and Sherrington, C. S. (1897). "A Text Book of Physiology. Part III. The Central Nervous System". Macmillan, New York.

Frank, K. and Raab, A. (1975). Male and female mice living in differently-sized groups. I. Noradrenaline metabolism in discrete brain areas, open-field activity and corticoid release. *J. comp. Physiol.* **99,** 153–164.

Fraumeni, J. F., Jr. (Ed.) (1975). "Persons at High Risk of Cancer". Academic Press, New York and London, 544 pp.

Freedman, D. X. (1968). A psychiatrist looks at LSD. *Fed. Probation,* June issue.

Freedman, D. X. (1970a). Drug dependence and its treatment. *Postgrad. Med.* **47,** 110–114.

Freedman, D. X. (1970b). Drug dependence and its treatment. *Postgrad. Med.* **47,** 150–154.

Frehan, F. A., Woodford, R. B. and Kety, S. S. (1951). Cerebral blood flow and metabolism in psychoses of senility. *J. Nerv. Ment. Dis.* **113,** 449–456.

Fried, P. A. (1972). Septum and behavior: A review. *Psychol. Bull.* **78,** 292–310.

Friedhoff, A. (Ed.) (1975a). "Catecholamines and Behavior. 1. Basic Neurobiology". Plenum Press, New York, 235 pp.

Friedhoff, A. (Ed.) (1975b). "Catecholamines and Behavior. 2. Neuropsychopharmacology". Plenum Press, New York, 225 pp.

Friedman, A. H. and Walker, C. A. (1972). The acute toxicity of drugs acting at cholinoceptive sites and twenty-four hour rhythms in brain acetylcholine. *Arch. Toxicol* **29,** 39–49.

Fuxe, K., Hökfelt, T. and Ungerstedt, U. (1970). Morphological and functional aspects of central monoamine neurons. *Int. Rev. Neurobiol.* **13,** 93–126.

Gadea-Ciria, M., Gervas-Camacho, J., Garcia-Gracia, M. and DeFeudis, F. V. (1975). Water contents of various regions of the feline nervous system. *Med. Biol.* **53,** 469–474.

Gaito, J. (1967). *In* "Chemistry of Learning" (W. C. Corning and S. C. Ratner, Eds). Plenum Press, New York.

Gaito, J. (Ed.) (1972). "Macromolecules and Behavior", 2nd Edn. Appleton-Century-Crofts, New York.

Gaitonde, M. K. and Festing, M. F. W. (1976). Brain glutamic acid decarboxylase and open field activity in ten inbred strains of mice. *Brain Res.* **103,** 617–621.

Gaitonde, M. K. and Richter, D. (1956). The metabolic activity of the proteins of the brain. *Proc. Roy. Soc. B.* **145,** 83–99.

Galambos, R. (1961). A glial-neural theory of brain function. *Proc. Nat. Acad. Sci., U.S.A.* **47,** 129–136.

Galambos, R. (1965). Introductory discussion on glial function. *Progr. Brain Res.* **15,** 267–277.

Galeano, C. (1972). Electrophysiological studies of learning in simplified nervous system preparations. *In* "The Structure and Function of Nervous Tissue" (G. H. Bourne, Ed.), Vol. V, pp. 507–549. Academic Press, New York and London.

Gallistel, C. R. (1973). Self-stimulation: The neurophysiology of reward and motivation. *In* "The Physiological Basis of Memory" (J. A. Deutsch, Ed.), pp. 175–267. Academic Press, New York and London.

Gambetti, P., Autilio-Gambetti, L., Rizzuto, N., Schafer, B. and Pfaff, L. (1974). Synapses and malnutrition: Quantitative ultrastructural study of rat cerebral cortex. *Exptl. Neurol.* **43,** 464–473.

Ganz, L. (1975). Orientation in visual space by neonates and its modification by visual deprivation. *In* "The Developmental Neuropsychology of Sensory Deprivation" (A. H. Riesen, Ed.), pp. 169–210. Academic Press, New York and London.

Ganz, L. and Haffner, M. E. (1974). Permanent perceptual and neurophysiological effects of visual deprivation in the cat. *Exptl. Brain Res.* **20,** 67–87.

Ganz, L., Hirsch, H. V. B. and Tieman, S. B. (1972). The nature of perceptual deficits in visually deprived cats. *Brain Res.* **44,** 547–568.

Garattini, S. and Valzelli, L. (1965). "Serotin". Elsevier Publ. Co., Amsterdam.

Garattini, S., Giacalone, E. and Valzelli, L. (1967). Isolation, aggressiveness and brain 5-hydroxytryptamine turnover. *J. Pharm. Pharmacol.* **19,** 338–339.

Garattini, S., Giacalone, E. and Valzelli, L. (1969). Biochemical changes during isolation-induced aggressiveness in mice. *In* "Aggressive Behaviour" (S. Garattini and E. B. Sigg, Eds), pp. 179–187. Excerpta Medica Fdn., Amsterdam.

Garcia, E. P., Denari, J. H. and Rosner, J. M. (1973). Uptake and metabolic effects of estradiol in the guinea pig cerebral cortex. *Steroids Lipids Res.* **4,** 248–256.

Gardner, E. (1968). "Fundamentals of Neurology", 5th Ed. W. B. Saunders Co., Philadelphia, 367 pp.

Gardner, E. L., Glick, S. D. and Jarvik, M. E. (1972). ECS dissociation of learning and one-way cross-dissociation with physostigmine and scopolamine. *Physiol. Behav.* **8,** 11–15.

Gardner, L. I. (1973). Deprivation dwarfism. *In* "The Nature and Nurture of Behavior. Developmental Psychobiology (Readings from Scientific American)", pp. 101–107. W. H. Freeman and Co., San Francisco.

Garey, L. J. and Pettigrew, J. D. (1974). Ultrastructural changes in kitten visual cortex after environmental modification. *Brain Res.* **66,** 165–172.

Garfinkel, D. (1973). Possible correlations between morphological structures in the brain and compartmentation indicated by simulation. *In* "Metabolic Compartmentation in Brain" (R. Balázs and J. E. Cremer, Eds), pp. 129–136. Macmillan, London.

Garfinkel, D., London, J. W., Dzubow, L. and Nicklas, W. J. (1975). Computer simulation of the metabolism of guinea pig brain slices, and how they differ from the intact brain. *Brain Res.* **92,** 207–218.

Gatfield, P. D., Lowry, O. H., Schulz, D. W. and Passonneau, J. V. (1966). Regional energy reserves in mouse brain and changes with ischaemia and anaesthesia. *J. Neurochem.* **13,** 185–195.

Gattoni, R. C. and Izquierdo, I. (1974). The effect of conditioning and pseudo-conditioning on RNA metabolism of rat hippocampus and neocortex. *Behav. Biol.* **12,** 67–80.

Gaull, G. E. (1972). Abnormal metabolism of sulfur-containing amino acids associated with brain dysfunction. *In* "Handbook of Neurochemistry" (A. Lajtha, Ed.), Vol. 7, pp. 169–190. Plenum Press, New York.

Gaze, R. M. (1970). "The Formation of Nerve Connections". Academic Press, London and New York, 288 pp.

Gazzaniga, M. S. and Blakemore, C. (Eds) (1975). "Handbook of Psychobiology", Academic Press, New York and London, 639 pp.

Geffen, L. B., Jessell, T. M., Cuello, A. C. and Iversen, L. L. (1976). Release of dopamine from dendrites in rat substantia nigra. *Nature* (*Lond.*) **260,** 258–260.

Geiger, A. (1958). Correlation of brain metabolism and function by the use of a brain perfusion method *in situ*. *Physiol. Rev.* **38,** 1–20.

Geiger, A., Yamasaki, S. and Lyons, R. (1956). Changes in nitrogenous components of brain produced by stimulation of short duration. *Amer. J. Physiol.* **184,** 239–247.

Geinsmann, Y. Y., Larima, V. N. and Mats, U. N. (1971). Changes in neuron dimensions as a possible morphological correlate of their increased functional activity. *Brain Res.* **26,** 247–257.

Geller, A. and Jarvick, M. E. (1968). The time relation of ECS induced amnesia. *Psychonom. Sci.* **12,** 169–170.

Geller, A., Robustelli, F., Barondes, S. H., Cohen, H. D. and Jarvik, M. E. (1969). Impaired performance by post-trial injections of cycloheximide in a passive avoidance. *Psychopharmacologia* **14,** 371–376.

Geller, E. (1971). Some observations on the effects of environmental complexity and isolation on biochemical ontogeny. *In* "Brain Development and Behavior" (D. J. McGinty and A. M. Adinolfi, Eds), pp. 277–296. Academic Press, New York and London.

Geller, E. and Yuwiler, A. (1968). Environmental effects on the biochemistry of the developing rat brain. *In* "Ontogenesis of the Brain" (L. Jilek and S. Trojan, Eds), pp. 277–284. Charles University, Prague.

Geller, E., Yuwiler, A. and Zolman, J. F. (1965). Effects of environmental complexity on constituents of brain and liver. *J. Neurochem.* **12,** 949–955.

Gendreau, P., Freedman, N. L., Wilde, G. J. S. and Scott, G. D. (1972). Changes in EEG alpha frequency and evoked response latency during solitary confinement. *J. Abnormal Psychol.* **79,** 54–59.

George, J. M. and Penrose, M. (1975). Increased incorporation of [^{3}H] uridine into RNA in the brain subfornical organ of ovariectomized mice. *Brain Res.* **97,** 167-170.

Gerard, R. W. (1949). Physiology and psychiatry. *Amer. J. Psychiat.* **106,** 161–173.

Gerard, R. W. (1955). Biological roots of psychiatry. *Science* **122,** 225–230.

Gerchman, L. B., Edgerton, V. R. and Carrow, R. E. (1975). Effects of physical training on the histochemistry and morphology of ventral motor neurons. *Exptl. Neurol.* **49,** 790–801.

Gerlach, J. L., McEwen, B. S., Pfaff, D. W., Moskovitz, S., Ferin, M., Carmel, P. W. and Zimmerman, E. A. (1976). Cells in regions of rhesus monkey brain and pituitary retain radioactive estradiol, corticosterone and cortisol differentially. *Brain Res.* **103,** 603–612.

Gershon, S. and Terry, R. D. (Eds) (1976). "Neurobiology of Aging; Aging, Vol. 3". Raven Press, New York, 480 pp.

Gervas-Camacho, J., Balfagón, G. and DeFeudis, F. V. (1976). Constancia del contenido total de proteinas en varias regiones del sistema nervioso central del gato. *Rev. Española Fisiol.* **32,** 111–114.

Gessa, G. L., Krishna, G., Forn, J., Tagliamonte, A. and Brodie, B. B. (1970). Behavioral and vegetative effects produced by dibutyryl cyclic AMP injected into different areas of the brain. *In* "Role of Cyclic AMP in Cell Function, Advances in Biochemical Psychopharmacology" (P. Greengard and E. Costa, Eds), Vol. 3, pp. 371–381. Raven Press, New York.

Ghazal, A.-E. M., Makar, A. B. and Daabees, T. T. (1976). Effect of oral contraceptives (Lyndiol) on rat brain gamma aminobutyric acid system. *Biochem. Pharmacol.* **25,** 115–118.

Ghent, L. (1951). The relation of experience to the development of hunger. *Can. J. Psychol.* **5** 77–

Giacalone, E., Tansella, M., Valzelli, L. and Garattini, S. (1968). Brain serotonin metabolism in isolated aggressive mice. *Biochem. Pharmacol.* **17,** 1315–1327.

Gibson, P. H., Stones, M. and Tomlinson, B. E. (1976). Senile changes in the human neocortex and hippocampus compared by the use of the electron and light microscopes. *J. Neurol. Sci.* **27,** 389–405.

Gilbert, J. M. (1972). Evidence of protein synthesis in synaptosomal membranes. *J. Biol. Chem.* **247,** 6541–6550.

Gilman, S., Dauth, G. W., Tennyson, V. M. and Kremzner, L. T. (1975). Chronic cerebellar stimulation in the monkey: Preliminary observations. *Arch. Neurol.* **33,** 474–477.

Giorgi, P. P., Karlsson, J. O., Sjöstrand, J. and Field, E. J. (1973). Axonal flow and myelin protein in the optic pathway. *Nature, New Biol.* **244,** 121–124.

Gisiger, V. (1971). Triggering of RNA synthesis by acetylcholine stimulation of the postsynaptic membrane in a mammalian ganglion. *Brain Res.* **33,** 139–147.

Gispen, W. H. and Schotman, P. (1973). Pituitary-adrenal system, learning and performance: Some neurochemical aspects. *Progr. Brain Res.* **39,** 453–458.

Gispen, W. H., DeWied, D., Schotman, P. and Jansz, H. S. (1971). Brain stem polysomes and avoidance performance of hypophysectomized rats subjected to peptide treatment. *Brain Res.* **31,** 341–351.

Gispen, W. H., Van Wimersma Greidanus, Tj. B., Bohus, B. and DeWied, D. (Eds) (1975a). "Hormones, Homeostasis and the Brain, Progress in Brain Research, Vol. 42". Elsevier, Amsterdam, 382 pp.

Gispen, W. H., Van Wimersma Greidanus, Tj. B., Waters-Ezrin, C., Zimmerman, E., Krivoy, W. A. and DeWied, D. (1975b). Influence of peptides on reduced response of rats to electric footshock after acute administration of morphine. *Eur. J. Pharmacol.* **33,** 99–105.

Gispen, W. H., Wiegant, V. M., Greven, H. M. and DeWied, D. (1975c). The induction of excessive grooming in the rat by intraventricular application of peptides derived from ACTH: Structure-activity studies. *Life Sci.* **17,** 645–652.

Glassman, E. (1969). The biochemistry of learning: An evaluation of the role of RNA and protein. *Ann. Rev. Biochem.* **38,** 605–646.

Glassman, E. (1974). Macromolecules and behavior: A commentary. *In* "The Neurosciences: Third Study Program" (F. O. Schmitt and F. G. Worden, Eds), pp. 667–678. M.I.T. Press, Cambridge, Mass.

Glassman, E. and Wilson, J. E. (1970). The incorporation of uridine into RNA during short experiences. *Brain Res.* **21,** 157–168.

Glassman, E. and Wilson, J. E. (1972). Brain function and RNA. *In* "Macromolecules and Behavior" (J. Gaito, Ed), 2nd Edn., pp. 39–46. Appleton-Century-Crofts, New York.

Glassman, E. and Wilson, J. E. (1973). RNA and brain function. *In* "Macromolecules and Behaviour" (G. B. Ansell and P. B. Bradley, Eds), pp. 81–92. University Park Press, London.

Glassman, E., Gispen, W., Perumal, R., Machlus, B. and Wilson, J. (1973). The effect of short experiences on the incorporation of radioactive phosphate into synaptosomal and non-histone acid-extractable nuclear proteins from rat and mouse brain. *Proc. 5th Int. Congr. Pharmacol.* **4,** 14–17.

Glick, S. M., Roth, J., Yalow, R. S. and Berson, S. A. (1965). The regulation of growth hormone secretion. *Recent Progr. Hormone Res.* **21,** 241–283.

Globus, A. (1975). Brain morphology as a function of presynaptic morphology and activity. *In* "The Developmental Neuropsychology of Sensory Deprivation" (A. Riesen, Ed.), pp. 9–91. Academic Press, London and New York.

Globus, A. and Scheibel, A. B. (1966). Loss of dendrite spines as an index of presynaptic terminal patterns. *Nature* (*Lond.*) **212,** 463–465.

Globus, A. and Scheibel, A. B. (1967a). The effect of visual deprivation on cortical neurons. *Exptl. Neurol.* **19,** 331–345.

Globus, A. and Scheibel, A. B. (1967b). Effects of visual deprivation on neurons of the visual cortex. *Anat. Rec.* **157,** 248.

Globus, A., Lux, H. D. and Schubert, P. (1968). Somadendritic spread of intracellularly injected tritiated glycine in cat spinal motoneurones. *Brain Res.* **11,** 440–445.

Globus, A., Rosenzweig, M. R., Bennett, E. L. and Diamond, M. C. (1972). Effects of differential experience on dendritic spine counts in rat cerebral cortex. *Int. J. Neurosci.* **4,** 124–129.

Globus, A., Lux, H. D. and Schubert, P. (1973a). Transfer of amino acids between neuroglial cells and neurons in the leech ganglion. *Exptl. Neurol.* **40,** 104–113.

Globus, A., Rosenzweig, M. R., Bennett, E. L. and Diamond, M. C. (1973b). Effects of differential experience on dendritic spine counts in rat cerebral cortex. *J. Comp. Physiol. Psychol.* **82,** 175–181.

Glowinski, J. and Axelrod, J. (1965). Effect of drugs on the uptake, release and metabolism of H^3-norepinephrine in the rat brain. *J. Pharmacol. Exptl. Therapeut.* **149,** 43–49.

Glowinski, J., Axelrod, J. and Iversen, L. L. (1966). Regional studies of catecholamines in the rat brain. IV. Effects of drugs on the disposition and metabolism of H^3-norepinephrine and H^3-dopamine. *J. Pharmacol. Exptl. Therapeut.* **153,** 30–41.

Gluck, J. P., Harlow, H. F. and Schlitz, K. A. (1973). Differential effect of early enrichment and deprivation on learning in the rhesus monkey (*Macaca mulatta*). *J. Comp. Physiol. Psychol.* **84,** 598–604.

Godin, Y. and Mandel, P. (1965). Distribution des acides aminés libres dans le système nerveux central du rat au cours du sommeil et de l'état de veille prolongée. *J. Neurochem.* **12,** 455–460.

Gold, P. E. and King, R. A. (1972). Caudate stimulation and retrograde amnesia: Amnesia threshold and gradient. *Behav. Biol.* **7,** 709–715.

Goldberg, A. M. and Hanin, I. (Eds) (1976). "Biology of Cholinergic Function". Raven Press, New York, 525 pp.

Goldberg, A. M. and Welch, B. L. (1972). Adaptation of the adrenal medulla: Sustained increase in choline acetyltransferase by psychosocial stimulation. *Science* **178,** 319–320.

Goldberg, M. L. (1975). The glucose effect: Carbohydrate repression of enzyme induction, RNA synthesis, and glucocorticoid activity—A role for cyclic AMP and cyclic GMP. *Life Sci.* **17,** 1747–1754.

Goldberg, N. D., Haddox, M. K., Hartle, D. K. and Hadden, J. W. (1972). The biological role of cyclic 3′, 5′-guanosine monophosphate. *In* "Pharmacology and the Future of Man, Fifth International Congress of Pharmacology" (R. A. Maxwell and G. H. Acheson, Eds), pp. 146–169. Karger, Basel.

Goldstein, D. B. (1975). Drug dependence as an adaptive response: Studies with ethanol in mice. *In* "Neurobiological Mechanisms of Adaptation and Behavior" (A. J. Mandell, Ed.), pp. 185–198. Raven Press, New York.

Golgi, C. (1873). Sulla sostanza grigia del cervello. *In* "Opera Omnia 1903", **1,** 91–98.

Gombos, G., Morgan, I. G., Waehneldt, T. V., Vincendon, G. and Breckenridge, W. C. (1972). Glycoproteins of the synaptosomal plasma membrane. *Adv. Exptl. Med. Biol.* **25,** 101–113.

Gomirato, N. and Hydén, H. (1963). A biochemical glia error in the Parkinson disease. *Brain* **86,** 773–780.

Gonatas, N. K. (1967). Axonic and synaptic lesions in neuropsychiatric disorders. *Nature* (*Lond.*) **214,** 352–355.

Goodman, D. B. P., Rasmussen, H., DiBella, F. and Guthrow, C. E., Jr. (1970). Cyclic adenosine 3′:5′-monophosphate-stimulated phosphorylation of isolated neurotubule subunits. *Proc. Nat. Acad. Sci., U.S.A.* **67,** 652–659.

Goodman, L. (1931). Effect of total absence of function on the optic system of rabbits. *Amer. J. Physiol.* **100,** 46–63.

Goodman, R., Otten, U. and Thoenen, H. (1975). Organ culture of the rat adrenal medulla: A model system for the study of trans-synaptic enzyme induction. *J. Neurochem.* **25,** 423–427.

Goodwin, F. K., Post, R. M. and Sack, R. L. (1975). Clinical evidence for neurochemical adaptation to psychotropic drugs. *In* "Neurobiological Mechanisms of Adaptation and Behavior" (A. J. Mandell, Ed.), pp. 33–45. Raven Press, New York.

Gordon, R., Spector, S., Sjoerdsma, A. and Udenfriend, S. (1966). Increased synthesis of norepinephrine and epinephrine in the intact rat during exercise and exposure to cold. *J. Pharmacol. Exptl. Therapeut.* **153,** 440–447.

Grafstein, B. (1967). Transport of protein by goldfish optic nerve fibers. *Science* **157,** 196–198.

Grafstein, B. (1969). Axonal transport: Communication between soma and synapse. *Adv. Biochem. Psychopharmacol.* **1,** 11–25.

Grafstein, B. (1971). Transneuronal transfer of radioactivity in the central nervous system. *Science* **172,** 177–179.

Grafstein, B. (1975). The nerve cell body response to axotomy. *Exptl. Neurol.* **48** (Part 2) 32–51.

Gramsbergen, A. (1976). EEG development in normal and undernourished rats. *Brain Res.* **105,** 287–308.

Gray, E. G. and Whittaker, V. P. (1962). The isolation of nerve endings from brain; an electron-microscopic study of cell fragments derived by homogenization and centrifugation. *J. Anat.* (*Lond.*) **96,** 79–88.

Greengard, P. (1975). Cyclic nucleotides, protein phosphorylation and neuronal function. *In* "Advances in Cyclic Nucleotide Research" (G. I. Drummond, P. Greengard and G. A. Robison, Eds), Vol. 5, pp. 585–601. Raven Press, New York.

Greengard, P. (1976). Possible role for cyclic nucleotides and phosphorylated membrane proteins in postsynaptic actions of neurotransmitters. *Nature* (*Lond.*) **260,** 101–108.

Greengard, P. and Costa, E. (Eds) (1970). "Role of Cyclic AMP in Cell Function". Raven Press, New York.

Greengard, P. and Kebabian, J. W. (1974). Role of cyclic AMP in synaptic transmission in the mammalian peripheral nervous system. *Fed. Proc.* **33,** 1059–1067.

Greengard, P., McAfee, D. A. and Kebabian, J. W. (1972). On the mechanism of action of cyclic AMP and its role in synaptic transmission. *In* "Advances in Cyclic Nucleotide Research" (P. Greengard, R. Paoletti, and G. A. Robison, Eds), Vol. 1, pp. 337–355. Raven Press, New York.

Greengard, P., Kebabian, J. W. and McAfee, D. A. (1973). Studies on the role of cyclic AMP in neural function. *In* "Pharmacology and the Future of Man, Proc. 5th Int. Congr. Pharmacology, San Francisco", Vol. 5, pp. 207–217. Karger, Basel.

Greenough, W. T. (1975). Experiential modification of the developing brain. *Amer. Scientist* **63,** 37–46.

Greenough, W. T. (1976). Enduring brain effects of differential experience and training. *In* "Neural Mechanisms of Memory and Learning" (M. R. Rosenzweig, and E. L. Bennett, Eds), pp. 255–277. M.I.T. Press, Cambridge, Mass.

Greenough, W. T. and Volkmar, F. R. (1973). Pattern of dendritic branching in occipital cortex of rats reared in complex environments. *Exptl. Neurol.* **40,** 491–504.

Greenough, W. T., Fulcher, J. K., Yuwiler, A. and Geller, E. (1970). Enriched rearing and chronic electroshock: Effects on brain and behavior in mice. *Physiol. Behav.* **5,** 371–373.

Greenough, W. T., Volkmar, F. R., and Juraska, J. M. (1973). Effects of rearing complexity on dendritic branching in frontolateral and temporal cortex of the rat. *Exptl. Neurol.* **41,** 371–378.

Greenwood, F. C. (1967). *In* "Hormones in Blood" (C. H. Gray, Ed.), 2nd Edn., pp. 195. Academic Press, London and New York. Cited by Kokka *et al.* (1971).

Gross, G. W. (1973). The effect of temperature on the rapid axoplasmic transport in C-fibers. *Brain Res.* **56,** 359–363.

Gross, G. W. and Beidler, L. M. (1975). A quantitative analysis of isotope concentration profiles and rapid transport velocities in the C-fibers of the garfish olfactory nerve. *J. Neurobiol.* **6,** 213–232.

Gross, N. J., Getz, G. S. and Rabinowitz, M. (1969). Apparent turnover of mitochondrial deoxyribonucleic acid and mitochondrial phospholipids in the tissue of rat. *J. Biol. Chem.* **244,** 1552–1562.

Groves, P. M., Wilson, C. J., Young, S. J. and Rebec, G. V. (1975). Self-inhibition by dopaminergic neurons. *Science* **190,** 522–529.

Guillery, R. W. (1969). A quantitative study of synaptic interconnections in the dorsal lateral geniculate nucleus of the cat. *Z. Zellforsch.* **96,** 39–48.

Gunn, J. A. and Gurd, M. R. (1940). The action of some amines related to adrenaline. Cyclohexylalkylamines. *J. Physiol.* (*Lond.*) **97,** 453–470.

Gunne, L. M. (1969). Brain catecholamines in the rage response evoked by intracerebral stimulation and ablation. *In* "Aggressive Behaviour" (S. Garattini and E. B. Sigg, Eds), pp. 238–242. Excerpta Medica Fdn., Amsterdam.

Guth, L. (1974). Axonal regeneration and functional plasticity in the central nervous system. *Exptl. Neurol.* **45,** 606–654.

Guth, L. (1975). History of central nervous system regeneration research. *Exptl. Neurol.* **48** (Part 2), 3–15.

Guth, L. and Watson, P. K. (1968). Correlated histochemical and quantitative study on cerebral glycogen after brain injury in the rat. *Exptl. Neurol.* **22,** 590–602.

Guth, L. and Windle, W. F. (1970). The enigma of central nervous regeneration. *Exptl. Neurol.* **28,** suppl. 5, 1–43.

Guth, L. and Windle, W. F. (1973). Physiological, molecular, and genetic aspects of central nervous system regeneration. *Exptl. Neurol.* **39,** iii–xvi.

Gyllensten, L. (1959). Postnatal development of the visual cortex in darkness (mice). *Acta Morphol. Neerl. Scand.* **2,** 331–345.

Gyllensten, L. and Lindberg, J. (1964). Development of the visual cortex in mice with inherited retinal dystrophy. *J. Comp. Neurol.* **122,** 79–90.

Gyllensten, L., Malmfors, T. and Norrlin, M. L. (1965). Effect of visual deprivation on the optic centers of growing and adult mice. *J. Comp. Neurol.* **124,** 149–160.

Hadfield, M. G. and Weber, N. E. (1975). Effect of fighting and diphenylhydantoin on the uptake of ^{3}H-*l*-norepinephrine *in vitro* in synaptosomes isolated from retired male breeding mice. *Biochem. Pharmacol.* **24,** 1538–1540.

Hager, H., Luh, S., Ruscakova, D. und Ruscak, M. (1967). Histochemie, elektronemikroskopische und biochemische Untersuchungen uber Glykogenanhäufung in reaktiv veränderten Astrozyten der traumatisch lädierten Saugergrosshirnrinde. *Z. Zellforsch.* **83,** 295–320.

Haglid, K. G. and Stavrou, D. (1973). Water soluble and pentanol extractable proteins in human brain normal tissue and human tumors, with special reference to S-100 protein. *J. Neurochem.* **20,** 1523–1532.

Hakkinen, H.-M. and Kulonen, E. (1959). Increase in the γ-aminobutyric acid content of rat brain after ingestion of ethanol. *Nature* (*Lond.*) **184,** 726–727.

Haldane, J. S. (1917). "Organism and Environment as Illustrated by the Physiology of Breathing". Yale University Press, New Haven, 138 pp.

Halstead, W. C. (1951). Brain and intelligence. *In* "Cerebral Mechanisms in Behavior" (L. A. Jeffress, Ed.), p. 244. Wiley, New York.

Hamberger, A., Hansson, H.-A. and Sjöstrand, J. (1970). Surface structure of isolated neurons: Detachment of nerve terminals during axon regeneration. *J. Cell Biol.* **47,** 319–331.

Hamburger, C.-A. and Hydén, H. (1945). Cytochemical changes in the cochlear ganglion caused by acoustic stimulation and trauma. *Acta Oto-Laryngol.* **61** (Suppl.) 5–89.

Hámori, J. (1973). The inductive role of presynaptic axons in the development of postsynaptic spines. *Brain Res.* **62,** 337–344.

Hámori, J., Pasik, P. and Pasik, T. (1975). Postnatal differentiation of "presynaptic dendrites" in the lateral geniculate nucleus of the Rhesus monkey. *In* "Advances in Neurology, Vol. 12, Physiology and Pathology of Dendrites" (G. W. Kreutzberg, Ed.), pp. 149–161. Raven Press, New York.

Hanbauer, I. and Costa, E. (1976). Induction of tyrosine hydroxylase in the superior cervical ganglia of rats: Opposite influence of muscarinic and nicotinic receptor agonists. *Neuropharmacology* **15,** 85–90.

Hanin, I. (1969). A specific gas chromatographic method for assaying tissue acetylcholine: Present status. *In* "Advances in Biochemistry and Psychopharmacology" (E. Costa and P. Greengard, Eds), Vol. 1, pp. 111–130. Raven Press, New York.

Hanley, T. (1974). "Neuronal fall-out" in the ageing brain: A critical review of the quantitative data. *Age and Ageing* **3,** 133–151.

Harding, C. F. and Leshner, A. I. (1972). The effects of adrenalectomy on the aggressiveness of differently housed mice. *Physiol. Behav.* **8,** 437–440.

Harlow, H. F. (1953). Motivation as a factor in the acquisition of new responses. *In* "Current Theory and Research in Motivation: A Symposium". University of Nebraska Press, Lincoln, Nebraska.

Harlow, H. F. (1962). The heterosexual affectional system in monkeys. *Amer. Psychologist* **17,** 1–9.

Harlow, H. F. and Suomi, S. J. (1974). Induced depression in monkeys. *Behav. Biol.* **12,** 273–296.

Harlow, H. F., Rowland, G. L. and Griffin, G. A. (1964). The effect of total social deprivation on the development of monkey behavior. *Psychiat. Res. Rept.* **19,** 116–135.

Harlow, H. F., Dodsworth, R. O. and Harlow, M. K. (1965). Total social isolation in monkeys. *Proc. Nat. Acad. Sci., U.S.A.* **54,** 90–96.

Harlow, H. F., Schlitz, K. A. and Harlow, M. K. (1969). Effects of social isolation on the learning performance of Rhesus monkeys. *In* "Proc. 2nd Int. Congr. Primatol", Vol. 1, pp. 178–185. Karger, Basel.

Harlow, H. F., Harlow, M. K. and Suomi, S. J. (1971). From thought to therapy: Lessons from a primate laboratory. *Amer. Scientist* **59,** 538–549.

Harlow, H. F., Plubell, P. E. and Baysinger, C. M. (1973). Induction of psychological death in Rhesus monkeys. *J. Autism and Childhood Schizophrenia* **3,** 299–307.

Harris, A. B. (1964). Cortical alterations due to methionine sulfoximine. *Arch. Neurol.* **11,** 388–407.

Harris, A. B. (1972). Degeneration in experimental epileptic foci. *Arch. Neurol.* **26,** 434–449.

Harris, A. B. (1975). Cortical neuroglia in experimental epilepsy. *Exptl. Neurol.* **49,** 691–715.

Hattori, T., McGeer, P. L., Fibiger, H. C. and McGeer, E. G. (1973). On the source of GABA-containing terminals in the substantia nigra. Electron microscopic autoradiographic and biochemical studies. *Brain Res.* **54,** 103–114.

Hayat, M. A. (Ed.) (1975a). "Principles and Techniques of Scanning Electron Microscopy", Vol. 3. Van Nostrand Reinhold Co., New York, 226 pp.

Hayat, M. A. (Ed.) (1975b). "Principles and Techniques of Scanning Electron Microscopy", Vol. 4. Van Nostrand Reinhold Co., New York, 230 pp.

Haywood, J., Rose, S. P. R. and Bateson, P. P. G. (1970). Effect of an imprinting procedure on RNA polymerase activity in the chick brain. *Nature* (*Lond.*) **228,** 373–374.

Haywood, J., Hambley, J. and Rose, S. P. R. (1973). Changes in acteylcholinesterase activity in chick brain following exposure to an imprinting stimulus. *4th Meeting, Int. Soc. for Neurochemistry, Tokyo* (Abstract).

Heald, P. J. (1962). Phosphoprotein metabolism and ion transport in nervous tissue; a suggested connexion. *Nature* (*Lond.*) **193,** 451–454.

Heath, R. G. and Krupp, I. M. (1967). Catatonia induced in monkeys by antibrain antibody. *Amer. J. Psychiat.* **123,** 1499–1504.

Hebb, D. O. (1947). The effects of early experience on problem-solving at maturity. *Amer. Psychologist* **2,** 306–307.

Hebb, D. O. (1949). "The Organization of Behavior". John Wiley and Sons, New York, 335 pp.

Hebb, D. O. (1955). Drives and the C. N. S. (conceptual nervous system) *Psychol. Rev.* **63,** 243–254.

Hebb, D. O. (1958). The motivating effects of exteroceptive stimulation. *Amer. Psychologist* **13,** 109–113.

Heilbronn, E., Mattson, C., Stalberg, E. and Hilton-Brown, P. (1975). Neurophysiological signs of myasthenia in rabbits after receptor antibody development. *J. Neurol. Sci.* **24,** 59–64.

Heimer, L. and Lohman, A. H. M. (1975). Anatomical methods for tracing connections in the central nervous system. *In* "Handbook of Psychobiology" (M. S. Gazzaniga and C. Blakemore, Eds), pp. 73–106. Academic Press, New York and London.

Heller, I. H. and Elliott, K. A. C. (1954). Desoxyribonucleic acid content and cell density in brain and human brain tumors. *Can J. Biochem. Physiol.* **32,** 584–592.

Henderson, N. D. (1973). Brain weight changes resulting from enriched rearing conditions: A diallel analysis. *Develop. Psychobiol.* **6,** 367–376.

Hendley, E. D., Moisset, B. and Welch, B. L. (1973). Catecholamine uptake in cerebral cortex: Adaptive change induced by fighting. *Science* **180,** 1050–1052.

Hendrickson, A. (1975). Technical modifications to facilitate tracing synapses by electron microscopic autoradiography. *Brain Res.* **85,** 241–247.

Hendry, I. A. and Thoenen, H. (1974). Changes of enzyme pattern in the sympathetic nervous system of adult mice after submaxillary gland removal; response to exogenous nerve growth factor. *J. Neurochem.* **22,** 999–1004.

Hendry, I. A., Stöckel, K., Thoenen, H. and Iversen, L. L. (1974). The retrograde axonal transport of nerve growth factor. *Brain Res.* **68,** 103–121.

Henneman, D. H., Altschule, M. D. and Goncz, R. M. (1954). Carbohydrate metabolism in brain disease. II. Glucose metabolism in schizophrenic, manic-depressive and involutional psychoses. *Arch. Int. Med.* **94,** 402–416.

Hermann, A., Rieske, E., Kreutzberg, G. W. and Lux, H. D. (1975). Transjunctional flux of radioactive precursors across electrotonic synapses between lateral giant axons of the crayfish. *Brain Res.* **95,** 125–131.

Hernández-Peón, R., Chavez-Ibarra, G., Morgane, P. and Timo-Iaria, C. (1963). Limbic cholinergic pathways involved in sleep and emotional behavior. *Exptl. Neurol.* **8,** 93–111.

Herndon, R. M. and Oster-Granite, M. L. (1975). Effect of granule cell destruction on development and maintenance of the Purkinje cell dendrite. *In* "Advances in Neurology; Physiology and Pathology of Dendrites" (G. W. Kreutzberg, Ed.), Vol. 12, pp. 361–371. Raven Press, New York.

Herndon, R. M., Margolis, G. and Kilham, L. (1971). The synaptic organization of malformed cerebellum induced by perinatal infection with the feline panleukopenia virus (PLV). *J. Neuropath. Exptl. Neurol.* **30,** 557–580.

Heuser, J. E. and Reese, T. S. (1973). Evidence for recycling of synaptic vesicle membrane during transmitter release at the frog neuromuscular junction. *J. Cell. Biol.* **57,** 315–344.

Hier, D. B., Arnason, B. G. W. and Young, M. (1972). Studies on the mechanism of action of nerve growth factor. *Proc. Nat. Acad. Sci., U.S.A.* **69,** 2268–2272.

Hilgard, E. R. and Marquis, D. G. (1940). "Conditioning and Learning". Appleton Century Co., New York.

Hillarp, N.-A., Fuxe, K. and Dahlström, A. (1966). Demonstration and mapping of central neurons containing dopamine, noradrenaline, and 5-hydroxytryptamine and their reactions to psychopharmaca. *Pharmacol. Rev.* **18,** 727–742.

Hille, B. (1976). Gating in sodium channels of nerve. *Ann. Rev. Physiol.* **38,** 139–152.

Hiller, J. M., Pearson, J. and Simon, E. J. (1973). Distribution of stereospecific binding of the potent narcotic analgesic etorphine in the human brain: Predominance in the limbic system. *Res. Commun. Chem. Pathol. Pharmacol.* **6,** 1052–1062.

Himmelsbach, C. K. (1943). Can the euphoric, analgetic and physical dependence effects of drugs be separated? *Fed. Proc.* **2,** 201–203.

Himwich, H. (Ed.) (1970). "Biochemistry, Schizophrenias, and Affective Illnesses". Williams and Wilkins Co., Baltimore, 500 pp.

Himwich, W. A. (1969). The effect of environment upon the developing brain. *In* "The Future of the Brain Sciences" (S. Bogoch, Ed.), pp. 237–255. Plenum Press, New York.

Himwich, W. A. (Ed.) (1970). "Developmental Neurobiology". C. C. Thomas Springfield, Ill.

Himwich, W. A. (1973). Problems in interpreting neurochemical changes occurring in developing and aging animals. *Progr. Brain Res.* **40,** 13–23.

Himwich, W. A. and Agrawal, H. C. (1969). Amino acids. *In* "Handbook of Neurochemistry" (A. Lajtha, Ed.), Vol. 1, pp. 33–52. Plenum Press, New York.

Himwich, W. A. and Himwich, H. E. (1959). *In* "Handbook of Aging and the Individual" (J. E. Birren, Ed.) pp. 187–215. University of Chicago Press, Chicago.

Himwich, W. A., Davis, J. M. and Agrawal, H. C. (1968). Effects of early weaning on some free amino acids and acetylcholinesterase activity of rat brain. *Rec. Adv. Biol. Psychiat.* **10,** 266–270.

Hirano, A. and Dembitzer, H. M. (1975). Aberrant development of the Purkinje cell dendritic spine. *In* "Advances in Neurology; Physiology and Pathology of Dendrites" (G. W. Kreutzberg, Ed.), Vol. 12, pp. 353–360. Raven Press, New York.

Hirsch, H. V. B. and Jacobson, M. (1975). The perfectible brain. *In* "Handbook of Psychobiology" (M. S. Gazzaniga and C. Blakemore, Eds), pp. 107–137. Academic Press, London and New York.

Hirsch, H. V. B. and Spinelli, D. N. (1970). Visual experience modifies distribution of horizontally and vertically oriented receptive fields in cats. *Science* **168,** 869–871.

Hirsch, H. V. B. and Spinelli, D. N. (1971). Modification of the distribution of receptive field orientation in cats by selective visual exposure during development. *Exptl. Brain Res.* **13,** 509–527.

Hodgkin, A. L. (1964). The ionic basis of nervous conduction. *Science* **145,** 1148–1154.

Hoffman, H. S., Boskoff, K. J., Eiserer, L. A. and Klein, S. H. (1975). Isolation-induced aggression in newly hatched ducklings. *J. Comp. Physiol. Psychol.* **89,** 447–456.

Hökfelt, T. and Ljungdahl, Å. (1975). Uptake mechanisms as a basis for the histochemical identification and tracing of transmitter-specific neuron populations. *In* "The Use of Axonal Transport for Studies of Neuronal Connectivity" (W. M. Cowan and M. Cuénod, Eds.) pp. 249–305. Elsevier, Amsterdam.

Hökfelt, T. and Ungerstedt, U. (1973). Specificity of 6-hydroxydopamine induced degeneration of central monoamine neurones: An electron and fluorescence microscopic study with special reference to intracerebral injection on the nigro-striatal dopamine system. *Brain Res.* **60,** 269–297.

Holloway, R. L. (1966). Dendritic branching: Some preliminary results of training and complexity in rat visual cortex. *Brain Res.* **2,** 393–396.

Holloway, R. L. (1968). The evolution of the primate brain: Some aspects of quantitative relations, *Brain Res.* **7,** 121–172.
Holloway, R. L. (1969). Some questions on parameters of neural evolution in primates. *Ann. N.Y. Acad. Sci.* **167,** 332–340.
Holm, M. (1972). Gangliosides of the optic pathway: Biosynthesis and biodegradation studied *in vivo*. *J. Neurochem.* **19,** 623–629.
Holmgren, E. (1902). Beiträge zur Morphologie der Zelle. I. Nervenzellen. *Anat. Hefte* **18,** 267–325.
Holtz, P. and Palm, D. (1964). Pharmacological aspects of vitamin B_6. *Pharmacol. Rev.* **16,** 113–178.
Holzbauer, M. and Sharman, D. F. (1972). The distribution of catecholamines in vertebrates. *In* "Catecholamines: Handbook of Experimental Pharmacology, XXXIII", (H. Blaschko and E. Muscholl, Eds), pp. 110–185. Springer-Verlag, Berlin.
Holtzman, E., Teichberg, S., Abrahams, S. J., Citkowitz, E., Crain, S. M., Kawai, N. and Peterson, E. (1973). Notes on synaptic vesicles and related structures, endoplasmic reticulum, lysosomes and peroxisomes in nervous tissue and the adrenal medulla. *J. Histochem. Cytochem.* **21,** 349–385.
Horn, G., Horn, A. L. D., Bateson, P. P. G. and Rose, S. P. R. (1971). Effects of imprinting on uracil incorporation into brain RNA in the "split-brain" chick. *Nature* (*Lond.*) **229,** 131–132.
Horn, G., Rose, S. P. R. and Bateson, P. P. G. (1973). Experiences and plasticity in the central nervous system. *Science* **181,** 506–514.
Hornykiewicz, O. (1966). Dopamine (3-hydroxytyramine) and brain function. *Pharmacol. Rev.* **18,** 925–964.
Hornykiewicz, O. (1973). Parkinson's disease: From brain homogenate to treatment. *Fed. Proc.* **32,** 183–190.
Horridge, G. A. (1962). Learning of leg position of the ventral nerve cord in headless insects. *Proc. Roy. Soc., B.* **157,** 33–52.
Horridge, G. A. (1965). The electrophysiological approach to learning in isolatable ganglia. *Animal Behaviour* (*Suppl.* **1**), 163–182.
Horst, W. D. and Spirt, N. (1974). A possible mechanism for the antidepressant activity of thyrotropin releasing hormone. *Life Sci.* **15,** 1073–1082.
Hörtnagl, M., Hörtnagl, H. and Winkler, H. (1969). Bovine splenic nerve: Characterization of noradrenaline-containing vesicles and other cell organelles by density gradient centrifugation. *J. Physiol.* (*Lond.*) **205,** 103–114.
Householder, A. S. and Landahl, H. D. (1945). "Mathematical Biophysics of the Central Nervous System". Principia Press, Bloomington, Indiana.
Howard, E. and Granoff, D. M. (1968). Effect of neonatal food restriction in mice on brain growth, DNA and cholesterol, and on adult delayed response learning. *J. Nutr.* **95,** 111–121.
Hoyle, G. (1975). Neural mechanisms underlying behavior of invertebrates. In "Handbook of Psychobiology" (M. S. Gazzaniga and C. Blakemore, Eds), pp. 3–48. Academic Press, New York and London.
Hsia, Y. E. (1972). Inherited disorders of amino acid, carbohydrate, and nucleic acid metabolism. In "Basic Neurochemistry" (R. W. Albers, G. J. Siegel, R. Katzman and B. W. Agranoff, Eds), pp. 451–484. Little, Brown and Co., Boston.
Hubbard, J. I. (1970). Mechanism of transmitter release. *Progr. Biophys. Mol. Biol.* **21,** 33–124.
Hubbard, J. I. (1973). Microphysiology of vertebrate neuromuscular transmission. *Physiol. Rev.* **53,** 674–723.
Hubel, D. H. and Wiesel, T. N. (1963). Receptive fields of cells in striate cortex of very young visually inexperienced kittens. *J. Neurophysiol.* **26,** 994–1002.

Hubel, D. H. and Wiesel, T. N. (1970). The period of susceptibility to the physiological effects of unilateral eye closure in kittens. *J. Physiol.* (*Lond.*) **206,** 419–436.

Hubscher, U., Knusel, A., Heidelberger, R. and Kuenzle, C. C. (1975). DNA polymerases in brain neuronal nuclei: Increased activity in rats required to perform a complex behavioral task. *First European Neurosciences Meeting, Munich,* 9–13 September, Abstr. No. 224.

Hudgens, G. A., Holloway, W. R., Suggs, D., Hanson, R. S., Hettler, N., Lagasse, L. and Bruhns, W. (1975). "Behavioral Modification and Changes in Central Nervous System Biochemistry: An Annotated Bibliography", U.S. Army Human Engineering Laboratory, Aberdeen Proving Ground, Maryland, 373 pp.

Hudson, D. B., Vernadakis, A. and Timiras, P. S. (1970). Regional changes in amino acid concentration in the developing brain and the effects of neonatal administration of estradiol. *Brain Res.* **23,** 213–222.

Hughes, J. (1975a). Isolation of an endogenous compound from the brain with pharmacological properties similar to morphine. *Brain Res.* **88,** 295–308.

Hughes, J. (1975b). Search for the endogenous ligand of the opiate receptor. *Neurosci. Res. Progr. Bull.* **13,** 55–58.

Hughes, J., Smith, T., Morgan, B. and Fothergill, L. (1975a). Purification and properties of enkephalin—the possible endogenous ligand for the morphine receptor. *Life Sci.* **16,** 1753–1758.

Hughes, J., Smith, T. W., Kosterlitz, H. W., Fothergill, L. A., Morgan, B. A. and Morris, H. R. (1975b). Identification of two related pentapeptides from the brain with potent opiate agonist activity. *Nature* (*Lond.*) **258,** 577–579.

Hughes, R. A. (1969). Retrograde amnesia in rats produced by hippocampal injections of potassium chloride: Gradient of effect and recovery. *J. Comp. Physiol. Psychol.* **68,** 637–644.

Hughes, W. L. (1959). *In* "The Kinetics of Cellular Proliferation" (F. Stohlman, Ed.), pp. 83–96. Grune and Stratton, New York.

Hull, E. M. and Homan, H. D. (1975). Olfactory bulbectomy, peripheral anosmia, and mouse killing and eating by rats. *Behav. Biol.* **14,** 481–488.

Hunter, W. S. A. (1930). A consideration of Lashley's theory of the equipotentiality of cerebral action. *J. Gen. Psychol.* **3,** 455–468.

Huston, J. P. (1975). Learning in the thalamic rat. *In* "Subcortical Mechanisms and Sensorimotor Activities" (T. L. Frigyesi, Ed.), pp. 217–227. Hans Huber, Bern.

Huston, J. P. and Borbély, A. A. (1973). Operant conditioning in forebrain ablated rats by use of rewarding hypothalamic stimulation. *Brain Res.* **50,** 467–472.

Huston, J. P. and Borbély, A. A. (1974). The thalamic rat: General behavior, operant learning with rewarding hypothalamic stimulation and effects of amphetamine. *Physiol. Behav.* **12,** 433–448.

Huston, J. P., Shibata, M. and Siegfried, B. (1975). Physiological and behavioral analysis of feeding induced by hippocampal and neocortical spreading depression in rats. *First European Neurosciences Meeting, Munich,* 9–13 September, Abstr. No. 190.

Huttunen, M. O. (1971). Persistent alteration of turnover of brain noradrenaline in the offspring of rats subjected to stress during pregnancy. *Nature* (*Lond.*) **230,** 53–55.

Huxley, A. F. (1964). Excitation and conduction in nerve: Quantitative analysis. *Science* **145,** 1154–1159.

Huxtable, R. and Barbeau, A. (Eds) (1976). "Taurine". Raven Press, New York, 416 pp.

Hydén, H. (1943). Protein metabolism in the nerve cell during growth and function. *Acta Physiol. Scand,* Suppl. **17,** 1–136.

Hydén, H. (1959a). Quantitative assay of compounds in isolated, fresh nerve cells and glial cells from control and stimulated animals. *Nature* (*Lond.*) **184,** 433–435.

Hydén, H. (1959b). Biochemical changes in glial cells and nerve cells at varying activity. *In* "Proc. Fourth Int. Congr. Biochemistry" (O. Hoffmann-Ostenhof, Ed.), Vol. 3, pp. 64–89. Pergamon Press, London.

Hydén, H. (1961). Satellite cells in the nervous system. *Sci. Amer.* **205,** 62–70.

Hydén, H. (1964a). RNA, a functional characteristic of the neuron and its glia. *In* "RNA and Brain Function, Memory and Learning" (M.A.B. Brazier, Ed.). University of California Press, Los Angeles, U.S.A.

Hydén, H. (1964b). Biochemical and functional interplay between neuron and glia. *In* "Recent Advances in Biological Psychiatry" (J. Wortis, Ed.), Vol. VI, pp. 31–54. Plenum Press, New York.

Hydén, H. (1966). Production of RNA in neurons and glia in Parkinson's disease indicating genic stimulation. *In* "Biochemistry and Pharmacology of the Basal Ganglia" (E. Costa, L. J. Côté and M. D. Yahr, Eds), pp. 195–204. Plenum Press, New York.

Hydén, H. (1967). Dynamic aspects on the neuron-glia relationship. *In* "The Neuron" pp. 179–219. Elsevier, Amsterdam.

Hydén, H. (1973a). Changes in brain protein during learning. *In* "Macromolecules and Behaviour" (G. B. Ansell and P. B. Bradley, Eds), pp. 3–26. University Park Press, London.

Hydén, H. (1973b). Nerve cells and their glia: Relationships and differences. *In* "Macromolecules and Behaviour" (G. B. Ansell and P. B. Bradley, Eds), pp. 27–50. University Park Press, London.

Hydén, H. (1973c). RNA changes in brain cells during changes in behaviour and function. *In* "Macromolecules and Behaviour" (G. B. Ansell and P. B. Bradley, Eds), pp. 51–75. University Park Press, London.

Hydén, H. and Egyhazi, E. (1962). Nuclear RNA changes of nerve cells during a learning experiment in rats. *Proc. Nat. Acad. Sci., U.S.A.* **48,** 1366–1372.

Hydén, H. and Egyhazi, E. (1963). Glial RNA changes during learning experiment in rats. *Proc. Nat. Acad. Sci., U.S.A.* **49,** 618–624.

Hydén, H. and Egyhazi, E. (1964). Changes in RNA content and base composition in cortical neurons of rats in a learning experiment involving transfer of handedness. *Proc. Nat. Acad. Sci., U.S.A.* **52,** 1030–1035.

Hydén, H. and Lange, P. W. (1968). Protein synthesis in the hippocampal pyramidal cells of rats during a behavioural test. *Science* **159,** 1370–1373.

Hydén, H. and Lange, P. W. (1970). Correlation of the S 100 brain protein with behavior. *Exptl. Cell Res.* **62,** 125–132.

Hydén, H. and Lange, P. W. (1972). Correlation of the S-100 brain protein with behavior. *In* "Macromolecules and Behavior" (J. Gaito, Ed.) 2nd Edn, pp. 131–143. Appleton-Century-Crofts, New York.

Hydén, H. and McEwen, B. (1966). A glial protein specific for the nervous system. *Proc. Nat. Acad. Sci., U.S.A.* **55,** 354–358.

Hydén, H. and Pigon, A. (1960). A cytophysiological study of the functional relationship between oligodendrial cells and nerve cells of Deiters' nucleus. *J. Neurochem.* **6,** 57–72.

Hydén, H. and Rönnbäck, L. (1975). Membrane-bound S-100 protein of nerve cells and its distribution. *Brain Res.* **100,** 615–628.

Hydén, H. Løvtrup, S. and Pigon, A. (1958). Cytochrome oxidase and succinoxidase activities in spinal ganglion cells and in glial capsule cells. *J. Neurochem.* **2,** 304–311.

Hymovich, B. (1952). The effects of environmental variations on problem solving in the rat. *J. Comp. Physiol. Psychol.* **45,** 313–321.

Ibuka, N. (1971). The differential effects of methamphetamine upon visual exploratory behavior and spontaneous motor activity in rhesus monkeys (*Macaca mulatta*). *Jap. Psychol. Res.* **13,** 26–33.

Ilyutchenok, R. Yu. (1969). Action of drugs on memory and learning. *Proc. IV Int. Congr. Pharmacol.*, Basel, pp. 55–78.

Ilyutchenok, R. Yu., Gilinsky, M. A. and Abuladze, G. V. (1973). Cholinergic mechanisms in the regulation of learning and memory. *In* "Current Biochemical Approaches to Learning and Memory" (W. B. Essman and S. Nakajima, Eds), pp. 111–131. Spectrum Publ., New York.

Im, H. S., Barnes, R. H. and Levitsky, D. A. (1971). Postnatal malnutrition and brain cholinesterase in rats. *Nature* (*Lond.*) **233,** 269–270.

Im, H. S., Barnes, R. H., Levitsky, D. A. and Pond, W. G. (1973). Postnatal malnutrition and regional cholinesterase activities in brain of pigs. *Brain Res.* **63,** 461–465.

Imura, H., Nakai, Y. and Yoshimi, T. (1973). Effect of 5-hydroxytryptophan (5-HTP) on growth hormone and ACTH release in man. *J. Clin. Endocrinol. Metab.* **36,** 204–206.

Ingoglia, N. A., Grafstein, B., McEwen, B. S. and McQuorrie, I. G. (1973). Axonal transport of radioactivity in the goldfish optic system following intraocular injection of labeled RNA precursors. *J. Neurochem.* **20,** 1605–1615.

Ingoglia, N. A., Weis, P. and Mycek, J. (1975). Axonal transport of RNA during regeneration of the optic nerves of goldfish. *J. Neurobiol.* **6,** 549–563.

Ingvar, D. H. (1973). Cerebral blood flow and metabolism in complete apallic syndromes, in states of severe dementia, and in akinetic mutism. *Acta Neurol. Scand.* **49,** 233–244.

Ingvar, D. H. and Gadea-Ciria, M. (1975). Assessment of severe damage to the brain by multiregional measurements of cerebral blood flow. *In* "Outcome of Severe Damage to the Central Nervous System", pp. 97–120, Ciba Fdn. Symp. 34 (new series). Elsevier-Excerpta Medica—North-Holland, Amsterdam.

Ingvar, D. H. and Gustafson, L. (1970). Regional cerebral blood flow in organic dementia with early onset. *Acta Neurol. Scand.* **46,** (Suppl. 43), 42–73.

Ingvar, D. H. and Schwartz, M. S. (1975). The cerebral blood flow in low pressure hydrocephalus. *In* "Intracranial Pressure II" (N. Landberg, U. Pontén and M. Brock, Eds), pp. 153–159. Springer-Verlag, Berlin.

Iqbal, K. and Tellez-Nagel, I. (1972). Isolation of neurons and glial cells from normal and pathological human brains. *Brain Res.* **45,** 296–301.

Iqbal, K., Tellez-Nagel, I. and Grundke-Iqbal, I. (1974a). Protein abnormalities in Huntington's chorea. *Brain Res.* **76,** 178–184.

Iqbal, K., Wisniewski, H. M., Shelanski, M. L., Brostoff, S., Liwnicz, B. H. and Terry, R. D. (1974b). Protein changes in senile dementia. *Brain Res.* **77,** 337–343.

Iqbal, K. Wisniewski, H. M., Grundke-Iqbal, I., Korthals, J. K. and Terry, R. D. (1975). Chemical pathology of neurofibrils: Neurofibrillary tangles of Alzheimer's presenile-senile dementia. *J. Histochem. Cytochem.* **23,** 563–569.

Irwin, L. N. (1969a). "Biochemical Changes in Stimulated Brain and Mechanisms of Information Processing". Doctoral Thesis, University of Kansas, 152 pp.

Irwin, L. N. (1969b). Protein and N-acetylneuraminic acid changes in subcellular fractions of brains of stimulated rats. *Brain Res.* **15,** 518–521.

Irwin, L. N. (1974). Glycolipids and glycoproteins in brain function. *In* "Reviews of Neuroscience" (S. Ehrenpreis and I. J. Kopin, Eds), Vol 1, pp. 137–179. Raven Press, New York.

Irwin, L. N. and Barraco, R. A. (1975). Metabolic profile of glucose incorporation into rat and pigeon brain: Time course and effect of training. *Fifth Int. Meeting, Int. Soc. Neurochem., Barcelona*, 2–6 September, Abstr. 385.

Irwin, L. N. and Samson, F. E. (1969). Ganglioside turnover in brains of stimulated and isolated rats. *Fed. Proc.* **28,** 277.

Irwin, L. N. and Samson, F. E., Jr. (1971). Content and turnover of gangliosides in rat brain following behavioural stimulation. *J. Neurochem.* **18,** 203–211.

Iversen, L. L. (1967). "The Uptake and Storage of Noradrenaline in Sympathetic Nerves". Cambridge University Press, Cambridge.

Izquierdo, I. (1974). Possible peripheral adrenergic and cholinergic mechanisms in pseudoconditioning. *Psychopharmacologia* **35,** 189–193.

Izquierdo, I. (1975a). The hippocampus and learning. *Progr. in Neurobiol.* **5,** 37–75.

Izquierdo, I. (1975b). Pharmacological observations on the role of hippocampal and autonomic pharmacology in performance and learning. *In* "Current Developments in Psychopharmacology" (W. B. Essman and L. Valzelli, Eds), Vol. 1, pp. 65–106. Spectrum Publ., Flushing, N.Y.

Izquierdo, I. and Thaddéu, R. (1975). The effect of adrenaline, tyramine and guanethidine on two-way avoidance conditioning and on pseudoconditioning. *Psychopharmacologia* **43,** 85–87.

Izquierdo, I., Marichich, E. S. and Nasello, A. G. (1969). Effect of potassium on hippocampal ribonucleic acid concentration. *Exptl. Neurol.* **25,** 259–266.

Jacobson, M. (1970a). "Developmental Neurobiology". Holt, Rinehart and Winston, New York.

Jacobson, M. (1970b). Development, specification, and diversification of neuronal connections. *In* "The Neurosciences. Second Study Program" (F. O. Schmitt, Ed.), pp. 116–129. Rockefeller University Press, New York.

Jacquet, Y. F. and Lajtha, A. (1973). Morphine action at central nervous sites in rat: Analgesia and hyperalgesia depending on site and dose. *Science* **182,** 490–492.

Jagenburg, O. R. (1965). Disturbances of amino acid metabolism in neurologic disorders. *Acta Neurol. Scand.* **41** (Suppl. 13), 339–355.

Jaffé, H. H. and Sharpless, S. K. (1968). Pharmacological denervation supersensitivity in the central nervous system: A theory of physical dependence. *In* "The Addictive States" (A. Winkler, Ed.), p. 226, Williams and Wilkins, Baltimore.

Jakoubec, B. and Semiginovsky, B. (1970). The effect of increased functional activity on the protein metabolism of the nervous system. *Int. Rev. Neurobiol.* **13,** 255–288.

Jankovic, B. D. (1972). Biological activity of antibrain antibody—an introduction to immunoneurology. *In* "Macromolecules and Behavior" (J. Gaito, Ed.) pp. 99–130. Appleton-Century-Crofts, New York.

Jankovic, B. D., Rakic, L., Veskov, R. and Horvat, J. (1968). Effect of intraventricular injection of anti-brain antibody on defensive conditioned reflexes. *Nature (Lond.)* **218,** 270–271.

Janssen, P. A. J. (1967). The pharmacology of haloperidol. *Int. J. Neuropsychiat.* **3,** (Suppl. 1) S10-S18.

Jasper, H. H., Proctor, L. D., Knighton, R. S., Noshay, W. C. and Costello, R. T. (Eds), (1958). "Reticular Formation of the Brain, Henry Ford Hospital International Symposium". Little, Brown and Co., Boston, 766 pp.

Jeffrey, P. L., James, K. A. C., Kidman, A. D., Richards, A. M. and Austin, L. (1972). The flow of mitochondria in chicken sciatic nerve. *J. Neurobiol.* **3,** 199–208.

Jelliffe, S. E. (1918). The epileptic attack in dynamic pathology. *New York Med. J.* **108,** 139–143.

John, E. R. (1967). "Mechanisms of Memory". Academic Press, New York and London, 468 pp.

Johnson, E. M., Ueda, T., Maeno, H. and Greengard, P. (1972). Adenosine 3′, 5′-monophosphate-dependent phosphorylation of a specific protein in synaptic membrane fractions from rat cerebrum. *J. Biol. Chem.* **247,** 5650–5652.

Jones, D. G. (1970). A further contribution to the study of the contact region of *Octopus* synaptosomes. *Z. Zellforsch.* **103,** 48–60.

Jones-Lecointe, A., Rose, S. P. R. and Sinha, A. K. (1976). Sub-cellular localisation of enhanced lysine incorporation into cerebral cortex proteins in dark-reared and light-exposed rats. *J. Neurochem.* **26,** 929–933.

Jouvet, M. (1972). The role of monoamines and acetylcholine-containing neurons in the regulation of the sleep-waking cycle. *Ergebn. Physiol.* **64,** 166–307.

Jung, R. and Hassler, R. (1960). The extrapyramidal motor system. *In* "Field's Handbook of Physiology, Sec. 1, Neurophysiology", Vol. II, pp. 863–927. American Physiological Society, Washington, D.C.

Jurna, I. and Grossman, W. (1976). The effect of morphine on the activity evoked in ventrolateral tract axons of the cat spinal cord. *Exptl. Brain Res.* **24,** 473–484.

Justesen, D. R., Sharp, J. C. and Porter, P. B. (1963). Self-stimulation of the caudate nucleus by instrumentally naive cats. *J. Comp. Physiol. Psychol.* **56,** 371–374.

Kahlenberg, A., Galsworthy, P. R. and Hokin, L. E. (1967). Sodium-potassium adenosine triphosphatase: Acyl phosphate "intermediate" shown to be L-glutamyl-γ-phosphate. *Science* **157,** 434–436.

Kahneman, D. (1973) "Attention and Effort". Prentice Hall, New York.

Kakiuchi, S., Rall, T. W. and McIlwain, H. (1969). The effect of electrical stimulation upon the accumulation of adenosine 3′, 5′-phosphate in isolated cerebral tissue. *J. Neurochem.* **16,** 485–491.

Kamback, M. C. (1973). The hippocampus and motivation: A re-examination. *J. Gen. Psychol.* **89,** 313–324.

Kandel, E. R. and Spencer, W. A. (1968). Cellular neurophysiological approaches in the study of learning. *Physiol. Rev.* **48,** 65–134.

Karadzić, V., Mičić, D. and Rakić, Lj. (1971). Alterations of free amino acid concentrations in cat brain induced by rapid eye movement sleep deprivation. *Experientia* **27,** 509–511.

Karczmar, A. G. and Scudder, C. L. (1967). Behavioral responses to drugs and brain catecholamine levels in mice of different strains and genera. *Fed. Proc.* **26,** 1186–1191.

Karlsson, J.-O. and Sjöstrand, J. (1968). Transport of labelled proteins in the optic nerve and tract of the rabbit. *Brain Res.* **11,** 431–439.

Karlsson, J.-O. and Sjöstrand, J. (1971a). Rapid intracellular transport of fucose-containing glycoproteins in retinal ganglion cells. *J. Neurochem.* **18,** 2209–2216.

Karlsson, J.-O. and Sjöstrand, J. (1971b). Synthesis, migration and turnover of protein in retinal ganglion cells. *J. Neurochem.* **18,** 749–767.

Karlsson, J.-O., Hansson, H.-A. and Sjöstrand, J. (1971). Effect of colchicine on axonal transport and morphology of retinal ganglion cells. *Z. Zellforsch. Mikrosk. Anat.* **115,** 265–283.

Karlsson, U. L. (1966). Three-dimensional studies of neurones in the lateral geniculate nucleus of the rat. II. Environment of perikarya and proximal parts of their branches. *J. Ultrastruct. Res.* **16,** 482–504.

Karpiak, S. E. and Rapport, M. M. (1975). Behavioral changes in 2-month-old rats following prenatal exposure to antibodies against synaptic membranes. *Brain Res.* **92,** 405–413.

Karpiak, S. E., Serokosz, M. and Rapport, M. M. (1976). Effects of antisera to S-100 protein and to synaptic membrane fraction on maze performance and EEG. *Brain Res.* **102,** 313–321.

Katz, B. (1966). "Nerve, Muscle and Synapse". McGraw-Hill, New York.

Katz, J. J. and Halstead, W.C. (1950). Protein organization and mental function. *Comp. Psychol. Monog.* **20,** 1–38.

Katzman, R., Björklund, A., Owman, Ch., Stenevi, U. and West, K. A. (1971).

Evidence for regenerative axon sprouting of central catecholamine neurons in the mesencephalon following electrolytic lesions. *Brain Res.* **25,** 579–596.

Kaufman, S. (1958). Phenylalanine hydroxylation cofactor in phenylketonuria. *Science* **128,** 1506–1507.

Kebabian, J. W., Steiner, A. L. and Greengard, P. (1975). Muscarinic cholinergic regulation of cyclic guanosine 3′, 5′-monophosphate in autonomic ganglia: Possible role in synaptic transmission. *J. Pharmacol. Exptl. Therapeut.* **193,** 474–488.

Keevil-Rogers, P. and Schnore, M. M. (1969). Short-term memory as a function of age in persons of above average intelligence. *J. Gerontol.* **24,** 184–188.

Kelleher, R. T. and Morse, W. H. (1968). Determinants of the specificity of behavioral effects of drugs. *Ergebn. Physiol.* **60,** 1–56.

Keller, H. H., Bartholini, G. and Pletscher, A. (1974). Enhancement of cerebral noradrenaline turnover by thyrotropin-releasing hormone. *Nature* (*Lond.*) **248,** 528–529.

Kellogg, C. (1976). Audiogenic seizures: Relation to age and mechanisms of monoamine neurotransmission. *Brain Res.* **106,** 87–103.

Kennedy, C., DesRosiers, M. H., Jehle, J. W., Reivich, M., Sharp, F. R. and Sokoloff, L. (1975). Mapping of functional neural pathways by autoradiographic survey of local metabolic rate with [^{14}C] deoxyglucose. *Science* **187,** 850–853.

Kenney, F. T., Hamkalo, B. A., Favelukes, G. and August, J. T. (Eds) (1973). "Gene Expression and its Regulation". Plenum Press, New York, 557 pp.

Kerkut, G. A., Emson, P. and Walker, R. J. (1973). Learning in lower animals. *In* "Macromolecules and Behaviour" (G. B. Ansell and P. B. Bradley, Eds), pp. 241–257. University Park Press, Baltimore.

Kernell, D. and Peterson, R. P. (1970). The effect of spike activity versus synaptic activity on the metabolism of ribonucleic acid in molluscan giant neurone. *J. Neurochem.* **17,** 1087–1094.

Kerr, F. W. L. (1975). Structural and functional evidence of plasticity in the central nervous system. *Exptl. Neurol.* **48,** (Part 2) 16–31.

Kesner, R. (1973). A neural system analysis of memory storage and retrieval. *Psychol. Bull.* **80,** 177–203.

Kesner, R. P. and Doty, R. W. (1968). Amnesia produced in cats by local seizure activity initiated from the amygdala. *Exptl. Neurol.* **21,** 58–68.

Kesner, R. P. and Wilburn, M. W. (1974). A review of electrical stimulation of the brain in context of learning and retention. *Behav. Biol.* **10,** 259–295.

Kety, S. S. (1955). Changes in cerebral circulation and oxygen consumption which accompany maturation and aging. *In* "Biochemistry of the Developing Nervous System" (H. Waelsch, Ed.), pp. 208–223. Academic Press, New York and London.

Kety, S. S., Javoy, F., Thierry, A. M., Julou, L. and Glowinski, J. (1967). A sustained effect of electroconvulsive shock on the turnover of norepinephrine in the central nervous system of the rat. *Proc. Nat. Acad. Sci., U.S.A.* **58,** 1249–1254.

Khan, M. A. and Ochs, S. (1975). Slow axoplasmic transport of mitochondria (MAO) and lactic dehydrogenase in mammalian nerve fibers. *Brain Res.* **96,** 267–277.

Kidd, M. (1963) Paired helical filaments in electron microscopy of Alzheimer's Disease. *Nature* (*Lond.*) **197,** 192–193.

Kim, H.-J., Miskit, D. and Routtenberg, A. (1975). Retention impairment of passive avoidance by post-trial injection of picrotoxin into the substantia nigra in rats. *Society for Neuroscience Meeting, New York,* November.

Kim, J.-S. and Hassler, R. (1975). Effects of acute haloperidol on the gamma-aminobutyric acid system in rat striatum and substantia nigra. *Brain Res.* **88,** 150–153.

Kim, J. S., Bak, I. J., Hassler, R. and Okada, Y. (1971). Role of γ-aminobutyric acid (GABA) in the extrapyramidal motor system. 2. Some evidence for the existence of a type of GABA-rich strionigral neurons. Exptl. *Brain Res.* **14,** 95–104.

Kiss, J. (1975). Dendritic protein transport in hippocampal pyramidal cells: An electron microscope autoradiography investigation. *First European Neurosciences Meeting, Munich,* 9–13 September, Abstr. No. 210.

Kissane, J. M. and Robins, E. (1958). The fluorometric measurement of DNA in animal tissues with special reference to the CNS. *J. Biol. Chem.* **233,** 184–188.

Kleihues, P. and Hossmann, K.-A. (1971). Protein synthesis in the cat brain after prolonged cerebral ischemia. *Brain Res.* **35,** 409–418.

Klein, D. C. and Weller, J. L. (1970). Indole metabolism in the pineal gland: A circadian rhythm in N-acetyltransferase. *Science* **169,** 1093–1095.

Knox, W. E. (1963). The adaptive control of tryptophan and tyrosine metabolism in animals. *Trans. N.Y. Acad. Sci.* **25,** 503–512.

Kobayashi, K. and van Hof, M. W. (1972). Visual evoked responses in rabbits deprived of light for seven months after birth. *Acta Soc. Ophthal. Jap.* **76,** 257–259.

Koe, B. K. and Weissman, A. (1966). p-Chlorophenylalanine: A specific depletor of brain serotonin. *J. Pharmacol. Exptl. Therapeut.* **154,** 499–516.

Koenig, E. (1972). Ribonucleic acid of nervous tissue. *In* "The Structure and Function of Nervous Tissue" (G. H. Bourne, Ed.), Vol. IV, pp. 179–214. Academic Press, New York and London.

Koenig, H. and Jibril, A. (1962). Acidic glycolipids and the role of ionic bonds in the structure-linked latency of lysosomal hydrolases. *Biochim. Biophys. Acta* **65,** 543–545.

Kohlenberg, R. and Trabasso, T. (1968). Recovery of a conditioned emotional response after one or two electroconvulsive shocks. *J. Comp. Physiol. Psychol.* **65,** 270–273.

Kohn, R. R. (1971). "Principles of Mammalian Aging". Prentice-Hall, Englewood Cliffs, New Jersey.

Koikegami, H. (1964). Amygdala and other related limbic structures; experimental studies on the anatomy and function. II. Functional experiments. *Acta Med. Biol.* **12,** 73–266.

Koizumi, J. and Shiraishi, H. (1970a). Ultrastructural appearance of glycogen in the hypothalamus of the rabbit following chlorpromazine administration. *Exptl. Brain Res.* **10,** 276–282.

Koizumi, J. and Shiraishi, H. (1970b). Glycogen accumulation in astrocytes of the striatum and pallidum of the rabbit following administration of psychotropic drugs. *J. Elect. Microscopy* **19,** 182–187.

Konorski, J. (1948). "Conditioned Reflexes and Neuron Organization". Cambridge University Press, Cambridge.

Konorski, J. (1961). The physiological approach to the problem of recent memory. *In* "Brain Mechanisms and Learning" (J. F. Delafresnaye, Ed.), pp. 115–132. Blackwell, Oxford.

Konrad, K. and Melzack, R. (1975). Novelty-enhancement effects associated with early sensory-social isolation. *In* "The Developmental Neuropsychology of Sensory Deprivation" (A. H. Riesen, Ed.), pp. 253–276. Academic Press, New York and London.

Koopowitz, H., Silver, D. and Rose, G. (1975). Neuronal plasticity and recovery of function in a polyclad flatworm. *Nature* (*Lond.*) **256,** 737–738.

Kopp, R., Bohdanecky, Z. and Jarvik, M. E. (1968). Proactive effect of a single electroconvulsive shock (ECS) on one-trial learning in mice. *J. Comp. Physiol. Psychol.* **65,** 514–517.

Korányi, L., Beyer, C. and Guzmán-Flores, C. (1971a). Effect of ACTH and hydrocortisone on multiple unit activity in the forebrain and thalamus in response to reticular stimulation. *Physiol. Behav.* **7,** 331–335.

Korányi, L., Beyer, C. and Guzmán-Flores, C. (1971b). Multiple unit activity during habituation, sleep-wakefulness cycle and the effect of ACTH and corticosteroid treatment. *Physiol. Behav.* **7,** 321–329.

Korf, J., Zieleman, M. and Westerink, B. H. C. (1976) Dopamine release in substantia nigra? *Nature* (*Lond.*) **260,** 257–258.

Korneliussen, H. (1972). Ultrastructure of normal and stimulated motor endplates with comments on the origin and fate of synaptic vesicles. *Z. Zellforsch.* **130,** 28–57.

Korsakoff, S. S. (1889). Étude médico-psychologique sur une forme des maladies de la mémoire. *Revue de Philosophie* **28,** 501–530.

Kosterlitz, H. W. and Hughes, J. (1975). Some thoughts on the significance of enkephalin, the endogenous ligand. *Life Sci.* **17,** 91–96.

Kramer, S. G. (1971). Dopamine: A retinal neurotransmitter. I. Retinal uptake, storage and light-stimulated release of H3-dopamine *in vivo*. *Invest. Ophthalmol.* **10,** 438–452.

Krech, D., Rosenzweig, M. R. and Bennett, E. L. (1960). Effects of environmental complexity and training on brain chemistry. *J. Comp. Physiol. Psychol.* **53,** 509–519.

Krech, D., Rosenzweig, M. R. and Bennett, E. L. (1962a). On the trail of the memory trace. *Psychologische Beitrage*, **6,** 472–483.

Krech, D., Rosenzweig, M. R. and Bennett, E. L. (1962b). Relations between brain chemistry and problem-solving among rats in enriched and impoverished environments. *J. Comp. Physiol. Psychol.* **55,** 801–807.

Krech, D., Rosenzweig, M. R. and Bennett, E. L. (1963). Effects of complex environment and blindness on rat brain. *Archs. Neurol.* **8,** 403–412.

Krech, D., Rosenzweig, M. R. and Bennett, E. L. (1966). Environmental impoverishment, social isolation and changes in brain chemistry and anatomy. *Physiol. Behav.* **1,** 99–104.

Kreutzberg, G. (1969). Neuronal dynamics and axonal flow. IV. Blockage of intra-axonal enzyme transport by colchicine. *Proc. Nat. Acad. Sci., U.S.A.* **62,** 722–728.

Kreutzberg, G. W. and Schubert, P. (1973). Neuronal activity and axonal flow. *In* "Metabolic Regulation and Functional Activity in the Central Nervous System" (H. Herken and E. Genazzani, Eds), pp. 84–93. Springer-Verlag, Berlin.

Kreutzberg, G. W. and Schubert, P. (1975). The cellular dynamics of intraneuronal transport. *In* "Use of Axonal Transport for Studies of Neuronal Connectivity" (W. M. Cowan and M. Cuénod, Eds), pp. 84–112. Elsevier, Amsterdam.

Kreutzberg, G. W., Schubert, P. and Lux, H. D. (1975a). Neuroplasmic transport in axons and dendrites. *In* "Golgi Centennial Symposium Proceedings" (M. Santini, Ed.), pp. 161–166. Raven Press, New York.

Kreutzberg, G. W., Schubert, P., Tóth, L. and Rieske, E. (1973). Intradendritic transport to postsynaptic sites. *Brain Res.* **62,** 399–404.

Kreutzberg, G. W., Tóth, L. and Kaiya, H. (1975b). Acetylcholinesterase as a marker for dendritic transport and dendritic secretion. *In* "Advances in Neurology", (G. W. Kreutzberg, Ed.), Vol. 12, pp. 269–281. Raven Press, New York.

Kripke, D. F. (1974). Ultradian rhythms in sleep and wakefulness. *In* "Advances in Sleep Research" (E. D. Weitzman, Ed.), Vol. VI, pp. 305–325. Spectrum Publ., New York.

Krivoy, W. A., Zimmerman, E. and Lande, S. (1974). Facilitation of development of resistance to morphine analgesia by desglycinamide^9-lysine-vasopressin. *Proc. Nat. Acad. Sci., U.S.A.* **71,** 1852–1856.

Kristensson, K. and Olsson, Y. (1971). Retrograde axonal transport of protein. *Brain Res.* **29,** 363–365.

Kristensson, K. and Olsson, Y. (1974). Retrograde transport of horseradish peroxidase in transected axons. I. Time relationships between transport and induction of chromatolysis. *Brain Res.* **79,** 101–109.

Krnjević, K. (1967). Chemical transmission and cortical arousal. *Anesthesiology* **28,** 100–105.

Krnjević, K. (1974). Chemical nature of synaptic transmission in vertebrates. *Physiol. Rev.* **54,** 418–540.

Krnjević, K. (1975). Coupling of neuronal metabolism and electrical activity. *In* "Brain Work, Alfred Benzon Symposium VIII", pp. 65–78. Munksgaard.

Krnjević, K. and Morris, M. E. (1975a). Correlation between extracellular focal potentials and K^+ potentials evoked by primary afferent activity. *Can. J. Physiol. Pharmacol.* **53,** 912–922.

Krnjević, K. and Morris, M. E. (1975b). Factors determining the decay of K^+ potentials and focal potentials in the central nervous system. *Can. J. Physiol. Pharmacol.* **53,** 923–934.

Krnjević, K. and Silver, A. (1965). A histochemical study of cholinergic fibres in the cerebral cortex. *J. Anat.* (*Lond.*) **99,** 711–759.

Krnjević, K., Puil, E. and Werman, R. (1975). Evidence for Ca^{2+}-activated K^+ conductance in cat spinal motoneurons from intracellular EGTA injections. *Can. J. Physiol. Pharmacol.* **53,** 1214–1218.

Kršiak, M. (1975). Timid singly-housed mice: Their value in prediction of psychotropic activity of drugs. *Brit. J. Pharmacol.* **55,** 141–150.

Krygier-Brévart, V., Weiss, D. G., Mehl, E., Schubert, P. and Kreutzberg, G. W. (1974). Maintenance of synaptic membranes by the fast axonal flow. *Brain Res.* **77,** 97–110.

Kuffler, S. W. and Nicholls, J. G. (1966). The physiology of neuroglial cells. *Ergebn. Physiol.* **57,** 1–90.

Kuhar, M. J., Pert, C. B. and Snyder, S. H. (1973). Regional distribution of opiate receptor binding in monkey and human brain. *Nature* (*Lond.*) **245,** 447–450.

Kulig, B. M. and Calhoun, W. H. (1972). Enhancement of successive discrimination reversal learning by methamphetamine. *Psychopharmacologia* **27,** 233–240.

Kuo, Z. Y. (1960) Studies on the basic factors in animal fighting: I-IV. *J. Genet. Psychol.* **96,** 201–239.

Kutner, S. J. and Brown, W. L. (1972). Types of oral contraceptives, depression, and premenstrual symptoms. *J. Nerv. Ment. Dis.* **155,** 153–162.

Kuypers, H. G. J. M. and Maisky, V. A. (1975). Retrograde axonal transport of horseradish peroxidase from spinal cord to brain stem cell groups in the cat. *Neurosci. Letters* **1,** 9–14.

Kvamme, E., Reichelt, K. L., Edminson, P. D., Sinichkin, A., Sterri, S. and Svenneby, G. (1975). N-Substituted peptides in brain. *Proceedings of the Tenth FEBS Meeting; Neurochemistry Symposium, Paris,* pp. 127–136.

Laemle, L. K. (1975). Cell populations of the lateral geniculate nucleus of the cat as determined with horseradish peroxidase. *Brain Res.* **100,** 650–656.

Lagerspetz, K. Y. H., Tirri, R. and Lagerspetz, K. M. J. (1967). Neurochemical and endocrinological studies of mice selectively bred for aggressiveness. *Rep. Inst. Psychol.* (*Turku*) **29,** 1–5.

Lagnado, J. R., Lyons, C. A., Weller, M. and Phillipson, O. (1972). The possible significance of adenosine 3′:5′-cyclic monophosphate-stimulated protein kinase activity associated with purified microtubular protein preparations from mammalian brain. *Biochem. J.* (*Lond.*) **128,** 95P.

Lahti, R. A. (1975). Alcohol, aldehydes and biogenic amines. *In* "Biochemical Pharmacology of Ethanol" (E. Majchrowicz, Ed.), pp. 239–253. Plenum Press, New York.

Lajtha, A. (1964). Alteration and pathology of cerebral protein metabolism. *Int. Rev. Neurobiol.* **7,** 1–40.

Lajtha, A. and Dunlop, D. (1974). *In* "Advances in Behavioral Biology" (A. Vernadakis and N. Weiner, Eds), Vol. 8, pp. 215–229. Plenum Press, New York.

Lajtha, A. and Sershen, H. (1975). Changes in the rates of protein synthesis in the brain of goldfish at various temperatures. *Life Sci.* **17,** 1861–1868.

Lal, H., DeFeo, J. J., Pitterman, A., Patel, G. and Baumel, I. (1972). Effects of prolonged social deprivation or enrichment on neuronal sensitivity for CNS depressants and stimulants. *In* "Drug Addiction: Experimental Pharmacology", Vol. 1, pp. 255–266. Futura Publ. Co., Mount Kisco, New York.

Lalley, P. M., Rossi, G. V. and Baker, W. W. (1970). Analysis of local cholinergic tremor mechanisms following selective neurochemical lesions. *Exptl. Neurol.* **27,** 258–275.

Lampert, P. W., Gajdusek, D. C. and Gibbs, C. J., Jr. (1972). Subacute spongiform virus encephalopathies. *Amer. J. Pathol.* **68,** 626–646.

Lampert, P. W., Gajdusek, D. C. and Gibbs, C. J., Jr. (1975). Pathology of dendrites in subacute spongiform virus encephalopathies. *In* "Advances in Neurology" (G. W. Kreutzberg, Ed.), Vol. 12, pp. 465–470. Raven Press, New York.

Lande, S., Flexner, J. B. and Flexner, L. B. (1972). Effect of corticotropin and desglycinamide^9-lysine vasopressin on suppression of memory by puromycin. *Proc. Nat. Acad. Sci., U.S.A.* **69,** 558–560.

Landis, S. C. and Bloom, F. E. (1975). Ultrastructural identification of noradrenergic boutons in mutant and normal mouse cerebellar cortex. *Brain Res.* **96,** 299–305.

Langeman, J., Shimada, M. and Rodier, P. (1972). Floxuridine and its influence on postnatal cerebellar development. *Pediatr. Res.* **6,** 758–764.

Langer, G., Heinze, G., Reim, B. and Matussek, N. (1975). Growth hormone response in affective disorders. *First European Neurosciences Meeting, Munich,* 9–13 September, Abstr. No. 233.

Lapetina, E. G., Soto, E. F. and DeRobertis, E. (1967). Gangliosides and acteylcholinesterase in isolated membranes of the rat-brain cortex. *Biochim. Biophys. Acta* **135,** 33–43.

Larochelle, L., Bédard, P., Poirier, L. J. and Sourkes, T. L. (1971). Correlative neuroanatomical and neuropharmacological study of tremor and catatonia in the monkey. *Neuropharmacology* **10,** 273–288.

Lasansky, A. (1971). Nervous function at the cellular level: Glia. *Ann. Rev. Physiol.* **33,** 241–256.

Lasek, R. J. (1967). Axoplasmic movement in cat dorsal root ganglion cells and rat ventral motoneurons. Ph.D. Thesis, State Univ. of New York, Upstate Medical Center, Syracuse.

Lasek, R. J. (1968). Axoplasmic transport in cat dorsal root ganglion cells: as studied with L-leucine-H^3. *Brain Res.* **7,** 360–377.

Lashley, K. S. (1929). "Brain Mechanisms and Intelligence: A Quantitative Study of Injuries to the Brain". Chicago University Press, Chicago.

Lashley, K. S. (1930a). Basic neural mechanisms in behavior. *Psychol. Rev.* **37,** 1–24.

Lashley, K. S. (1930b). The mechanism of vision: II. The influence of cerebral lesions upon the threshold of discrimination for brightness. *J. Genet. Psychol.* **37,** 461–480.

Lashley, K. S. (1931). Mass action in cerebral function. *Science* **73,** 245–254.

Lashley, K. S. (1935). The mechanism of vision: XII. Nervous structures concerned in habits based on reactions to light. *Comp. Psychol. Monogr.* **11,** 43–79.

Lashley, K. S. (1938). The mechanism of vision: XV. Preliminary studies of the rat's capacity for detail vision. *J. Genet. Psychol.* **18,** 123–193.

Lashley, K. S. (1942). The problem of cerebral organization in vision. *In* "Visual Mechanisms. Biol. Symp." (H. Kluver, Ed.), Vol. 7, pp. 301–322.

Lashley, K. S. (1950). In search of the engram. *In* "Symp. Soc. Exptl. Biol. No. 4", pp. 454–482. Cambridge University Press, Cambridge, England.

Lashley, K. S. (1951). Paper presented at the symposium on "The Brain and the Mind", American Neurological Association; cited by S. Cobb in Beach *et al.* (1960).

Lashley, K. S. (1958). Cerebral organization and behavior. *In* "The Brain and Behavior, Proc. Ass. Res. Nerv. Ment. Dis." **36,** 1–18.

Laursen, A. and Mosfeldt, A. (1963). Corpus striatum. *Acta Physiol. Scand.* **59,** suppl. **211,** 13–106.

LaVail, J. H. and LaVail, M. M. (1972). Retrograde axonal transport in the central nervous system. *Science* **176,** 1416–1417.

LaVail, J. H. and LaVail, M. M. (1974). The retrograde intraaxonal transport of horseradish peroxidase in the chick visual system: A light and electron microscopic study. *J. Comp. Neurol.* **157,** 303–357.

LaVail, M. M. and LaVail, J. H. (1975). Retrograde intraaxonal transport of horseradish peroxidase in retinal ganglion cells of the chick. *Brain Res.* **85,** 273–280.

LaVail, J. M., Winston, K. R. and Tish, A. (1973). A method based on retrograde intraaxonal transport of protein for identification of cell bodies terminating within the CNS. *Brain Res.* **58,** 470–477.

Laverty, R. (1974). On the roles of dopamine and noradrenaline in animal behaviour. *Progr. Neurobiol.* **3,** 31–70.

Leathwood, P., Bush, M., Berent, C. and Mauron, J. (1974). Effects of early malnutrition on Swiss white mice: Avoidance learning after rearing in large litters. *Life Sci.* **14,** 157–162.

Ledeen, R. W., Skrivanek, J. A., Tirri, L. J., Margolis, R. K. and Margolis, R. U. (1976). Gangliosides of the neuron: Localization and origin. *In press.*

Leestma, J. E. (1976). Velocity measurements of particulate neuroplasmic flow in organized mammalian CNS tissue cultures. *J. Neurobiol.* **7,** 173–183.

Lefresne, P., Guyenet, P. and Glowinski, J. (1973). Acetylcholine synthesis from [2-^{14}C]pyruvate in rat striatal slices. *J. Neurochem.* **20,** 1083–1097.

Legein, C. P. J. J. M. M. and van Hof, M. W. (1970). The effect of light deprivation on the electroretinogram of the guinea pig. *Pflüg. Arch.* **318,** 1–6.

Legrand, J. (1967). Analyse de l'action morphogénétique des hormones thyroidiennes sur le cervelet du jeune rat. *Arch. Anat. Micr. Morph. Exp.* **56,** 205–244.

Leigh, H. and Hofer, M. (1973). Behavioral and physiologic effects of littermate removal on the remaining single pup and mother during the preweaning period in rats. *Psychosomat. Med.* **35,** 497–508.

Leigh, H. and Hofer, M. A. (1975). Long-term effects of preweaning isolation from littermates in rats. *Behav. Biol.* **15,** 173–181.

Leslie, I. and Davidson, J. N. (1951). The chemical composition of the chick embryonic cell. *Biochim. Biophys. Acta* **7,** 413–428.

LeVay, S. (1975). Polyribosomes in visual cortex: Effects of deprivation and anesthesia. *5th Ann. Meeting, Soc. for Neuroscience, New York,* Vol. 1, p. 497.

Levi, G. (1970). Cerebral amino acid transport *in vitro* during development: A kinetic analysis. *Archs. Biochem. Biophys.* **138,** 347–349.

Levi, G. and Morisi, G. (1971). Free amino acids and related compounds in chick brain during development. *Brain Res.* **26,** 131–140.

Levi, L. (Ed.) (1975). *Emotions: Their Parameters and Measurement,* Raven Press, New York, 814 pp.

Levi-Montalcini, R. and Angeletti, P. U. (1968). The nerve growth factor. *Physiol. Rev.* **48,** 534–569.

Levine, S. (Ed.) (1972). "Hormones and Behavior", Academic Press, New York and London, 363 pp.

Levitan, I. B., Mushynski, W. E. and Ramirez, G. (1972a). Effects of an enriched environment on amino acid incorporation into rat brain sub-cellular fractions *in vivo*. *Brain Res.* **41,** 498–502.

Levitan, I. B., Mushynski, W. E. and Ramirez, G. (1972b). Effects of environmental complexity on amino acid incorporation into rat cortex and hippocampus *in vivo*. *J. Neurochem.* **19,** 2621–2630.

Levitan, I. B., Ramirez, G. and Mushynski, W. E. (1972c). Amino acid incorporation in the brains of rats trained to use the non-preferred paw in retrieving food. *Brain Res.* **47,** 147–156.

Levitsky, D. A. and Barnes, R. H. (1970). Effect of early malnutrition on the reaction of adult rats to aversive stimuli. *Nature* (*Lond.*) **225,** 468–469.

Levitsky, D. A. and Barnes, R. H. (1972). Nutritional and environmental interactions in the behavioral development of the rat: Long term effects. *Science* **176,** 68–71.

Lewin, B. (1974a). "Gene Expression", Vol. 1. John Wiley and Sons, London, 642 pp.

Lewin, B. (1974b). "Gene Expression", Vol. 2. John Wiley and Sons, London, 467 pp.

Lewin, E., Golden, T. and Walker, J. E. (1976). Stimulation of cyclic AMP accumulation by pentylenetetrazol. *Exptl. Neurol.* **50,** 422–426.

Lewis, P. D., Patel, A. J., Johnson, A. L. and Balázs, R. (1976). Effect of thyroid deficiency on cell acquisition in the postnatal rat brain: A quantitative histological study. *Brain Res.* **104,** 49–62.

Lewy, A. J. and Seiden, L. S. (1972). Operant behavior changes norepinephrine metabolism in rat brain. *Science* **175,** 454–456.

Lichtshtein, D. and Dobkin, J. (1976). Reduction of γ-aminobutyric acid level in the cat cerebral cortex by afferent electrical stimulation. *J. Neurochem.* **26,** 135–139.

Liddle, G. W., Island, S. and Meador, C. K. (1962). Normal and abnormal regulation of corticotropin secretion in man. *Recent Progr. Hormone Res.* **18,** 125–166.

Lidsky, A. and Slotnick, B. M. (1970). Electrical stimulation of the hippocampus and electroconvulsive shock produce similar amnestic effects in mice. *Neuropsychologia* **8,** 363–369.

Lindsley, D. B. (1951). Emotion. *In* "Handbook of Experimental Psychology" (S. S. Stevens, Ed.), pp. 473–516. Wiley, New York.

Litteria, M. and Thorner, M. W. (1976). Inhibitory action of neonatal estrogenization on the incorporation of [^{3}H] lysine into cortical neuroproteins. *Brain Res.* **103,** 584–587.

Litteria, M. and Timiras, P. S. (1970). *In vivo* inhibition of protein synthesis in specific hypothalamic nuclei by 17β-estradiol. *Proc. Soc. Exptl. Biol. Med.* **134,** 256–261.

Little, W. A. and Shaw, G. L. (1975). A statistical theory of short and long term memory. *Behav. Biol.* **14,** 115–133.

Livingston, R. B., Fulton, J. F., Delgado, J. M. R., Sachs, E., Brendler, S. J. and Davis, G. D. (1947). Functions of the orbital surface. *Res. Publ. Assoc. Nerv. Ment. Dis.* **27,** 405–420.

Llinás, R. and Nicholson, C. (1975). Calcium role in depolarization-secretion coupling: An aequorin study in squid giant synapse. *Proc. Nat. Acad. Sci., U.S.A.* **72,** 187–190.

Lloyd, K. G., Davidson, L. and Hornykiewicz, O. (1975a). The neurochemistry of Parkinson's disease: Effect of *l*-DOPA therapy. *J. Pharmacol. Exptl. Therapeut.* **195,** 453–464.

Lloyd, K. G., Möhler, H., Heitz, Ph. and Bartholini, G. (1975b). Distribution of choline acetyltransferase and glutamate decarboxylase within the substantia nigra and in other brain regions from control and Parkinsonian patients. *J. Neurochem.* **25,** 789–795.

Lorenz, K. (1966). "On Aggression" (M. K. Wilson, Trans.) Harcourt, Brace and World, Inc., New York (Originally published in 1963).

Lovell, R. A. (1971). Some neurochemical aspects of convulsions. *In* "Handbook of Neurochemistry" (A. Lajtha, Ed.), Vol. 6, pp. 63–102. Plenum Press, New York.

Lovenberg, W., Goodwin, J. S. and Wallace, E. F. (1975). Molecular properties and regulation of dopamine-β-hydroxylase. *In* "Neurobiological Mechanisms of Adaptation and Behavior" (A. J. Mandell, Ed.), pp. 77–93. Raven Press, New York.

Lowden, J. A. and Wolfe, L. S. (1964). Studies on brain gangliosides. III. Evidence for the location of gangliosides specifically in neurones. *Can. J. Biochem.* **42,** 1587–1594.

Lowe, C. R. and Dean, P. D. G. (1974). "Affinity Chromatography". John Wiley and Sons, London, 272 pp.

Lowry, O. H., Rosebrough, N. J., Farr, A. L. and Randall, R. J. (1951). Protein measurement with the folin phenol reagent. *J. Biol. Chem.* **193,** 265–275.

Lubar, J. F. and Numan, R. (1973). Behavioral and physiological studies of septal function and related medial cortical structures. *Behav. Biol.* **8,** 1–25.

Lubinska, L. (1964). Axoplasmic streaming in regenerating and in normal nerve fibers. *In* "Progress in Brain Research, Mechanisms of Neural Regeneration" (M. Singer and J. P. Schadé, Eds), Vol. 13, pp. 56–66. Elsevier, Amsterdam.

Lubinska, L. (1975). On axoplasmic flow. *Int. Rev. Neurobiol.* **17,** 241–296.

Luco, J. V. and Eyzaguirre, C. (1955). Fibrillation and hypersensitivity to ACh in denervated muscle: Effect of length of degenerated nerve fibers. *J. Neurophysiol.* **18,** 65–73.

Lundberg, N., Pontén, U. and Brock, M. (Eds) (1975). "Intracranial Pressure II". Springer-Verlag, Berlin, 527 pp.

Lupo, C. and Dessi-Fulgheri, F. (1975). Effect of long-term isolation on *in vitro* steroidogenesis by the brain of aged rat. *First European Neurosciences Meeting, Munich*, 9–13 September, Abstr. No. 257.

Lupo, C., Gobbetti, A., Dessi-Fulgheri, F. and Valeri, P. (1975). *In vivo* metabolism of corticosterone in rat brain. *Fed. Eur. Biochem. Soc., 10th Meeting, Pargi*, Abstr. No. 1371.

Lux, H. D., Schubert, P., Kreutzberg, G. W. and Globus, A. (1970). Excitation and axonal flow: Autoradiographic study on motoneurons intracellularly injected with ^{3}H-amino acid. *Exptl. Brain Res.* **10,** 197–204.

Lynch, G., Brecha, N., Cotman, C. W. and Globus, A. (1974). Spine loss and regrowth in hippocampus following deafferentation. *Nature* (*Lond.*) **248,** 71–73.

Lynch, G., Smith, R. L., Browning, M. D. and Deadwyler, S. (1975). Evidence for bidirectional dendritic transport of horseradish peroxidase. *In* "Advances in Neurology; Physiology and Pathology of Dendrites" (G. W. Kreutzberg, Ed.), Vol. 12, pp. 297–313. Raven Press, New York.

Maas, J. W. (1962). Neurochemical differences between two strains of mice. *Science* **137,** 621–622.

Maccioni, A. H. R., Gimenez, M. S. and Caputto, R. (1971). The labelling of the gangliosidic fraction from brains of rats exposed to different levels of stimulation after injection of [6-^{3}H]-glucosamine. *J. Neurochem.* **18,** 2363–2370.

Machiyama, Y., Utena, H. and Kikuchi, M. (1970). Behavioral disorders in Japanese monkeys produced by the long-term administration of methamphetamine. *Proc. Japan Acad.* **46,** 738–743.

Machiyama, Y., Hsu, S. C., Utena, H., Katagiri, M. and Hirata, A. (1974). Aberrant social behaviour induced in monkeys by the chronic methamphetamine administration as a model for schizophrenia. *In* "Biological Mechanisms of Schizophrenia and Schizophrenia-like Psychoses" (H. Mitsuda and T. Fukuda, Eds), pp. 97–105. Igaku Shoin Ltd., Tokyo.

Machlus, B. (1971). Phosphorylation of Nuclear Proteins During Behavior of Rats. Ph.D. Thesis, Univ. of North Carolina, Chapel Hill.

Machlus, B. J., Wilson, J. E. and Glassman, E. (1974a). Brain phosphoproteins: The effect of short experiences on the phosphorylation of nuclear proteins of rat brain. *Behav. Biol.* **10,** 43–62.

Machlus, B. J., Entingh, D., Wilson, J. E. and Glassman, E. (1974b). Brain phosphoproteins and the effect of various behaviors and reminding experiences on the incorporation of radioactive phosphate into nuclear proteins. *Behav. Biol.* **10,** 63–73.

MacInnes, J. W., McConkey, E. H. and Schlesinger, K. (1970). Changes in brain polyribosomes following an electro-convulsive seizure. *J. Neurochem.* **17,** 457–460.

MacLean, P. D. (1958). Contrasting functions of limbic and neocortical systems of the brain and their relevance to psychophysiological aspects of medicine. *Amer. J. Med.* **26,** 611–626.

MacPherson, F. C. and Chinerman, J. (1970). Effect of intraventricular injections of brain iso-antibodies on learning. *Exptl. Neurol.* **31,** 45–52.

MacPherson, F. C. and Shek, R. (1970). Effect of brain iso-antibodies on learning and memory in the rat. *Exptl. Neurol.* **29,** 1–15.

Maengwyn-Davies, G. D., Johnson, D. G., Thoa, N. B., Weise, V. K. and Kopin, I. J. (1973). Influence of isolating and of fighting on adrenal tyrosine hydroxylase and phenylethanolamine-N-methyltransferase activities in three strains of mice. *Psychopharmacologia* **28,** 339–350.

Maeno, H. and Greengard, P. (1972). Phosphoprotein phosphatases from rat cerebral cortex: Subcellular distribution and characterization. *J. Biol. Chem.* **247,** 3269–3277.

Maeno, H., Johnson, E. M. and Greengard, P. (1971). Subcellular distribution of adenosine 3′, 5′-monophosphate-dependent protein kinase in rat brain. *J. Biol. Chem.* **246,** 134–142.

Maeno, H., Ueda, T. and Greengard, P. (1975). Adenosine 3′: 5′-monophosphate-dependent protein phosphatase activity in synaptic membrane fractions. *J. Cyclic Nucl. Res.* **1,** 37–48.

Mahler, H. R. (1972). Nucleic acids and brain function. *In* "Basic Neurochemistry" (R. W. Albers, G. J. Siegel, R. Katzman and B. W. Agranoff, Eds), pp. 245–265. Little, Brown and Co., Boston.

Maizel, J. V., Jr. (1971). Polyacrylamide gel electrophoresis of viral proteins. *In* "Methods in Virology" (K. Maramorosch and H. Koprowski, Eds), Vol. V, pp. 179–246. Academic Press, New York, and London.

Maj, J., Baran, L., Grabowska, M. and Sowinska, H. (1973). Effect of clonidine on the 5-hydroxytryptamine and 5-hydroxyindoleacetic acid brain levels. *Biochem. Pharmacol.* **22,** 2679–2683.

Maker, H. S. and Lehrer, G. M. (1972). Carbohydrate chemistry of brain. *In* "Basic Neurochemistry" (R. W. Albers, G. J. Siegel, R. Katzman and B. W. Agranoff, Eds), pp. 169–189. Little, Brown and Company, Boston.

Malin, D. H. (1974). Synthetic scotophobin: Analysis of behavioral effects on mice. *Pharmacol. Biochem. Behav.* **2,** 147–153.

Malitz, S. (Ed.) (1972). "L-DOPA and Behavior". Raven Press, New York, 144 pp.

Malkasian, D. R. and Diamond, M. C. (1971). The effects of environmental manipulation on the morphology of the neonate rat brain. *Int. J. Neurosci.* **2,** 161–170.

Malkinson, A. M., Krueger, B. K., Rudolph, S. A., Casnellie, J. E., Haley, B. E. and Greengard, P. (1975). Widespread occurrence of a specific protein in vertebrate tissues and regulation by cyclic AMP of its endogenous phosphorylation and dephosphorylation. *Metabolism* **24,** 331–341.

Mandel, P., Bieth, R. and Stoll, E. (1948). Biochemical evolution of the brain of a chicken embryo. The nucleic acids. *Compt. Rend. Soc. Biol.* **142,** 1020–1022.

Mandel, P., Rein, H., Harth-Edel, S. and Mardell, R. (1964). Distribution and metabolism of ribonucleic acid in the vertebrate central nervous system. *In* "Comparative Neurochemistry" (D. Richter, Ed.), pp. 149–163. Pergamon Press, New York.

Mandel, P., DiCarlo, R., Simler, S. and Randrianarisoa, H. (1973). Specific and nonspecific RNA synthesis in relation to behavioral changes. *In* "Macromolecules and Behaviour" (G. B. Ansell and P. B. Bradley, Eds), pp. 163–179. University Park Press, London.

Mandell, A. J. (1970). Drug induced alterations in brain biosynthetic enzyme activity—a model for adaptation to the environment by the central nervous system. *In* "Biochemistry of Brain and Behavior", pp. 97–121. Plenum Press, New York.

Mandell, A. J. (Ed.) (1975a). "Neurobiological Mechanisms of Adaptation and Behavior". Raven Press, New York.

Mandell, A. J. (1975b). Neurobiological mechanisms of presynaptic metabolic adaptation and their organization: Implications for a pathophysiology of the affective disorders. *In* "Neurobiological Mechanisms of Adaptation and Behavior" (A. J. Mandell, Ed.), pp. 1–32. Raven Press, New York.

Mandell, A. J. and Morgan, M. (1970). Amphetamine induced increase in tyrosine hydroxylase activity. *Nature (Lond.)* **227,** 75–76.

Mann, G. (1894). Histological changes induced in sympathetic, motor, and sensory nerve cells by functional activity. (Preliminary note.) Presented at the Scottish Micr. Soc. under the title: "What alterations are produced in nerve cells by work?" May 18.

Manocha, S. L. and Perachio, A. A. (1973). Histochemical correlates of changes in the primate brain associated with varying environmental light conditions. *Histochemie* **34,** 217–225.

Mao, C. C., Guidotti, A. and Costa, E. (1974). Interactions between γ-aminobutyric acid and guanosine cyclic 3′, 5′-monophosphate in rat cerebellum. *Molec. Pharmacol.* **10,** 736–745.

Marangos, P., Zomzely-Neurath, C., York, C. and Bondy, S. C. (1975). Axoplasmic transport of a brain-specific soluble protein. *Biochim. Biophys. Acta* **392,** 75–81.

Marcucci, F., Mussini, E., Valzelli, L. and Garattini, S. (1968). Decrease in N-acetyl-L-aspartic acid in brain of aggressive mice. *J. Neurochem.* **15,** 53–54.

Marin-Padilla, M. (1967). Number and distribution of the apical dendritic spines of the layer V pyramidal cells in man. *J. Comp. Neurol.* **131,** 475–489.

Marin-Padilla, M. (1972). Structural abnormalities of the cerebral cortex in human chromosomal aberrations. A Golgi study. *Brain Res.* **44,** 625–629.

Marin-Padilla, M. (1974). Structural organization of the cerebral cortex (motor area) in human chromosomal aberrations. A Golgi study. 1. D_1 (13–15) trisomy, Patau syndrome. *Brain Res.* **66,** 373–391.

Mark, R. F. and Watts, M. E. (1971). Drug inhibition of memory formation in chickens. I. Long-term memory. *Proc. Roy. Soc. B.* **178,** 439–454.

Martin, J. B. (1973). Neural regulation of growth hormone secretion. *New Engl. J. Med.* **288,** 1384–1393.

Martini, L. and Ganong, W. F. (Eds) (1976). "Frontiers in Neuroendocrinology", Vol. 4. Raven Press, New York, 304 pp.

Mason, W. A. (1963). Social development of rhesus monkeys with restricted social experience. *Perceptual and Motor Skills* **16,** 263–270.

Mason, W. A. (1968). Early social deprivation in the nonhuman primates: Implications for human behavior. *In* "Environmental Influences" (D. C. Glass, Ed.), pp. 70–101. Rockefeller University Press, New York.

Massey, P. F., Cummins, R. A., Walsh, R. N., Budtz-Olsen, O. E. and Vine, P. (1973). Relative increases in total cortex volume in environmentally stimulated rats. *Proc. Aust. Soc. Med. Res.* **3,** 173.

Massieu, G. H., Ortega, B. G., Syrguin, A. and Tuena, M. (1962). Free amino acids in brains and liver of deoxypyridoxine-treated mice subject to insulin shock. *J. Neurochem.* **9,** 143–151.

Mattsson, C. and Heilbronn, E. (1975). The nicotinic acetylcholine receptor; a glycoprotein. *J. Neurochem.* **25,** 899–901.

Maxson, S. C., Sze, P. Y. and Towle, A. (1974). Protein synthesis and the sensory induction of susceptibility to audiogenic seizures in C57BL/6 mice. *Soc. for Neurosci., 4th Ann. Meeting, St. Louis,* October, 20–24, p. 328.

Mayer, D. J. (1975). Pain inhibition by electrical brain stimulation: Comparison to morphine. *Neurosci. Res. Progr. Bull.* **13,** 94–100.

Mehraein, P., Yamada, M. and Tarnowska-Dziduszko, E. (1975). Quantitative study on dendrites and dendritic spines in Alzheimer's disease and senile dementia. *In* "Advances in Neurology; Physiology and Pathology of Dendrites" (G. W. Kreutzberg, Ed.), Vol. 12, pp. 453–458. Raven Press, New York.

Meier-Ruge, W. (1976). Experimental pathology and pharmacology in brain research and aging. *Life Sci.* **17,** 1627–1636.

Meissner, W. W. (1966). Special review. Hippocampal functions in learning. *J. Psychiat. Res.* **4,** 235–304.

Meister, A. (1973). On the enzymology of amino acid transport. *Science* **180,** 33–39.

Melzack, R. (1954). "The Effects of Early Experience on the Emotional Responses to Pain". Ph.D. Thesis, McGill University.

Melzack, R. (1959). The role of early experience in emotional arousal. *Ann. N.Y. Acad. Sci.* **159,** 721–730.

Meyer, V. and Knobil, E. (1967). Growth hormone secretion in the unanesthetized rhesus monkey in response to noxious stimuli. *Endocrinology* **80,** 163–171.

Miani, N., DiGirolamo, A. and DiGirolamo, M. (1966). Sedimentation characteristics of axonal RNA in rabbit. *J. Neurochem.* **13,** 755–759.

Miani, N., DeRenzis, G., Michetti, F., Correr, S., Oliveri Sangiacomo, C. and Caniglia, A. (1972). Axonal transport of S-100 protein in mammalian nerve fibres. *J. Neurochem.* **19,** 1387–1394.

Miani, N., Michetti, F., DeRenzis, G. and Caniglia, A. (1973). Effect of a brain-specific protein (S-100 protein) on the nucleolar RNA-polymerase activity in isolated brain nuclei. *Experientia* **29,** 1499–1501.

Michaelis, E. K. (1975). Partial purification and characterization of a glutamate-binding membrane glycoprotein from rat brain. *Biochem. Biophys. Res. Commun.* **65,** 1004–1012.

Michaelis, E. K., Michaelis, M. L. and Boyarsky, L. L. (1974). High-affinity glutamic acid binding to brain synaptic membranes. *Biochim. Biophys. Acta* **367,** 338–348.

Michetti, F., Miani, N., DeRenzis, G., Caniglia, A. and Correr, S. (1974). Nuclear localization of S-100 protein. *J. Neurochem.* **22,** 239–244.

Michetti, F., DeRenzis, G., Donato, R. and Miani, N. (1976). Brain-specific effect of S-100 protein on the RNA-polymerase I activity in isolated nuclei. *Brain Res.* **105,** 372–375.

Mihailovic, L. and Curpic, D. (1971). Epileptiform activity evoked by intracerebral injection of antibrain antibodies. *Brain Res.* **32,** 72–97.

Mihailovic, L. and Hydén, H. (1969). On antigenic differences between nerve cells and glia. *Brain Res.* **16,** 243–256.

Mihailovic, L. and Jankovic, B. D. (1961). Effects of intraventricularly injected anti-n. caudatus antibody on the electrical activity of the cat brain. *Nature* (*Lond.*) **192,** 665–666.

Miller, L. H., Kastin, A. J., Sandman, C. A., Fink, M. and Van Veen, W. J. (1974). Polypeptide influences on attention, memory and anxiety in man. *Pharmacol. Biochem. Behav.* **2,** 663–668.

Miller, N. E. (1968). Les périodes critiques du cerveau. *Impact: science et société,* vol. XVIII, 175–186.

Miller, N. E. (1972). Interactions between learned and physical factors in mental illness. *Seminars in Psychiatry* **4,** 239–254.

Miller, N. E. (1975). Applications of learning and biofeedback to psychiatry and medicine. *In* "Comprehensive Textbook of Psychiatry/II" (A. M. Freedman. H. I. Kaplan and B. J. Sadock, Eds), pp. 349–365. Academic Press, New York and London.

Miller, N. E. and Kessen, M. L. (1952). Reward effects of food via stomach fistula compared with those via mouth. *J. Comp. Physiol. Psychol.* **45,** 555–564.

Milner, B. (1970). Memory and the medial temporal regions of the brain. *In* "Biology of Memory" (K. H. Pribram, and D. E. Broadbent, Eds), pp. 29–50. Academic Press, New York and London.

Milner, B. and Teuber, H. L. (1968). Alteration of perception and memory in man: Reflections on methods. *In* "Analysis of Behavioral Change" (L. Weiskrantz, Ed.), pp. 268–375. Harper, New York.

Milner, P. M. (1970). "Physiological Psychology". Holt, Rinehart and Winston, Inc., New York, 531 pp.

Miquel, J. and Haymaker, W. (1965). Astroglial reactions to ionizing radiation with emphasis on glycogen accumulation. *In* "Biology of Neuroglia, Progress in Brain Research" (E. D. P. DeRobertis and A. Carrea, Eds), Vol. 15, pp. 89–114. Elsevier, Amsterdam.

Miyamoto, E., Kuo, J. F. and Greengard, P. (1969a). Adenosine 3′, 5′-monophosphate-dependent protein kinase from brain. *Science* **165,** 63–65.

Miyamoto, E., Kuo, J. F. and Greengard, P. (1969b). Cyclic nucleotide-dependent protein kinases. III. Purification and properties of adenosine 3′, 5′-monophosphate-dependent protein kinase from bovine brain. *J. Biol. Chem.* **244,** 6395–6402,

Model, P. G., Highstein, S. M. and Bennett, M. V. L. (1975). Depletion of vesicles and fatigue of transmission at a vertebrate central synapse. *Brain Res.* **98,** 209–228.

Modigh, K. (1974). Functional aspects of 5-hydroxytryptamine turnover in the central nervous system. *Acta Physiol. Scand. Suppl.* **403,** 1–56.

Moisset, B., Hendley, E. D. and Welch, B. L. (1975). Norepinephrine uptake by cerebral synaptosomes of mouse: Strain differences. *Brain Res.* **92,** 157–164.

Molinoff, P. B. and Axelrod, J. (1971). Biochemistry of catecholamines. *Ann. Rev. Biochem.* **40,** 465–500.

Møllgard, K., Diamond, M. C., Bennett, E. L., Rosenzweig, M. R. and Lindner, B. (1971). Quantitative synaptic changes with differential experience in rat brain. *Int. J. Neurosci.* **2,** 113–127.

Monniér, M., Sauer, R. and Hatt, A. M. (1970). The activating effect of histamine on the central nervous system. *Int. Rev. Neurobiol.* **12,** 265–305.

Moore, B. W. (1972). Chemistry and biology of two proteins, S-100 and 14-3-2, specific to the nervous system. *Int. Rev. Neurobiol.* **15,** 215–225.

Moore, B. W. (1973). Brain-specific proteins. *In* "Proteins of the Nervous System" (D. J. Schneider, R. H. Angeletti, R. A. Bradshaw, A. Grasso and B. W. Moore, Eds), pp. 1–12. Raven Press, New York.

Moore, B. W. and McGregor, D. (1965). Chromatographic and electrophoretic fractionation of soluble proteins of brain and liver. *J. Biol. Chem.* **240,** 1647–1653.

Moore, K. E. (1968). Studies with chronically isolated rats: Tissue levels and urinary excretion of catecholamines and plasma levels of corticosterone. *Can. J. Physiol. Pharmacol.* **46,** 553–558.

Moore, K. E., Sawdy, L. C. and Shaul, S. R. (1965). Effects of d-amphetamine on blood glucose and tissue glycogen levels of isolated and aggregated mice. *Biochem. Pharmacol.* **14,** 197–204.

Morden, B., Conner, R., Mitchell, G., Dement, W. and Levine, S. (1968). Effects of rapid eye movement (REM) sleep deprivation on shock-induced fighting. *Physiol. Behav.* **3,** 425–432.

Morest, D. K. (1971). Dendrodendritic synapses of cells that have axons: The fine structure of the Golgi type II cell in the medial geniculate body of the cat. *Z. Anat. Entwicklungsgesch.* **133,** 216–246.

Morgan, C. T. (1943). "Physiological Psychology". McGraw-Hill, New York.

Morgan, I. (1970). Protein synthesis in brain mitochondrial and synaptosomal preparations. *FEBS letters* **10,** 273–275.

Morgan, I. G., Breckenridge, W. C., Vincendon, G. and Gombos, G. (1973). The proteins of nerve-ending membranes. *In* "Proteins of the Nervous System" (D. J. Schneider, Ed.), pp. 171–192. Raven Press, New York.

Morgan, M. J. (1975). The effects of isolation rearing in the rat upon inhibitory phenomena. *First European Neurosciences Meeting, Munich,* 9–13 September, Abstr. No. 286.

Morgan, M. J. and Einon, D. F. (1976). Activity and exploration in thyroid-deficient and socially-isolated rats. *Physiol. Behav.* **16,** 107–110.

Morgane, P. J. and Stern, W. C. (1974). Chemical anatomy of brain circuits in relation to sleep and wakefulness. *In* "Advances in Sleep Research" (E. Weitzman, Ed.), Vol. 1, pp. 1–131. Spectrum Publ., Inc., New York.

Morrell, F. (1969). Physiology and histochemistry of the mirror focus. *In* "Basic Mechanisms of the Epilepsies" (H. H. Jasper, A. A. Ward and A. Pope, Eds), pp. 357–370. Little, Brown and Co., Boston.

Moss, R. L. and McCann, S. M. (1973). Induction of mating behavior in rats by luteinizing hormone releasing factor. *Science* **181,** 177–179.

Motta, M., Crosignani, P. G. and Martini, L. (Eds) (1975). "Hypothalamic Hormones". Academic Press, New York and London, 352 pp.

Mountcastle, V. B. and Powell, T. P. S. (1959). Neural mechanisms subserving cutaneous sensibility, with special reference to the role of afferent inhibition in sensory perception and discrimination. *Bull. Johns Hopkins Hospital* **105,** 201–232.

Mouren-Mathieu, A.-M. and Colonnier, M. (1969). The molecular layer of the adult cat cerebellar cortex after lesions of the parallel fibres: An optic and electron microscope study. *Brain Res.* **16,** 307–323.

Mueller, P. S., Heninger, G. R. and McDonald, R. K. (1969). Insulin tolerance test in depression. *Archs. Gen. Psychiat.* **21,** 587–594.

Mueller, R. A., Thoenen, H. and Axelrod, J. (1969a). Increase in tyrosine hydroxylase activity after reserpine administration. *J. Pharmacol. Exptl. Therapeut.* **169,** 74–79.

Mueller, R. A., Thoenen, H. and Axelrod, J. (1969b). Inhibition of trans-synaptically increased tyrosine hydroxylase activity by cycloheximide and actinomycin D. *Molec. Pharmacol.* **5,** 463–469.

Mueller, R. A., Thoenen, H. and Axelrod, J. (1970). Inhibition of neuronally induced tyrosine hydroxylase by nicotinic receptor blockade. *Eur. J. Pharmacol.* **10,** 51–56.

Müller, J. (1826). "Zut verleichenden Physiologie des Gesichtsinnes des Menschen und der Tiere". Knobloch, Leipzig.

Muller, P. B. and Langemann, H. (1962). Distribution of glutamic acid decarboxylase activity in human brain. *J. Neurochem.* **9,** 399–401.

Munkvad, I., Pakkenberg, H. and Randrup, A. (1968). Aminergic systems in basal ganglia associated with stereotyped hyperactive behaviour and catalepsy. *Brain, Behav. Evolut.* **1,** 89–100.

Murphey, R. K. and Matsumoto, S. G. (1976). Experience modifies the plastic properties of identified neurons. *Science* **191**, 564–566.

Murthy, M. R. V., Roux, H., Bharucha, A. D. and Charbonneau, R. (1976). Alterations in the structure and function of free and membrane bound polysomes and m-RNA of neuronal and glial enriched fractions of rat cerebral cortex during light deprivation and exposure to lights of different wavelengths. *In* "Behavioral Neurochemistry" (J. M. R. Delgado and F. V. DeFeudis, Eds), pp. 85–111. Spectrum Publ., Holliswood, N.Y. (In press).

Mushynski, W. E., Levitan, I. B. and Ramirez, G. (1973). Competition hybridization studies on brain ribonucleic acid from rats reared in enriched and deprived environments. *J. Neurochem.* **20,** 309–317.

McAfee, D. A. and Greengard, P. (1972). Adenosine 3′, 5′-monophosphate: Electrophysiological evidence for a role in synaptic transmission. *Science* **178,** 310–312.

McCouch, G. P., Austin, G. M., Liu, C. N. and Liu, C. Y. (1958). Sprouting as a cause of spasticity. *J. Neurophysiol.* **21,** 205–216.

McEwen, B. S. and Grafstein, B. (1968). Fast and slow components in axonal transport of protein. *J. Cell Biol.* **38,** 494–508.

McGaugh, J. L. (1966). Time dependent processes in memory storage. *Science* **153,** 1351–1358.

McGaugh, J. L. (1973). Drug facilitation of learning and memory. *Ann. Rev. Pharmacol.* **13,** 229–241.

McGaugh, J. L. and Dawson, R. G. (1971). Modification of memory storage processes. *Behav. Sci.* **16,** 45–63.

McGaugh, J. L. and Gold, P. E. (1974). The effect of drugs and electrical stimulation of the brain on memory storage processes. *Adv. Behav. Biol.* **10,** 189–206.

McGaugh, J. L. and Herz, M. J. (1972). "Memory Consolidation". Albion Publ. Co., San Francisco.

McGeer, P. L. and McGeer, E. G. (1976). Enzymes associated with the metabolism of catecholamines, acetycholine and GABA in human controls and patients with Parkinson's disease and Huntington's chorea. *J. Neurochem.* **26,** 65–76.

McGeer, E. G., Fibiger, H. C., McGeer, P. L. and Brooke, S. (1973). Temporal changes in amine synthesizing enzymes of rat extrapyramidal structures after hemitransections or 6-hydroxydopamine administration. *Brain Res.* **52,** 289–300.

McGeer, P. L., McGeer, E. G., Wada, J. A. and Jung, E. (1971). Effect of globus pallidus lesions and Parkinson's disease on brain glutamic acid decarboxylase. *Brain Res.* **32,** 425–431.

McGee-Russell, S. M., Brown, A. W. and Brierley, J. B. (1970). A combined light and electron microscope study of early anoxic-ischaemic cell change in rat brain. *Brain Res.* **20,** 193–200.

McIlwain, H. (1963). "Chemical Exploration of the Brain". Elsevier, Amsterdam.

McIlwain, H. (1966). "Biochemistry and the Central Nervous System". Churchill, London.

McIlwain, H. (1974). Regulatory significance of the release and action of adenine derivatives in cerebral systems. *Biochem. Soc. Symp.* **36,** 69–85.

McIlwain, H. and Bachelard, H. S. (1971). "Biochemistry and the Central Nervous System". Churchill Livingstone, Edinburgh.

McInnes, J. W. and Luttges, M. W. (1973). Interactive effects of cycloheximide and puromycin in altering brain polyribosomes and neural and behavioral responses to electroshock in mice. *J. Neurochem.* **21,** 775–781.

McKenzie, G. M. (1971). Apomorphine-induced aggression in the rat. *Brain Res.* **34,** 323–330.

McKenzie, G. M. and Viik, K. (1975). Chemically induced choreiform activity: Antagonism by GABA and EEG patterns. *Exptl. Neurol.* **46,** 229–243.

McKinney, T. D. (1970). Brain cholinesterase in grouped and singly caged adrenal-demedullated rats. *Amer. J. Physiol.* **219,** 331–334.

McKinney, W. T. (1974). Primate social isolation. *Archs. Gen. Psychiat.* **31,** 422–426.

McKinney, W. T., Suomi, S. J. and Harlow, H. F. (1972). Vertical-chamber confinement of juvenile-age Rhesus monkeys. *Arch. Gen. Psychiat.* **26,** 223–228.

McKinney, W. T., Jr., Young, L. D., Suomi, S. J. and Davis, J. M. (1973). Chlorpromazine treatment of disturbed monkeys. *Arch. Gen. Psychiat.* **29,** 490–494.

McLennan, H. (1970). "Synaptic Transmission", 2nd Edn. W. B. Saunders, Philadelphia, 178 pp.

Nadler, J. V., Vaca, K. W., White, W. F., Lynch, G. S. and Cotman, C. W. (1976). Aspartate and glutamate as possible transmitters of excitatory hippocampal afferents. *Nature* (*Lond.*) **260,** 538–540.

Nakajima, S. (1969). Interference with relearning in the rat after hippocampal injection of actinomycin D. *J. Comp. Physiol. Psychol.* **67,** 457–461.

Nakajima, S. (1972). Proactive effect of actinomycin D on maze performance in the rat. *Physiol. Behav.* **8,** 1063–1067.

Nakajima, S. (1973). Biochemical disruption of memory: A re-examination. *In* "Current Biochemical Approaches to Learning and Memory" (W. B. Essman and S. Nakajima, Eds), pp. 133–146. Spectrum Publ., Flushing, New York.

Nakajima, S. and Essman, W. B. (1973). Biochemical studies of learning and memory: An historical overview. *In* "Current Biochemical Approaches to Learning and Memory" (W. B. Essman and S. Nakajima, Eds), pp. 1–27. Spectrum Publ., Flushing, New York.

Nakamura, R., Cheng, S.-C. and Naruse, H. (1970). A study on the precursors of the acetyl moiety of acetylcholine in brain slices. Observations on the compartmentalization of the acetyl-coenzyme A pool. *Biochem. J.* (*Lond.*) **118,** 443–450.

Nakamura, Y. (1975). An electron microscope study of the red nucleus in the cat, with special reference to the quantitative analysis of the axosomatic synapses. *Brain Res.* **94,** 1–17.

Nakamura, Y., Mizuno, N., Konishi, A. and Sato, M. (1974). Synaptic reorganization of the red nucleus after chronic deafferentation from cerebellorubral fibers: An electron microscopic study in the cat. *Brain Res.* **82,** 298–301.

Nasello, A. G. and Izquierdo, I. (1969). Effect of learning and of drugs on the ribonucleic acid concentration of brain structures of the rat. *Exptl. Neurol.* **23,** 521–528.

Nathanson, N., Cole, G. H. and Van der Loos, H. (1969). Heterotopic cerebellar granule cells following administration of cyclophosphamide to suckling rats. *Brain Res.* **15,** 532–536.

Neale, J. H., Klinger, P. D. and Agranoff, B. W. (1973a). Temperature-dependent consolidation of puromycin-susceptible memory in the goldfish. *Behav. Biol.* **9,** 267–278.

Neale, J. H., Klinger, P. D. and Agranoff, B. W. (1973b). Camptothecin blocks memory of conditioned avoidance in the goldfish. *Science* **179,** 1243–1246.

Nicholson, A. N., Freeland, S. A. and Brierley, J. B. (1970). A behavioral and neuro-pathological study of the sequelae of profound hypoxia. *Brain Res.* **22,** 327–345.

Nishioka, H. (1960). *J. Okayama Med. Soc.*, **72,** 1293, cited by Ghazal *et al.*, (1976).

Nissl, F. (1891). Ueber die Veränderungen der Ganglienzellen am Facialiskern des Kaninchens nach Ausreissung der Nerven. *Allg. Z. Psychiat.* **48,** 197–198.

Nissl, F. (1894). Ueber eine neue Untersuchungsmethode des Centralorgans speciell zur Feststellung der Localisation der Nervenzellen. *Cbl. Nervenheilk. Psychiat.* **17,** 337–344.

Nobin, A., Baumgarten, H. G., Bjorklund, A., Lachenmayer, L. and Stenevi, U. (1973). Axonal degeneration and regeneration of the bulbospinal indolamine neurons after 5,6-dihydroxytryptamine treatment. *Brain Res.* **56,** 1–24.

Norton, W. T. and Poduslo, S. E. (1970). Neuronal soma and whole neuroglia of rat brain: A new isolation technique. *Science* **167,** 1144–1146.

Nyhan, W. L. (Ed.) (1967). "Amino Acid Metabolism and Genetic Variation". McGraw-Hill, New York.

Obrig, T. G., Culp, W. J., McKeehan, W. L. and Hardesty, B. (1971). The mechanism by which cycloheximide and related glutarimide antibiotics inhibit peptide synthesis on reticulocyte ribosomes. *J. Biol. Chem.* **246,** 174–181.

Ochs, S. (1962). The nature of spreading depression in neuronal networks. *Int. Rev. Neurobiol.* **4,** 1–69.

Ochs, S. (1971). Characteristics and a model for the fast axoplasmic transport in nerve. *J. Neurobiol.* **2,** 331–345.

Ochs, S. (1972a). Fast transport of materials in mammalian nerve fibers. *Science* **176,** 252–260.

Ochs, S. (1972b). Axoplasmic flow—the fast transport system in mammalian nerve fibers. *In* "Macromolecules and Behavior" (J. Gaito, Ed.), 2nd Edn, pp. 147–166. Appleton-Century-Crofts, New York.

Ochs, S. and Hunt, K. (1960). Apical dendrites and propagation of spreading depression in cerebral cortex. *J. Neurophysiol.* **23,** 432–444.

Ochs, S. and Ranish, N. (1970). Metabolic dependence of fast axoplasmic transport in nerve. *Science* **167,** 878–879.

Ochs, S., Sabri, M. I. and Ranish, N. (1970). Somal site of synthesis of fast transported materials in mammalian nerve fibers. *J. Neurobiol.* **1,** 329–344.

Oey, J. (1975). Noradrenaline induces morphological alterations in nucleated and enucleated rat C_6 glioma cells. *Nature* (*Lond.*) **257,** 317–319.

Oja, S. S. (1966). Postnatal changes in the concentration of nucleic acids, nucleotides and amino acids in the rat brain. *Ann. Acad. Sci. Fennicae*, Series A (V. Medica), 7–69.

Oja, S. S., Uusitalo, A. J., Vahvelainen, M.-L. and Piha, R. S. (1968a). Changes in cerebral and hepatic amino acids in the rat and guinea pig during development. *Brain Res.* **11,** 655–661.

Oja, S. S., Oja, H. and Sarajas, H. S. S. (1968b). Cerebral free amino acids in rats undergoing varying periods of anesthesia and hypothermia. *Ann. Acad. Sci. Fennicae*, **142,** Series A (IV. Biologica) 3–14.

Okada, Y., Nitsch-Hassler, C., Kim, J. S., Bak, I. J. and Hassler, R. (1971). The role of γ-aminobutyric acid (GABA) in the extrapyramidal motor system. 1. Regional distribution of GABA in rabbit, rat and guinea pig brain. *Exptl. Brain Res.* **13,** 514–518.

Olds, J. (1956). A preliminary mapping of electrical reinforcing effects in the rat brain. *J. Comp. Physiol. Psychol.* **49,** 281–285.

Olds, J. and Milner, P. (1954). Positive reinforcement produced by electrical stimulation of septal area and other regions of rat brain. *J. Comp. Physiol. Psychol.* **47,** 419–427.

Olds, M. E. and Olds, J. (1963). Approach-avoidance analysis of rat diencephalon. *J. Comp. Neurol.* **120,** 259–295.

Oliver, G. W. O. (1973). Neurochemical aspects of shock-avoidance learning in cockroaches. *In* "Macromolecules and Behaviour" (G. B. Ansell and P. B. Bradley, Eds), pp. 113–131. University Park Press, London.

Olney, J. W. (1969). Brain lesions, obesity, and other disturbances in mice treated with monosodium glutamate. *Science* **164,** 719–721.

Olney, J. W. (1971). Glutamate-induced neuronal necrosis in the infant mouse hypothalamus. *J. Neuropathol. Exptl. Neurol.* **30,** 75–90.

Olney, J. W. and Sharpe, L. G. (1969). Brain lesions in an infant Rhesus monkey treated with monosodium glutamate. *Science* **166,** 386–388.

Olsen, L. and Fuxe, K. (1972). Further mapping out of central noradrenaline neuron systems: Projections of the subcoeruleus area. *Brain Res.* **43,** 289–295.

Olson, L. and Malmfors, T. (1970). Growth characteristics of adrenergic nerves in the adult rat. *Acta Physiol. Scand. Suppl.* 348, 112 pp.

Op Den Velde, W. and Hooghwinkel, G. J. M. (1975). The brain ganglioside pattern in presenile and senile dementia. *J. Amer. Geriat. Soc.* **23,** 301–303.

Ordy, J. M. and Brizzee, K. R. (Eds) (1975). "Neurobiology of Aging; Vol. 16, Advances in Behavioral Biology". Plenum Press, New York, 599 pp.

Ordy, J. M. and Schjeide, O. A. (1973). Univariate and multivariate models for evaluating long-term changes in neurobiological development, maturity and aging. *In* "Progress in Brain Research, Vol. 40: Neurobiological Aspects of Maturation and Aging", pp. 25–51. Elsevier, Amsterdam.

Ordy, J. M., Kaack, B. and Brizzee, K. R. (1975). Life-span neurochemical changes in the human and nonhuman primate brain. *In* "Aging" (H. Brody, D. Harman and J. M. Ordy, Eds), Vol. 1, pp. 133–189. Raven Press, New York.

Orgel, L. E. (1963). The maintenance of the accuracy of protein synthesis and its relevance to aging. *Proc. Nat. Acad. Sci., U.S.A.* **49,** 517–521.

Orgel, L. E. (1970). The maintenance of the accuracy of protein synthesis and its relevance to ageing: A correction. *Proc. Nat. Acad. Sci., U.S.A.* **67,** 1476.

Orsingher, O. A. and Fulginiti, S. (1973). Influence of peripheral mechanisms on the facilitatory learning action of amphetamine and nicotine in rats. *Pharmacology* **9,** 138–144.

Oswald, I. (1969). Human brain protein, drugs and dreams. *Nature* (*Lond.*) **223,** 893–897.

Otsuka, M., Konishi, S. and Takahashi, T. (1972). The presence of a motoneuron-depolarizing peptide in bovine dorsal roots of spinal nerves. *Proc. Jap. Acad.* **48,** 342–346.

Otten, U., Paravicini, U., Oesch, F. and Thoenen, H. (1973). Time requirements for the single steps of trans-synaptic induction of tyrosine hydroxylase in the peripheral sympathetic nervous system. *Naunyn-Schmiedeberg's Arch. Exp. Path. Pharmacol.* **280,** 117–127.

Owen, O. E., Morgan, A. P., Kemp, H. G., Sullivan, J. M., Herrera, M. G. and Cahill, G. F., Jr. (1967). Brain metabolism during fasting. *J. Clin. Invest.* **46,** 1589–1595.

Pagel, J., Pegram, V., Vaughn, S., Donaldson, P. and Bridgers, W. (1973). The relationship of REM sleep with learning and memory in mice. *Behav. Biol.* **9,** 383–388.

Papeschi, R. (1972). Dopamine, extrapyramidal system and psychomotor function. *Psychiatria, Neurologia, Neurochirurgia,* **75,** 13–48.

Papez, J. W. (1937). A proposed mechanism of emotion. *Arch. Neurol Psychiat.* **38,** 725–743.

Pappas, G. D. and Purpura, D. P. (Eds) (1972). "Structure and Function of Synapses". Raven Press, New York, 308 pp.

Pappas, G. D. and Waxman, S. G. (1972). Synaptic fine structure—Morphological correlates of chemical and electrotonic transmission. *In* "Structure and Function of Synapses" (G. D. Pappas and D. P. Purpura, Eds), pp. 1–43. Raven Press, New York.

Pappas, G. D., Cohen, E. B. and Purpura, D. P. (1966). Fine structure of synaptic and non-synaptic relations in the thalamus of the cat. *In* "The Thalamus" (D. P. Purpura and M. D. Yahr, Eds), pp. 47–75. Columbia University Press, New York.

Pappius, H. M. (1969). Water spaces. *In* "Handbook of Neurochemistry" (A. Lajtha, Ed.), Vol. 2, pp. 1–10. Plenum Press, New York.

Párducz, A. and Fehér, O. (1970). Fine structural alterations of presynaptic endings in the superior cervical ganglion of the cat after exhausting preganglionic stimulation. *Experientia* **26,** 629–630.

Parker, G. H. (1932). On the trophic impulse so-called, its rate and nature. *Amer. Natur.* **67,** 147–158.

Parnavelas, J. G. and Uylings, H. B. M. (1975). Continuous illumination from birth affects the morphology of neurons in the visual cortex of rats. *First European Neurosciences Meeting, Munich,* 9–13 September, Abstr. No. 304.

Parnavelas, J. G., Globus, A. and Kaups, P. (1973a). Changes in lateral geniculate neurones of rats as a result of continuous exposure to light. *Nature New Biol.* (*Lond.*) **245,** 287–288.

Parnavelas, J. G., Globus, A. and Kaups, P. (1973). Continuous illumination from birth affects spine density of neurons in the visual cortex of the rat. *Exptl. Neurol.* **40,** 742–747.

Paroli, E. (1967). Indagini sull'effetto antimorfinico dell' ACTH. I° Relazioni con il corticosurrene ed i livelli ematici degli 11-OH steroidi. *Arch. Ital. Sci. Farmacol.* **13,** 234.

Pasquini, J. M., Kaplún, B., Garcia Argiz, C. A. and Gomez, C. J. (1967). Hormonal regulation of brain development. I. The effect of neonatal thyroidectomy upon nucleic acids, protein and two enzymes in developing cerebral cortex and cerebellum of the rat. *Brain Res.* **6,** 621–634.

Passingham, R. E. (1975). Changes in the size and organisation of the brain in man and his ancestors. *Brain Behav. Evolut.* **11,** 73–90.

Pasternak, G. W. and Snyder, S. H. (1974). Opiate receptor binding: Effects of enzymatic treatments. *Mol. Pharmacol.* **10,** 183–193.

Pasternak, G. W. and Snyder, S. H. (1975). Identification of a novel high affinity opiate receptor binding in rat brain. *Nature* (*Lond.*) **253,** 563–565.

Pasternak, G. W., Goodman, R. and Snyder, S. H. (1975). An endogenous morphine-like factor in mammalian brain. *Life Sci.* **16,** 1765–1769.

Patel, A. J., Rabié, A., Lewis, P. D. and Balázs, R. (1976). Effect of thyroid deficiency on postnatal cell formation in the rat brain: A biochemical investigation. *Brain Res.* **104,** 33–48.

Patel, G. J. and Lal, H. (1973). Reduction in brain γ-aminobutyric acid and in barbital narcosis during ethanol withdrawal. *J. Pharmacol. Exptl. Therapeut.* **186,** 625–629.

Patel, G. J., Schatz, R. P., Constantinides, S. M. and Lal, H. (1975). Effect of desipramine and pargyline on brain gamma-aminobutyric acid. *Biochem. Pharmacol.* **24,** 57–60.

Patton, H. D. and Amassian, V. E. (1950). Cortical projection zone of the chorda tympani. *J. Neurophysiol.* **15,** 245–250.

Pavlov, I. P. (1927). "Conditioned Reflexes" (trans. by G. V. Anrep). Oxford University Press, London.

Pease, D. C. (1966). Polysaccharides associated with the exterior surface of epithelial cells: Kidney, intestine, brain. *J. Ultrastruct. Res.* **15,** 555–588.

Pederson, T. and Robbins, E. (1970). Absence of translational control of histone synthesis during the HeLa cell life cycle. *J. Cell. Biol.* **45,** 509–513.

Peeke, H. V. S. and Herz, M. J. (1967). Permanence of electroconvulsive shock produced retrograde amnesia. *Proceedings 75th Annual Convention of the American Psychological Association* **2,** 85–86.

Peeke, H. V. S. and Herz, M. J. (Eds) (1973). "Habituation: Behavioral Studies", Vol. 1, Academic Press, New York and London, 300 pp.

Pegram. V., Hammond, D. and Bridgers, W. (1973). The effects of protein synthesis inhibition on sleep in mice. *Behav. Biol.* **9,** 377–382.

Pelc, S. R. (1970). Metabolic DNA and the problem of ageing. *Exptl. Gerontol.* **5,** 217–226.

Penfield, W. (1927). The mechanism of cicatricial contraction in the brain. *Brain* **50,** 499–517.

Penfield, W. and Humphreys, S. (1940). Epileptogenic lesions of the brain. *Arch. Neurol. Psychiat.* **43,** 240–261.

Penfield, W. and Jasper, H. H. (1954). "Epilepsy and the Functional Anatomy of the Human Brain", Little, Brown and Co., Boston.

Perez, V. J. and Moore, B. W. (1968). Wallerian degeneration in rabbit tibial nerve: Changes in amounts of the S-100 protein. *J. Neurochem.* **15,** 971–977.

Perez, V. J. and Moore, B. W. (1970). Biochemistry of the nervous system in aging. *Interdisc. Top. Geront.* **7,** 22–45.

Perez, V. J., Olney, J. W., Cicero, T. J., Moore, B. W. and Bahn, B. A. (1970). Wallerian degeneration in rabbit optic nerve: Cellular localization in the central nervous system of the S-100 and 14.3.2 proteins. *J. Neurochem.* **17,** 511–519.

Perry, T. L., Hansen, S., Stedman, D. and Love, D. (1968). Homocarnosine in human cerebrospinal fluid: An age-dependent phenomenon. *J. Neurochem.* **15,** 1203–1206.

Perry, T. L., Bratty, P. J. A., Hansen, S., Kennedy, J., Urquhart, N. and Dolman, C. L. (1975). Hereditary mental depression and Parkinsonism with taurine deficiency. *Arch. Neurol.* **32,** 108–113.

Pert, C. B. and Snyder, S. H. (1974). Opiate receptor binding of agonists and antagonists affected differentially by sodium. *Mol. Pharmacol.* **10,** 868–879.

Pert, A. and Yaksh, T. (1974). Sites of morphine induced analgesia in the primate brain: Relation to pain pathways. *Brain Res.* **80,** 135–140.

Pert, A. and Yaksh, T. (1975). Localization of the antinociceptive action of morphine in primate brain. *Pharmacol. Biochem. Behav.* **3,** 133–138.

Peters, A. and Palay, S. L. (1965). An electron microscope study of the distribution and patterns of astroglial processes in the central nervous system. *J. Anat.* (*Lond.*) **99,** 419.

Peterson, R. P. and Erulkar, S. D. (1973). Parameters of stimulation of RNA synthesis and characterization by hybridization in a molluscan neuron. *Brain Res.* **60,** 177–190.

Peterson, J. A., Bray, J. J. and Austin, L. (1968). An autoradiographic study of the flow of protein and RNA along peripheral nerve. *J. Neurochem.* **15,** 741–745.

Pettigrew, J. D. and Garey, L. J. (1974). Selective modification of single neuron properties in the visual cortex of kittens. *Brain Res.* **66,** 160–164.

Pevsner, L. Z. (1966). Nucleic acid changes during behavioural events. *In* "Macromolecules and Behavior" (J. Gaito, Ed.), 1st Edn, pp. 43–70. Appleton-Century-Crofts, New York.

Pevsner, L. Z. (1972). Macromolecular changes within neuron-neuroglia unit during behavioral events. *In* "Macromolecules and Behavior" (J. Gaito, Ed.), 2nd Edn., pp. 335–358. Appleton-Century-Crofts, New York.

Pfaff, D. W. (1966). Morphological changes in the brains of adult male rats after neonatal castration. *J. Endocrinol.* **36,** 415–416.

Pfaff, D. W. (1973). Luteinizing hormone-releasing factor potentiates lordosis behavior in hypophysectomized ovariectomized female rats. *Science* **182,** 1148–1149.

Phelps, C. H. (1972). Barbiturate-induced glycogen accumulation in brain. An electron microscopic study. *Brain Res.* **39,** 225–234.

Phillis, J. W. (1970). "The Pharmacology of Synapses". Pergamon Press, Oxford, 358 pp.

Piepho, R. W. and Friedman, A. H. (1971). Twenty-four hour rhythms in the glycine content of rat hindbrain and spinal cord. *Life Sci.* **10** (Part I) 1355–1362.

Piezzi, R. S. and Wurtman, R. J. (1970). Pituitary serotonin content: Effects of melatonin or deprivation of water. *Science* **169,** 285–286.

Piha, R. S., Bergström, R. M., Bergström, L., Uusitalo, A. J. and Oja, S. S. (1963). Studies on the metabolism of brain proteins. II. Effect of chlorpromazine and lysergic acid diethylamide on turnover rate of rat brain proteins. *Ann. Med. Exp. Fenn.* **41,** 498–515.

Piha, R. S., Oja, S. S. and Uusitalo, A. J. (1962). The effect of chlorpromazine on free amino acids in the rat brain. *Ann. Med. Exp. Biol. Fennicae* **40,** (Suppl. 5), 1–28.

Pilcher, C. W. T. (1976). Inducing dependence by a single administration of morphine in rats. *Proc. Brit. Pharmacol. Soc., Chelsea College Meeting,* 1–2 April, Abstract C.14.

Pilcher, C. W. T. and Booth, D. A. (1975). Effect of cycloheximide on the long-term retention of reversed paw-preference in the rat. *First European Neurosciences Meeting, Munich,* 9–13 September, Abstr. No. 318.

Plaut, S. M. and Grota, L. J. (1971). Effects of differential housing on adrenocortical reactivity. *Neuroendocrinology* **7,** 348–360.

Plotnikoff, N. P., Prange, A. J., Jr., Breese, G.R., Anderson, M. S. and Wilson, I. C. (1972). Thyrotropin releasing hormone: Enhancement of DOPA activity by a hypothalamic hormone. *Science* **178,** 417–418.

Plotnikoff, N. P., Prange, A. J., Jr., Breese, G. R. Anderson, M. S. and Wilson, I. C. (1974a). The effects of thyrotropin-releasing hormone on DOPA response in normal, hypophysectomized, and thyroidectomized animals. *In* "The Thyroid Axis, Drugs, and Behavior" (A. J. Prange, Jr., Ed.), pp. 103–113. Raven Press, New York.

Plotnikoff, N. P., Prange, A. J., Jr. Breese, G. R. and Wilson, I. C. (1974b). Thyrotropin releasing hormone: Enhancement of DOPA activity in thyroidectomized rats. *Life Sci.* **14,** 1271–1278.

Plum, F. and Posner, J. B. (1966). "Diagnosis of Stupor and Coma." F. A. Davis, Philadelphia.

Pohle, W. and Matthies, H. (1974a) Incorporation of RNA-precursors into neuronal and glial cells of rat brain during a learning experiment. *Brain Res.* **65,** 231–237.

Pohle, W. and Matthies, H. (1974b). Incorporation of ^{3}H-leucine into brain cells after learning. *Pharmacol. Biochem. Behav.* **2,** 573–577.

Poirier, L., Filion, M., Langelier, P. and Larochelle, L. (1975). Brain nervous mechanisms involved in the so-called extra pyramidal motor and psychomotor distrubances. *Progr. Neurobiol* **5,** 197–244.

Pollen, D. A. and Trachtenberg, M. C. (1970). Neuroglia: Gliosis and focal epilepsy. *Science* **167,** 1252–1253.

Pomerat, C. M., Hendelman, W. J., Raiborn, C. W., Jr. and Massey, J. F. (1967). Dynamic activities of nervous tissue *in vitro*. *In* "The Neuron" (H. Hydén, Ed.), pp. 119–178. Elsevier, Amsterdam.

Popov, N., Schulzeck, S., Schmidt, S. and Matthies, H. (1975). Changes in labeling of soluble and solubilized rat brain proteins using [^{3}H] leucine as precursor during a learning experiment. *Acta Biol. Med. Germ.* **34,** 583–592.

Popov, N., Rüthrich, H.-L., Pohle, W., Schulzeck, S. and Matthies, H. (1976). Increased fucose incorporation into rat hippocampus during learning. A biochemical and microautoradiographic study. *Brain Res.* **101,** 295–304.

Posner, J. B., Plum, F. and Poznak, A. V. (1969). Cerebral metabolism during electrically induced seizures in man. *Arch. Neurol.* **20,** 388–395.

Posner, M. I. (1975). Psychobiology of attention. *In* "Handbook of Psychobiology" (M. S. Gazzaniga and C. Blakemore, Eds), pp. 441–480. Academic Press, New York and London.

Pradhan, S. N. (1975). Aggression and central neurotransmitters. *Int. Rev. Neurobiol.* **18,** 213–262.

Pradhan, S. N. and Dutta, S. N. (1971). Central cholinergic mechanisms and behavior. *Int. Rev. Neurobiol.* **14,** 173–231.

Prange, A. J., Jr., Breese, G. R., Cott, J. M., Martin, B. R., Cooper, B. R., Wilson, I. C. and Plotnikoff, N. P. (1974). Antagonism of pentobarbital in rodents. *Life Sci.* **14,** 447–455.

Prange, A. J., Jr., Breese, G. R., Wilson, I. C. and Lipton, M. A. (1975). Pituitary and suprapituitary hormones: Brain-behavioral effects. *In* "Topics in Psychoendocrinology" (E. J. Sachar, Ed.), pp. 105–119. Grune and Stratton, Inc., New York.

Prescott, J. W. (1971). Early somatosensory deprivation as an ontogenetic process in the abnormal development of the brain and behavior. *In* "Proc. 2nd Conf. Exp. Med. Surg. Primates", pp. 356–375. Karger, Basel.

Prestige, M. C. (1970). Differentiation, degeneration and the role of the periphery: Quantitative considerations. *In* "The Neurosciences, Second Study Program" (F. O. Schmitt, Ed.), pp. 73–82. Rockefeller University Press, New York.

Pribram, K. H. (1966). Some dimensions of remembering: Steps toward a neuropsychological model of memory. *In* "Macromolecules and Behavior" (J. Gaito, Ed.), 1st Edn, pp. 165–187. Appleton-Century-Crofts, New York.

Pribram, K. H. (1972). Some dimensions of remembering: Steps toward a neuropsychological model of memory. *In* "Macromolecules and Behavior" (J. Gaito, Ed.), 2nd Edn, pp. 367–393. Appleton-Century-Crofts, New York.

Prokhorova, M. I., Romanova, S. L. and Sokolova, G. P. (1964). *Chem. Abstr.* **61,** 1098 g.

Prokhorova, M. I., Romanova, S. L. and Tumonova, S. Y. (1967). *Chem. Abstr.* **68,** 5784 j.

Puerto, A., Molina, F., Rogers, J. and Moss, D. E. (1976). Physostigmine-induced amnesia for an escape response 12 to 72 hours after training. *Behav. Biol.* **16,** 85–90.

Pull, I. and McIlwain, H. (1972). Adenine derivatives as neurohumoral agents in the brain. *Biochem. J.* (*Lond.*), **130,** 975–981.

Purpura, D. P. (1974). Dendritic spine "dysgenesis" and mental retardation. *Science* **186,** 1126–1128.

Purpura, D. P. (1975). Dendritic differentiation in human cerebral cortex: Normal and aberrant developmental patterns. *In* "Advances in Neurology, Vol. 12, Physiology and Pathology of Dendrites" (G. W. Kreutzberg, Ed.), pp. 91–116. Raven Press, New York.

Purpura, D. P. and Schadé, J. P. (Eds) (1964). "Growth and Maturation of the Brain, Progress in Brain Research", Vol. 4. Elsevier, Amsterdam, 289 pp.

Pysh, J. J. and Wiley, R. G. (1974). Synaptic vesicle depletion and recovery in cat sympathetic ganglia electrically stimulated *in vivo*. *J. Cell. Biol.* **60,** 365–374.

Quamina, A. and Bogoch, S. (1966). *In* "Proc. XIII Colloq., Protides of Biological Fluids, Brugge" (H. Peeters, Ed.), p. 211. Elsevier, Amsterdam.

Quartermain, D., McEwen, B. S. and Azmitia, E. C. (1970). Amnesia produced by electroconvulsive shock or cycloheximide: Conditions for recovery. *Science* **169,** 683–686.

Quastel, J. H. (1969). Carbohydrate metabolism in the nervous system. *In* "The Structure and Function of Nervous Tissue" (G. H. Bourne, Ed.), Vol. III, pp. 61–107. Academic Press, London and New York.

Quastel, J. H. (1974). Fifty years of biochemistry. A personal account. *Can. J. Biochem.* **52**, 71–82.

Quastel, J. H. (1975). Effects of drugs on energy metabolism of the brain and on cerebral transport. In "Handbook of Psychopharmacology" (L. L. Iversen, S. D Iversen and S. H. Snyder, Eds), Vol. 5, pp. 1–46. Plenum Press, New York.

Quastel, J. H. and Quastel, D. M. J. (1961). "The Chemistry of Brain Metabolism in Health and Disease". C. C. Thomas, Springfield, Ill., 170 pp.

Quastel, J. H. and Wheatley, A. H. M. (1932a). Oxidations by the brain. *Biochem. J.* (*Lond.*), **26,** 725–744.

Quastel, J. H. and Wheatley, A. H. M. (1932b). Narcosis and oxidations of the brain. *Proc. Roy. Soc., Ser. B,* **112,** 60–79.

Quay, W. B. (1963). Circadian rhythms in rat pineal serotonin and its modifications by estrous cycle and photoperiod. *J. Comp. Endocr.* **3,** 473–479.

Quay, W. B. (1966). 24-Hour rhythms in pineal 5-hydroxytryptamine and hydroxy indole-O-methyl transferase activity in the macaque. *Proc. Soc. Exptl. Biol. Med.* **121,** 946–948.

Quenzer, L. F., Feldman, R. S. and Moore, J. W. (1974). Toward a mechanism of the anti-aggression effects of chlordiazepoxide in rats. *Psychopharmacologia* **34,** 81–94.

Quinton, E. E. (1966). Retrograde amnesia induced by carbon dioxide inhalation. *Psychonom. Sci.* **5,** 417–418.

Quinton, E. (1971). The cycloheximide-induced amnesia of a passive avoidance task. *Psychonom. Sci.* **25,** 295–296.

Quinton, E. (1972). Memory retrieval, reactivation, and protein synthesis. *Paper presented at the meetings of the Rocky Mountain Psychological Assn.*

Raab, A. (1971). Der Serotoninstoffwechsel in einzelnen Hirnteilen vom Tupaia (*Tupaia Belangeri*) bei soziopsychischen stress. *Z. Vergl. Physiol.* **72,** 54–66.

Raab, A. and Deisz, R. (1975). Male and female mice living in differently-sized groups. II. Serotonin metabolism in discrete brain areas, open-field activity and corticoid release. *J. Comp. Physiol.* **99,** 165–175.

Raabe, W. and Ayala, G. F. (1976). Diphenylhydantoin increases cortical postsynaptic inhibition. *Brain Res.* **105,** 597–601.

Rachevsky, N. (1938). "Mathematical Biophysics: Physicomathematical Foundations of Biology", 1st Edn. University of Chicago Press, Chicago.

Radcliffe, G. J. and Shelton, J. W. (1974). Molecular coding of maze learning; demonstration by bioassay. *Experientia* **30,** 1284–1286.

Raichle, M. E., Eichling, J. O., Gado, M. H., Grubb, R. L. Jr. and Ter-Pogossian, M. M. (1975). Cerebral blood volume in dementia. *In* "Intracranial Pressure II" (N. Lundberg, U. Pontén and M. Brock, Eds), pp. 150–152. Springer-Verlag, Berlin.

Raisman, G. (1969). Neuronal plasticity in the septal nuclei of the adult rat. *Brain Res.* **14,** 25–48.

Raisman, G. (1973). Anatomical evidence for plasticity in the central nervous system in the adult rat. *In* "Macromolecules and Behaviour" (G. B. Ansell and P. B. Bradley, Eds), pp. 183–191. University Park Press, London.

Raisman, G. and Field, P. M. (1973). Sexual dimorphism in the neuropil of the preoptic area of the rat and its dependence on neonatal androgen. *Brain Res.* **54,** 1–29.

Rakic, P. (1974). Intrinsic and extrinsic factors influencing the shape of neurons and their assembly into neuronal circuits. *In* "Frontiers in Neurology and Neuro-science Research" (P. Seeman and G. M. Brown, Eds) University of Toronto Press, Toronto.

Rakic, P. (1975). Role of cell interaction in development of dendritic patterns. *In* "Advances in Neurology" (G. W. Kreutzberg, Ed.) Vol. 12, pp. 117–134. Raven Press, New York.

Rall, W. (1970). Cable properties of dendrites and effects of synaptic location. *In* "Excitatory Synaptic Mechanisms" (P. Andersen and J. K. S. Jansen, Eds), pp. 175–187. Universitetsforlaget, Oslo.

Rall, T. W. and Gilman, A. G. (1970). The role of cyclic AMP in the nervous system. *Neurosci. Res. Progr. Bull.* **8,** 221–323.

Ralston, H. J., III and Chow, K. L. (1973). Synaptic reorganization in the degenerating lateral geniculate nucleus of the rabbit. *J. Comp. Neurol.* **147,** 321–350.

Ralston, H. J. and Herman, M. D. (1969). The fine structure of neurons and synapses in the ventrobasal thalamus of the cat. *Brain Res.* **14,** 77–97.

Ramirez, G. (1974). Nonspecific incorporation of glucosamine into rat brain synaptosomes and endoplasmic reticulum. *Biochim. Biophys. Acta* **338,** 337–344.

Ramirez, G., Levitan, I. B. and Mushynski, W. E. (1972). Highly purified synaptosomal membranes from rat brain; incorporation of amino acids into membrane proteins *in vitro*. *J. Biol. Chem.* **247,** 5382–5390.

Ramirez, G., Karlsson, B. and Perrin, C. L. (1974). Effects of environmental complexity on glucosamine incorporation into rat cerebral cortex and hippocampus *in vivo*. *Brain Res.* **79,** 296–300.

Ramón-Moliner, E. (1975). Can dendrites be presynaptic? *In* "Advances in Neurology; Physiology and Pathology of Dendrites" (G. W. Kreutzberg, Ed.) Vol. 12, pp. 135–147. Raven Press, New York.

Ramón y Cajal, S. (1888). Sobre las fibras nerviosas de la capa molecular del cerebelo. *Revista Trimestral de Histologia Normal y Patologica*, Barcelona.

Ramón y Cajal, S. (1891). Sur la structure de l'écorce cérébrale de quelques mammifères. *La Cellule*, **7,** 1–54.

Ramón y Cajal, S. (1895). "Les Nouvelles Idées sur la Structure du Système Nerveux chez l'Homme et chez les Vertébrés". Edition Française revue et augmentée par l'auteur; traduite de l'espagnol par L. Azoulay; C. Reinwald et Cie., Paris, quote from p. 79.

Ramón y Cajal, S. (1896). Les épines collaterales des cellules du cerveau colorée au bleu de méthyléne. *Rev. Trim. Micrgr.* **1,** 5–19.

Ramón y Cajal, S. (1899). "Textura del Sistema Nervioso del Hombre y de los Vertebrados", Tomo 1. Imprenta y Librería de Nicolás Moya, Madrid, pp. 161–173.

Ramón y Cajal, S. (1909). "Histologie du Système Nerveux de l'Homme et des Vertébrés" (translated by L. Azoulay) pp. 248–249. Maloine, Paris.

Ramón y Cajal, S. (1910). "Histologie du Système Nerveux de l'Homme et des Vertébrés". Maloine, Paris.

Ramón y Cajal, S. (1911). "Histologie du Système Nerveux de l'Homme et des Vertébrés". Trans by L. Azoulay, Paris, Maloine, Vol. II.

Ramón y Cajal, S. (1929). "Études sur la Neurogenèse de Quelques Vertébrés". Tipografía Artística, Madrid, 393 pp.

Ramón y Cajal, S. (1934). Les preuves objectives de l'unité anatomique des cellules nerveuses. *Trab. Lab. Rech. Biol., Univ. Madrid* **29,** 1–137.

Ramón y Cajal, S. (1952a). "Histologie du Système Nerveux de l'Homme et des Vertébrés". Vol. I., Trans. by L. Azoulay, Inst. Ramón y Cajal, Madrid.

Ramón y Cajal, S. (1952b) "Histologie du Système Nerveux de l'Homme et des Vertébrés". Vol. II., Trans. by L. Azoulay, Inst. Ramón y Cajal, Madrid.

Randrup, A. and Munkvad, I. (1968). Behavioral stereotypies induced by pharmacological agents. *In* "Pharmakopsychiat. Neuro-Psychopharmakol", Vol. 1, pp. 18–26. Thieme, Stuttgart.

Randrup, A. and Munkvad, I. (1970). Biochemical, anatomical and psychological investigations of stereotyped behaviour induced by amphetamines. *In* "Symp. on Amphetamine and Related Drugs", Milan, 1969 (E. Costa and S. Garattini, Eds.) pp. 695–713. Raven Press, New York.

Rappoport, D. A., Fritz, R. R. and Myers, J. L. (1969). Nucleic acids. *In* "Handbook of Neurochemistry" (A. Lajtha, Ed.), Vol. 1, pp. 101–119. Plenum Press, New York.

Reading, H. W. (1972). Effects of some adrenocorticotrophin analogues on protein synthesis in brain. *Biochem. J.* (*Lond.*) **127,** 7 P.

Rees, H. D., Brogan, L. L., Entingh, D. J., Dunn, A. J., Shinkman, P. G., Damstra-Entingh, T., Wilson, J. E. and Glassman, E. (1974). Effect of sensory stimulation on the uptake and incorporation of radioactive lysine into protein of mouse brain and liver. *Brain Res.* **68,** 143–156.

Regan, D. (1972). "Evoked Potentials". Wiley, New York.

Reichelt, K. L. (1970). The isolation of gamma-glutamyl peptides from monkey brain. *J. Neurochem.* **17,** 19–25.

Reichelt, K. L. and Edminson, P. D. (1974). Biogenic amine specificity of cortical peptide synthesis in monkey brain. *FEBS Letters* **47,** 185–189.

Reichelt, K. L. and Kvamme, E. (1967). Acetylated and peptide bound glutamate and aspartate in brain. *J. Neurochem* **14,** 987–996.

Reichelt, K. L., Edminson, P. D. and Kvamme, E. (1976). The formation of peptido-amines from constituent amino acids and histamine in hypothalamic tissue. *J. Neurochem* **26,** 811–815.

Reichlin, S., Saperstein, R., Jackson, I. M. D., Boyd, A. E., III and Patel, Y. (1976). Hypothalamic hormones. *Ann. Rev. Physiol.* **38,** 389–424.

Reinis, S. (1972). Autoradiographic study of ^{3}H-thymidine incorporation into brain DNA during learning. *Physiol. Chem. Physics* **4,** 391–397.

Reinis, S. and Lamble, R. W. (1972). Labelling of brain DNA by ^{3}H-thymidine during learning. *Physiol. Chem. Physics* **4,** 335–338.

Reinis, S., Abbott, J. and Clark, J. J. (1972). Brain DNA changes during learning studies by administration of 5-iodo-2′-deoxyuridine. *Physiol. Chem. Physics* **4,** 440–448.

Reis, D. and Fuxe, K. (1969). Brain norepinephrine: Evidence that neuronal release is essential for sham rage behaviour following brainstem transection in cat. *Proc. Nat. Acad. Sci., U.S.A.* **64,** 108–112.

Reis, D. J. (1971). Brain monoamines in aggression and sleep. *Clin. Neurosurg.* **18,** 471–502.

Reith, M. E. A., Schotman, P. and Gispen, W. H. (1975). The neurotropic action of ACTH: Effects of ACTH-like peptides on the incorporation of leucine into protein of brain stem slices from hypophysectomized rats. *Neurosci. Letters* **1,** 55–59.

Reuter, J. H., Legein, C. P. J. J. M. M., Van der Mark, F. and van Hof, M. W. (1971). The electroretinogram in normal and light-deprived rabbits. *Doc. Ophthal.* **30,** 349–361.

Reynolds, E. H. (1970). Water, electrolytes and epilepsy. *J. Neurol. Sci.* **11,** 327–358.

Richardson, K. and Rose, S. P. R. (1971). A diurnal rhythmicity in incorporation of lysine into rat brain regions. *Nature* (*Lond.*) **233,** 182–184.

Richardson, K. and Rose, S. P. R. (1972). Changes in [^{3}H] lysine incorporation following first exposure to light. *Brain Res.* **44,** 299–303.

Richardson, K. and Rose, S. P. R. (1973a). Differential incorporation of lysine into retinal protein fractions following first exposure to light. *J. Neurochem.* **21,** 521–530.

Richardson, K. and Rose, S. P. R. (1973b) Differential incorporation of [^{3}H] lysine into visual cortex protein fractions during first exposure to light. *J. Neurochem.* **21,** 531–537.

Richardson, K. and Rose, S. P. R. (1975). Differential incorporation of lysine into retinal protein fractions following first exposure to light (In press).

Richter, C. P. (1927). Animal behavior and internal drives. *Quart. Rev. Biol.* **2,** 307–343.

Richter, C. P. (1942). Total self regulatory functions in animals and human beings. *Harvey Lectures Series,* **38,** 63–103.

Rick, J. T., Tunnicliff, G., Kerkut, G. A., Fulker, D. W., Wilcock, J. and Broadbent, P. L. (1971). GABA production in brain cortex related to activity and avoidance behaviour in eight strains of rat. *Brain Res.* **32,** 234–238.

Riege, W. H. and Morimoto, H. (1970). Effects of chronic stress and differential environments upon brain weights and biogenic amine levels in rats. *J. Comp. Physiol. Psychol.* **71,** 396–404.

Riesen, A. H. (1960). Effects of stimulus deprivation on the development and atrophy of the visual sensory system. *Amer. J. Orthopsychiat.* **30,** 23–36.

Rieske, E., Schubert, P. and Kreutzberg, G. W. (1975). Transneuronal transfer of radioactive material between electrically coupled neurons of the leech central nervous system. *Brain Res.* **84,** 365–382.

Rigter, H. and van Riezen, H. (1975a). Attenuation of amnesia: Distinct effects of ACTH- and vasopressin-analogues. *First European Neurosciences Meeting, Munich,* 9–13 September, Abstr. No. 346.

Rigter, H. and Van Riezen, H. (1975b). Anti-amnesic effect of $ACTH_{4-10}$: Its independence of the nature of the amnesic agent and the behavioral test. *Physiol. Behav.* **14,** 563–566.

Rigter, H., Van Riezen, H. and De Wied, D. (1974). The effects of ACTH- and vasopressin-analogues on CO_2-induced retrograde amnesia in rats. *Physiol. Behav.* **13,** 381–388.

Rigter, H., Elbertse, R. and Van Riezen, H. (1975). Time-dependent anti-amnesic effect of $ACTH_{4-10}$ and desglycinamide-lysine vasopressin. *In* "Progress in Brain Research; Hormones, Homeostasis and the Brain" (W. H. Gispen, Tj.B. Van Wimersma Greidanus, B. Bohus and D. DeWied, Eds), Vol. 42, pp. 163–171. Elsevier, Amsterdam.

Rindi, G. and Ventura, U. (1961). *Biochem.* **10,** 135; cited by Ghazal *et al.* (1976).

Robbins, E. and Borun, T. W. (1967). The cytoplasmic synthesis of histones in HeLa cells and its temporal relationship to DNA replication. *Proc. Nat. Acad. Sci., U.S.A.* **57,** 409–416.

Roberts, E. and Matthysse, S. (1970). Neurochemistry: At the crossroads of neurobiology. *Ann. Rev. Biochem.* **39,** 777–820.

Roberts, E., Wein, J. and Simonsen, D. G. (1964). *In* "Vitamins and Hormones, Advances in Research and Applications (R. S. Harris, I. G. Wool and J. A. Loraine, Eds), Vol. 22, pp. 503–559. Academic Press, New York and London.

Roberts, E., Chase, T. N. and Tower, D. B. (eds) (1976). "GABA in Nervou. System Function", Kroc Foundation Series, Vol. 5, Raven Press, New York, 554 pps

Roberts, S. and Morelos, B. S. (1976). Role of ribonuclease action in phenylalanine-induced disaggregation of rat cerebral polyribosomes. *J. Neurochem.* **26,** 387–400.

Robins, E. and Hartman, B. K. (1972). Biochemical theories of mental disorders. *In* "Basic Neurochemistry" (R. W. Albers, G. J. Siegel, R. Katzman and B. W. Agranoff, Eds), pp. 607–644. Little, Brown and Co., Boston.

Robinson, N. and Wells, F. (1973). Distribution and localization of sites of gamma-aminobutyric acid metabolism in the adult rat brain. *J. Anat.* (*Lond.*) **114,** 365–378·

Rodnight, R. (1966). *In* "Protides of the Biological Fluids", Proc. XIIIth Colloq., Brugge (Peeters, H., Ed.) Vol. 13, pp. 39–52. Elsevier, Amsterdam.

Rodnight, R. (1968). The metabolism of amines and amino acids in mental disorders. *In* "Applied Neurochemistry" (A. N. Davison and J. Dobbing, Eds) pp. 377–400. F. A. Davis Co., Philadelphia.

Rogers, M., Dismukes, K. and Daly. J. W. (1975). Histamine-elicited accumulations of cyclic adenosine 3′, 5′-monophosphate in guinea-pig brain slices: Effect of H_1- and H_2-antagonists. *J. Neurochem.* **25,** 531–534.

Roisen, F. J., Murphy, R. A. and Braden, W. G. (1972a). Dibutyryl cyclic adenosine monophosphate stimulation of colcemid-inhibited axonal elongation. *Science* **177,** 809–811.

Roisen, F. J., Murphy, R. A., Pichichero, M. E. and Braden, W. G. (1972b). Cyclic adenosine monophosphate stimulation of axonal elongation. *Science* **175,** 73–74.

Rolls, E. T. (1975). "The Brain and Reward". Pergamon Press, Oxford, 124 pp.

Rose, S. P. R. (1967). Changes in visual cortex on first exposure of rats to light. *Nature* (*Lond.*) **215,** 253–255.

Rose, S. P. R. (1969a). Neurochemical correlates of learning and environmental change. *FEBS Letters,* **5,** 305–312.

Rose, S. P. R. (1969b). Neurones and glia: Separation techniques and biochemical interrelationships. *In* "Handbook of Neurochemistry" (A. Lajtha, Ed.) Vol. 2, pp. 183–193. Plenum Press, New York.

Rose, S. P. R. and Sinha, A. K. (1974). Incorporation of ^{3}H-lysine into a rapidly labelling neuronal protein fraction in visual cortex is suppressed in dark reared rats. *Life Sci.* **15,** 223–230.

Rose, S. P. R., Bateson, P. P. G., Horn, A. L. D. and Horn, G. (1970a). Effects of an imprinting procedure on regional incorporation of tritiated uracil into chick brain RNA. *Nature* (*Lond.*) **225,** 650–651.

Rose, S. P. R., Bateson, P., Horn, A. and Horn, D. (1970b). Effects of an imprinting procedure on regional incorporation of tritiated uracil into chick brain RNA. *Nature* (*Lond.*) **225,** 650–651.

Rose, S. P. R., Bateson, P. P. G. and Horn, G. (1973a). Biochemistry and behaviour in the chick. *In* "Macromolecules and Behaviour" (G. B. Ansell and P. B. Bradley, Eds), pp. 93–104. University Park Press, London.

Rose, S. P. R., Sinha, A. K. and Broomhead, S. (1973b). Precursor incorporation into cortical protein during first exposure of rats to light; cellular localization of effects. *J. Neurochem.* **21,** 539–546.

Rose, S. P. R., Hambley, J. and Haywood, J. (1976). Neurochemical approaches to developmental plasticity and learning. *In* "Neural Mechanisms of Learning and Memory" (M. R. Rosenzweig and E. L. Bennett, Eds). The M.I.T. Press, Cambridge, Mass.

Rosenberg, L. E. (1969). *In* "Duncan's Diseases of Metabolism" (P. K. Bondy, Ed.) 6th Edn, Vol. 1, pp. 23–45. W. B. Saunders, Philadelphia.

Rosenzweig, M. R. (1967). Effects of experience on brain chemistry and brain anatomy. *Accad. Naz. Lincei, Roma* **109,** 43–63.

Rosenzweig, M. R. and Bennett, E. L. (1969). Effects of differential environments on brain weights and enzyme activities in gerbils, rats, and mice. *Develop. Psychobiol.* **2,** 87–95.

Rosenzweig, M. R. and Bennett, E. L. (1972). Cerebral changes in rats exposed individually to an enriched environment. *J. Comp. Physiol. Psychol.* **80,** 304–313.

Rosenzweig, M. R. and Bennett, E. L. (1976a). Enriched environments: Facts, factors, and fantasies. *In* "Knowing, Thinking and Behaving" (J. L. McGaugh and L. Petrinovich, Eds). Plenum Press, New York (In press).

Rosenzweig, M. R. and Bennett, E. L. (Eds) (1976b). "Neural Mechanisms of Learning and Memory". The M.I.T. Press, Cambridge, Mass, 488 pp.

Rosenzweig, M. R., Love, W. and Bennett, E. L. (1968). Effects of a few hours a day of enriched experience on brain chemistry and brain weights. *Physiol. Behav.* **3,** 819–825.

Rosenzweig, M. R., Bennett, E. L., Diamond, M. C., Wu, S., Slagle, R. W. and Saffran, E. (1969). Influences of environmental complexity and visual stimulation on development of occipital cortex in rat. *Brain Res.* **14,** 427–445.

Rosenzweig, M. R., Bennett, E. L. and Diamond, M. C. (1972a). Chemical and anatomical plasticity of brain: Replications and extensions, 1970. *In* "Macromolecules and Behavior" (J. Gaito, Ed.) 2nd Edn, pp. 205–277. Appleton-Century-Crofts, New York.

Rosenzweig, M. R., Bennett, E. L. and Diamond, M. C. (1972b). Cerebral effects of differential experience in hypophysectomized rats. *J. Comp. Physiol. Psychol.* **79,** 56–66.

Rosenzweig, M. R., Møllgaard, K., Diamond, M. C. and Bennett, E. L. (1972c). Theoretical note—Negative as well as positive synaptic changes may store memory. *Psychol. Rev.* **79,** 93–96.

Rösner, H. (1975). Incorporation of sialic acid into gangliosides and glycoproteins of the optic pathway following an intraocular injection of [N-^{3}H] acetylmannosamine in the chicken. *Brain Res.* **97,** 107–116.

Rösner, H. Wiegandt, H. and Rahmann, H. (1973). Sialic acid incorporation into gangliosides and glycoproteins of the fish brain. *J. Neurochem.* **21,** 655–665.

Ross, S. B. and Renyi, A. L. (1967). Inhibition of the uptake of tritiated catecholamines by antidepressant and related drugs. *Eur. J. Pharmacol.* **2,** 181–186.

Routtenberg, A. (1975). Significance of intracranial self-stimulation pathways for memory consolidation. *In* "Methods in Brain Research" (P. B. Bradley, Ed.) pp. 453–474, John Wiley and Sons. London.

Routtenberg, A. and Ehrlich, Y. H. (1975). Endogenous phosphorylation of four cerebral cortical membrane proteins: Role of cyclic nucleotides, ATP and divalent cations. *Brain Res.* **92,** 415–430.

Routtenberg, A. and Holzman, N. (1973). Memory disruption by electrical stimulation of substantia nigra, pars compacta. *Science* **181,** 83–86.

Routtenberg, A. and Kramis, R. C. (1967). "Foot-stomping" in the gerbil: Rewarding brain stimulation, sexual behavior, and foot shock. *Nature* (*Lond.*) **214,** 173–174.

Routtenberg, A. and Malsbury, C. (1969). Brainstem pathways of reward. *J. Comp. Physiol. Psychol.* **68,** 22–30.

Routtenberg, A., George, D. R., Davis, L. G. and Brunngraber, E. G. (1974). Memory consolidation and fucosylation of crude synaptosomal glycoproteins resolved by gel electrophoresis: A regional study. *Behav. Biol.* **12,** 461–475.

Routtenberg, A., Ehrlich, Y. H. and Rabjohns, R. R. (1975). Effect of a training experience on phosphorylation of a specific protein in neocortical and subcortical membrane preparations. *Fed. Proc.* **34,** 293 (Abstr. No. 426).

Rubin, E., Lieber, C. S., Altman, K., Gordon, G. G. and Southern, A. L. (1976). Prolonged ethanol consumption increases testosterone metabolism in the liver. *Science* **191,** 563–564.

Ruch, T. C. and Shenkin, H. A. (1943). The relation of area 13 to hyperactivity in monkeys. *J. Neurophysiol.* **6,** 349–360.

Rudman, D., Scott, J. W., Del Rio, A. E., Houser, D. H. and Sheen, S. (1974). Effect of melanotrophic peptides on protein synthesis in mouse brain. *Amer. J. Physiol.* **226,** 687–692.

Rusca, G., Calissano, P. and Alemà, S. (1972). Identification of a membrane-bound fraction of the S-100 protein. *Brain Res.* **49,** 223–227.

Russell, W. R. (1959). "Brain: Memory: Learning". Clarendon Press, Oxford.

Sachar, E. J. (Ed.) (1976). "Hormones, Behavior, and Psychopathology", American Psychopathological Series. Raven Press, New York, 325 pp.

Sachar, E. J., Finkelstein, J. and Hellman, L. (1971). Growth hormone response in depressive illness. *Archs. Gen. Psychiat.* **25,** 263–269.

Sackett, G. P. (1972). Isolation-rearing in primates. *Neurosci. Res. Progr. Bull.* **10,** 388–392.

Sackett, G. P. (1974). Sex differences in rhesus monkeys following rearing experiences. *In* "Sex Differences in Behavior" (R. C. Friedman, R. M. Richart and R. L. Vande Wiele, Eds), pp. 99–122. John Wiley and Sons, New York.

Sacks, W. (1973). Disorders of glucose metabolism in brain dysfunction. In "Biology of Brain Dysfunction" (G. E. Gaull, Ed.) Vol. 1, pp. 143–189, Plenum Press, New York.

Sacktor, B., Wilson, J. E. and Tiekert, C. G. (1966). Regulation of glycolysis in brain, *in situ,* during convulsions. *J. Biol. Chem.* **241,** 5071–5075.

Sahakian, B. J., Robbins, T. W., Morgan, M. J. and Iversen, S. D. (1975). The effects of psychomotor stimulants on stereotypy and locomotor activity in socially-deprived and control rats. *Brain Res.* **84,** 195–205.

Samorajski, T., Rolsten, C. and Ordy, J. M. (1971). Changes in behavior, brain, and neuroendocrine chemistry with age and stress in C57BL/10 male mice. *J. Gerontol.* **26,** 168–175.

Sandler, M. (1972). Catecholamine synthesis and metabolism in man: Clinical implications (with special reference to Parkinsonism) *In* "Handbook of Experimental Pharmacology" (H. Blaschko and E. Muscholl, Eds), XXXIII, pp. 845–899. Springer-Verlag, Berlin.

Sandoval, M.-E. and Tapia, R. (1975). GABA metabolism and cerebral protein synthesis. *Brain Res.* **96,** 279–286.

Sano, I., Kakimoto, Y., Kanazawa, A., Nakajima, T. and Shimizu, H. (1966). Identifizierung einiger glutamylpeptide aus gehirn. *J. Neurochem.* **13,** 711–719.

Sara, S. J. (1973). Recovery from hypoxia and ECS-induced amnesia after a single exposure to training environment. *Physiol. Behav.* **10,** 85–89.

Sassin, J. F., Parker, D. C., Mace, J. W., Gotlin, R. W., Johnson, L. C. and Rossman, L. G. (1969). Human growth hormone release: Relation to slow-wave sleep and sleep-walking cycles. *Science* **165,** 513–515.

Satoh, M., Zeiglgansberger, W. and Herz, A. (1976). Supersensitivity of cortical neurones of the rat to acetylcholine and L-glutamate following chronic morphine treatment. *Naunyn-Schmeideberg's Arch. Pharmacol.* **293,** 101–103.

Schadé, J. P., Van Backer, H. and Colon, E. (1964). Quantitative analysis of neuronal parameters in the maturing cerebral cortex. *In* "Growth and Maturation of the Brain; Progress in Brain Research" (D. P. Purpura and J. P. Schadé, Eds), Vol. 4, pp. 150–175. Elsevier, Amsterdam.

Schain, R. J. and Watanabe, K. (1975). Effect of chronic phenobarbital administration upon brain growth of the infant rat. *Exptl. Neurol.* **47,** 509–515.

Schain, R. J. and Watanabe, K. (1976). Origin of brain growth retardation in young rats treated with phenobarbital. *Exptl. Neurol.* **50,** 806–809.

Schalch, D. S. (1967). *J. Lab. Clin. Med.* **69,** 256. Cited by Kokka *et al.* (1971).

Schally, A. V., Arimura, A. and Kastin, A. J. (1973). Hypothalamic regulatory hormones. *Science* **179,** 341–350.

Schapiro, S. (1971). Hormonal and environmental influences on rat brain development and behavior. *In* "Brain Development and Behavior" (M. B. Sterman, D. J. McGinty and A. M. Adinolfi, Eds), pp. 307–334, Academic Press, New York and London.

Schapiro, S. and Vukovich, K. R. (1970). Early experience effects upon cortical dendrites; A proposed model for development. *Science* **167,** 292–294.

Schapiro, S., Vukovich, K. and Globus, A. (1973). Effects of neonatal thyroxine and hydrocortisone administration on the development of dendritic spines in the visual cortex of rats. *Exptl. Neurol.* **40,** 286–296.

Schaumberg, H. H., Byck, R., Gerstl, R. and Mashman, J. H. (1969). Monosodium L-glutamate: Its pharmacology and role in the Chinese restaurant syndrome. *Science* **163,** 826–828.

Scheel-Kruger, J. and Hasselager, E. (1974). Studies of various amphetamines, apomorphine and clonidine on body temperature and brain 5-hydroxytryptamine metabolism in rats. *Psychopharmacologia* **36,** 189–202.

Scheel-Kruger, J. and Randrup, A. (1967). Stereotype hyperactive behaviour produced by dopamine in the absence of noradrenaline. *Life Sci.* **6,** 1389–1398.

Scheibel, M. E. and Scheibel, A. B. (1964). Some structural and functional substrates of development in young cats. *In* "The Developing Brain; Progress in Brain Research" (W. Himwich and H. Himwich, Eds), pp. 6–25. Elsevier, Amsterdam.

Scheibel, M. E., Lindsay, R. D., Tomiyasu, U. and Scheibel, A. B. (1975). Progressive dendritic changes in aging human cortex. *Exptl. Neurol.* **47,** 392–403.

Scheinberg, P., Blackburn, I., Rich, M. and Saslaw, M. (1953). Effects of aging on cerebral circulation and metabolism. *Arch. Neurol. Psychiat.* **70,** 77–85.

Schildkraut, J. J. (1965). The catecholamine hypothesis of affective disorders: A review of supporting evidence. *Amer. J. Psychiat.* **122,** 509–522.

Schildkraut, J. J. and Kety, S. J. (1967). Biogenic amines and emotions. *Science* **156,** 21–30.

Schilkrut, R., Chandra, O., Osswald, M.. Rüther, E., Baarfüsser, B. and Matussek, N. (1975). Growth hormone release during sleep and with thermal stimulation in depressed patients. *Neuropsychobiology* **1,** 70–79.

Schlesinger, K., Boggan, W. O. and Freedman, D. X. (1965). Genetics of audiogenic seizures: I. Relation to brain serotonin and norepinephrine in mice. *Life Sci.* **4,** 2345–2351.

Schlesinger, K., Boggan, W. O. and Freedman, D. X. (1970). Genetics of audiogenic seizures: III. Time response relationships between drug administration and seizure susceptibility. *Life Sci.*, **9** (Part I): 721–729.

Schlosberg, H. (1954). Three dimensions of emotion. *Psychol. Rev.* **61,** 81–88.

Schlote, W., Betz, E. and Nguyen-Duong, H. (1975). Reversible apical swelling of dendrites in the cerebral cortex of cats during respiratory acidosis. *In* "Advances in Neurology; Physiology and Pathology of Dendrites" (G. W. Kreutzberg, Ed.), Vol. 12, pp. 483–495. Raven Press, New York.

Schmitt, F. O. (1968). The molecular biology of neuronal fibrous proteins. *Neurosci. Res. Progr. Bull.* **6,** 119–144.

Schneider, A. M. (1973). Spreading depression: A behavioral analysis. *In* "The Physiological Basis of Memory" (J. A. Deutsch, Ed.) pp. 269–303. Academic Press, New York and London.

Schubert, P. (1975). Transport in Dendriten einzelner Motoneurone. *Bull. Schweiz. Akad. Med. Wiss.* **30,** 56–65.

Schubert, P. and Krentzberg, G. W. (1974). Axonal transport of adenosine and uridine derivatives and transfer to postsynaptic neurons. *Brain Res.* **76,** 526–530.

Schubert, P. and Kreutzberg, G. W. (1975a). Dendritic and axonal transport of nucleoside derivatives in single motoneurones and release from dendrites. *Brain Res.* **90,** 319–323.

Schubert, P. and Kreutzberg, G. W. (1975b). [^{3}H] Adenosine, a tracer for neuronal connectivity. *Brain Res.* **85,** 317–319.

Schubert, P. and Kreutzberg, G. W. (1975c). Parameters of dendritic transport. *In* "Advances in Neurology; Physiology and Pathology of Dendrites" (G. W. Kreutzberg, Ed.), Vol. 12, pp. 255–268. Raven Press, New York.

Schubert, P., Lux, H. D. and Kreutzberg, G. W. (1971). Single cell isotope injection technique, a tool for studing axonal and dendritic transport. *Acta Neuropathol.* **5,** 179–186.

Schubert, P., Kreutzberg, G. W. and Lux, H. D. (1972). Neuroplasmic transport in dendrites: Effect of colchicine on morphology and physiology of motoneurones in the cat. *Brain Res.* **62,** 331–343.

Schubert, P., Lee, K., West, M., Deadwyler, S. and Lynch, G. (1976). Stimulation-dependent release of ^{3}H-adenosine derivatives from central axon terminals to target neurones. *Nature* (*Lond.*) **260,** 541–542.

Schultz, D. P. (1965). "Sensory Restriction; Effects on Behavior". Academic Press, New York and London, 216 pp.

Schwab, M. E. and Thoenen, H. (1975). Early effects of nerve growth factor on adrenergic neurons: An electron microscopic morphometric study of the rat superior cervical ganglion. *Cell Tiss. Res.* **158,** 543–553.

Schwartz, J. H., Castellucci, V. F. and Kandel, E. R. (1971). The functioning of identified neurons and synapses in the absence of protein synthesis. *J. Neurophysiol.* **34,** 939–953.

Schwartzkroin, R. A. and Wester, K. (1975). Long-lasting facilitation of a synaptic potential following tetanization in the *in vitro* hippocampal slice. *Brain Res.* **89,** 107–119.

Scott, J. P. (1964). The effects of early experience on social behavior and organization. *In* "Social Behavior and Organization Among Vertebrates" (W. Etkin, Ed.). University of Chicago Press, Chicago.

Scott, W. W., Scott, C. C. and Luckhardt, A. B. (1938). Observations on the blood sugar level before, during, and after hunger periods in humans, *Amer. J. Physiol.* **123,** 243–247.

Sechenov, I. M. (1952). "Collected Works", Vol. 1. Acad. Nauk, SSSR.

Segal, D. S. and Mandell, A. J. (1974). Differential behavioral effects of hypothalamic polypeptides. *In* "The Thyroid Axis, Drugs, and Behavior" (A. J. Prange, Jr., Ed.) pp. 129–133. Raven Press, New York.

Segal, D. S., Knapp, S., Kuczenski, R. T. and Mandell, A. J. (1973). The effects of environmental isolation on behavior and regional rat brain tyrosine hydroxylase and tryptophan hydroxylase activities. *Behav. Biol.* **8,** 47–53.

Sellinger, O. Z. and Azcurra, J. M. (1970). The breakdown of polysomes and the stimulation of protein synthesis as cerebral mechanisms of defense against seizures. *In* "Protein Metabolism of the Nervous System", pp. 517–543. Plenum Press, New York.

Sellinger, O. Z. and Nordrum, L. M. (1969). A regional study of some osmotic, ionic and age factors affecting the stability of cerebral lysosomes. *J. Neurochem.* **16,** 1219–1229.

Sellström, A., Sjoberg, L.-B. and Hamberger, A. (1975). Neuronal and glial systems for γ-aminobutyric acid metabolism. *J. Neurochem.* **25,** 393–398.

Selye, H. (1936). Thymus and adrenals in the response of the organism to injuries and intoxication. *Brit. J. Pathol.* **17,** 234.

Selye, H. (1956). "The Stress of Life". McGraw-Hill, New York.

Semiginovsky, B. and Jakoubec, B. (1971). Effects of ACTH on the incorporation of L-[U-^{14}C] leucine into the brain and spinal cord in mice. *Brain Res.* **35,** 319–323.

Seminario, L. M., Hren, N. and Gomez, C. J. (1964). Lipid distribution in subcellular fractions of the rat brain. *J. Neurochem.* **11,** 197–207.

Sereni, F., Principi, N., Perletti, L. and Sereni, L. P. (1966). Undernutrition and the developing rat brain: I. Influence on acetylcholinesterase and succinic acid dehydrogenase activities and on norepinephrine and 5-OH-tryptamine tissue concentrations. *Biol. Neonat.* **10,** 254–265.

Serota, H. M. and Gerard, R. W. (1938). Localized thermal changes in the cat's brain. *J. Neurophysiol.* **1,** 115–124.

Serota, R. G., Roberts, R. B. and Flexner, L. B. (1972). Acetoxycycloheximide-induced transient amnesia: Protective effects of adrenergic stimulants. *Proc. Nat. Acad. Sci., U.S.A.* **69,** 340–342.

Shamoo, A. E. (Ed.) (1975). "Carriers and Channels in Biological Systems". Annals of the New York Academy of Sciences, Vol. 264, 485 pp.

Shapiro, A. L., Vinuela, E. and Maizel, J. V., Jr. (1967). Molecular weight estimation of polypeptide chains by electrophoresis in SDS-polyacrylamide gels. *Biochem. Biophys. Res. Commun.* **28,** 815–820.

Shapiro, D., Barber, T. X., DiCara, L. V., Kamiya, J., Miller, N. E. and Stoyva, J. (Eds) (1973). "Biofeedback and Self-Control". Aldine Press, Chicago, 534 pp.

Sharp, F. R., Kauer, J. S. and Shepherd, G. M. (1975). Local sites of activity-related glucose metabolism in rat olfactory bulb during olfactory stimulation. *Brain Res.* **98,** 596–600.

Shashoua, V. E. (1974). RNA metabolism in the brain. *Int. Rev. Neurobiol.* **16,** 183–231.

Shashoua, V. E. (1975). Brain biochemistry and learning: Evidence for the synthesis of three specific proteins in goldfish brain. *Fifth Int. Meeting, Int. Soc. for Neurochemistry, Barcelona,* 2–6 September, Abstr, No. 379.

Sheard, M. H. (1969). The effect of p-chlorophenylalanine on behavior in rats: Relation to brain serotonin and 5-hydroxyindoleacetic acid. *Brain Res.* **15,** 524–528.

Sheer, D. E. (Ed.) (1961). "Electrical Stimulation of the Brain". University of Texas Press, Austin, Texas.

Shelanski, M. L., and Taylor, E. W. (1970). Biochemistry of neurofilaments and neurotubules. *In* "Alzheimer's Disease and Related Conditions" (G. E. W. Wolstenholme and M. O'Connor, Eds), pp. 249–266. J. and A. Churchill, Ltd., London.

Shelanski, M. L., Albert, S., De Vries, G. H. and Norton, W. T. (1971). Isolation o filaments from brain. *Science* **174,** 1242–1245.

Shelanski, M. L., Gaskin, F. and Cantor, C. R. (1973). Microtubule assembly in the absence of added nucleotides. *Proc. Nat. Acad. Sci., U.S.A.* **70,** 765–768.

Sherman, S. M. (1972). Visual development in cats. *Invest. Ophthalmol.* **11,** 394–402.

Sherrington, C. S. (1906). "Integrative Action of the Nervous System", Yale University Press, New Haven.

Shimizu, A. and Himwich, H. E. (1968). The effects of amphetamine on the sleep-wakefulness cycle of developing kittens. *Psychopharmacologia* **13,** 161–169.

Shimizu, H. and Ishii, S. (1964). Electron microscopic observation of catecholamine containing granules in the hypothalamus and the area postrema and their changes following reserpine injection. *Arch. Histol. Jap.* **24,** 489–497.

Shoemaker, W. J. and Wurtman, R. J. (1971). Perinatal undernutrition: Accumulation of catecholamines in rat brain. *Science* **171,** 1017–1019.

Shopsin, B., Stern, S. and Gershon, S. (1972). Altered carbohydrate metabolism during treatment with lithium carbonate. *Arch. Gen. Psychiat.* **26,** 566–571.

Shute, C. C. D. and Lewis, P. R. (1967). The ascending cholinergic reticular system: Neocortical, olfactory and subcortical projections. *Brain.* **90,** 497–520.

Siegel, S. (1975). Evidence from rats that morphine tolerance is a learned response. *J. Comp. Physiol. Psychol.* **89,** 498–506.

Sigg, E. B. (1969). Relationship of aggressive behaviour to adrenal and gonadal function in male mice. *In* "Aggressive Behaviour" (S. Garattini and E. B. Sigg, Eds), pp. 143–149. Excerpta Medica Fdn., Amsterdam.

Sigg, E. B., Day, C. and Colombo, C. (1966). Endocrine factors in isolation-induced aggressiveness in rodents. *Endocrinology* **78,** 679–684.

Silbergeld, E. K. and Goldberg, A. M. (1974). Hyperactivity: A lead-induced behavior disorder. Environmental Health Perspect., May issue.

Siler, T. M., Van den Berg, G. and Yen, S. S. C. (1973). Inhibition of growth hormone release in human by somatostatin. *J. Clin. Endocrinol. Metab.* **37,** 632–634.

Silver, A. (1967). Cholinesterases of the central nervous system with special reference to the cerebellum. *Int. Rev. Neurobiol.* **10,** 57–109.

Simantov, R., Kuhar, M. J., Pasternak, G. W. and Snyder, S. H. (1976a). The regional distribution of a morphine-like factor enkephalin in monkey brain. *Brain Res.* **106,** 189–197.

Simantov, R., Oster-Granite, M. L., Herndon, R. M. and Snyder, S. H. (1976b). Gamma-aminobutyric acid (GABA) receptor binding selectively depleted by viral induced granule cell loss in hamster cerebellum. *Brain Res.* **105,** 365–371.

Sims, K. L. and Pitts, F. N., Jr. (1970). Brain glutamate decarboxylase: Changes in the developing rat brain. *J. Neurochem.* **17,** 1607–1612.

Sinclair, D. C. (1955). Cutaneous sensation and the doctrine of specific energy. *Brain* **78,** 584–614.

Singh, V. B. and Talwar, G. P. (1969). Identification of a protein fraction in the occipital cortex of the monkey rapidly labelled during the exposure of the animal to rhythmically flickering light. *J. Neurochem.* **16,** 951–959.

Sinha, A. K. and Rose, S. P. R. (1972). Neuronal locus of enhanced incorporation of ^3H-lysine into acid insoluble material in visual cortex on first exposure of rats to light. *Life Sci.* **11,** (Part II), 663–668.

Sisken, B., Sano, K. and Roberts, E. (1961). γ-Aminobutyric acid content and glutamic decarboxylase and γ-aminobutyrate transaminase activities in the optic lobe of the developing chick. *J. Biol. Chem.* **236,** 503–507.

Sjöstrand, J. (1965). Proliferative changes in glial cells during nerve regeneration. *Z. Zellforsch.* **68,** 481–493.

Sjöstrand, J. (1970). Fast and slow components of axoplasmic transport in the hypoglossal and vagus nerves of the rabbit. *Brain Res.* **18,** 461–467.

Skinner, B. F. (1938). "The Behavior of Organisms". Appleton-Century-Crofts, New York, 457 pp.

Skinner, B. F. (1953). "Science and Human Behavior". Macmillan, New York, 461 pp.

Sklenovsky, A. (1967). *Acta Univ. Palacki Olomuc., Fac. Med.* **44,** 151; cited by Ghazal *et al.* (1976).

Sloboda, R. D., Rudolph, S. A., Rosenbaum, J. L. and Greengard, P. (1975). Cyclic AMP-dependent endogenous phosphorylation of a microtubule-associated protein. *Proc. Nat. Acad. Sci., U.S.A.* **72,** 177–181.

Smart, J. L. (1975). Lowered threshold of behavioural arousal after early growth restriction in rats. *First European Neurosciences Meeting, Munich,* 9–13 September, Abstr. No. 378.

Smith, C. E. (1962). Is memory a matter of enzyme induction? *Science* **138,** 889–890.

Snyder, S. H. and Axelrod, J. (1965). Circadian rhythm pineal serotonin: Effect of monomaine oxidase inhibition and reserpine. *Science* **149,** 562.

Sokoloff, L. (1961). Local cerebral circulation at rest and during altered cerebral activity induced by anesthesia or visual stimulation. *In* "Regional Neurochemistry" (S. S. Kety and J. Elkes, Eds), pp. 107–117. Pergamon Press, London.

Sokoloff, L. (1975). Measurement of local glucose utilization and its use in mapping local functional activity in the central nervous system. *Proc. IV World Congress of Psychiatric Surgery,* Madrid, 7–12 September, (Abstr. No. 25).

Sokoloff, L., Reivich, M., Patlak, C. S., Pettigrew, K. D., Des Rosiers, M. and Kennedy, C. (1974). The [^{14}C] deoxyglucose method for the quantitative determination oflocal cerebral glucose consumption. *Proc. 5th Meeting Amer. Soc. Neurochem.,* p. 86. Pergamon Press, New York.

Sollberger, A. (1971). Biological rhythms and their control in neurobehavioral perspective. *In* "Neurosciences Research" (S. Ehrenpreis and O. C. Solnitzky, Eds), Vol. 4, pp. 101–163. Academic Press, New York and London.

Sols, A. and Crane, R. K. (1954). Substrate specificity of brain hexokinase. *J. Biol. Chem.* **210,** 581–595.

Sols, A. and Marco, R. (1970). Concentrations of metabolites and binding sites. Implications in metabolic regulation. *In* "Current Topics in Cellular Regulation" (B. L. Horecker and E. R. Stadtman, Eds), Vol. 2, pp. 227–273. Academic Press, New York and London.

Sosula, L. and Glow, P. H. (1971). Increase in number of synapses in the inner plexiform layer in light deprived rat retinae: Quantitative electron microscopy. *J. Comp. Neurol.* **141,** 427–452.

Sotelo, C. (1975a) Anatomical, physiological and biochemical studies of the cerebellum from mutant mice. II. Morphological study of cerebellar cortical neurons and circuits in the *weaver* mouse. *Brain Res.* **94**, 19–44.

Sotelo, C. (1975b) Dendritic abnormalities of Purkinje cells in the cerebellum of neurologic mutant mice (weaver and staggerer). *In* "Advances in Neurology; Physiology and Pathology of Dendrites" (G. W. Kreutzberg, Ed.), Vol. 12, pp. 335–351. Raven Press, New York.

Sourkes, T. L. (1962). "Biochemistry of Mental Disease". Harper and Row, New York, 402 pp.

Spector, S., Gordon, R., Sjoerdsma, A. and Udenfriend, S. (1967). End-product inhibition of tyrosine hydroxylase as a possible mechanism for regulating norepinephrine synthesis. *Mol. Pharmacol.* **3,** 549–555.

Spehlmann, R. (1975). The effects of acetylcholine and dopamine on the caudate nucleus depleted of biogenic amines. *Brain* **98,** 219–230.

Spitz, R. A. (1945). Hospitalism, an inquiry into the genesis of psychiatric conditions in early childhood. *Psychoanal. Stud. Childh.* **1,** 53–74.

Spitz, R. A. (1946). Anaclitic depression. *Psychoanal. Stud. Childh.* **2,** 313–347.

Spragg, S. D. S. (1940). Morphine addiction in chimpanzees. *Comp. Psychol. Monogr.* **15,** No. 7.

Springer, A. D., Schoel, W. M., Klinger, P. D. and Agranoff, B. W. (1975). Anterograde and retrograde effects of electroconvulsive shock and of puromycin on memory formation in the goldfish. *Behav. Biol.* **13,** 467–481.

Sprott, R. L. and Staats, J. (1975). Behavioral studies using genetically defined mice —A bibliography. *Behav. Genet.* **5,** 27–82.

Spurzheim, J. G. (1815). "The Physiognomical System of Drs. Gall and Spurzheim". Baldwin, Cradock and Joy, London, 2nd Edn, pp. 554–555.

Squire, L. (1974). Remote memory as affected by aging. *Neuropsychologia* **12,** 429–435.

Squire, L. R. and Barondes, S. H. (1970). Actinomycin-D: Effects on memory at different times after training. *Nature* (*Lond.*) **225,** 649–650.

Squire, L. R. and Barondes, S. H. (1972). Inhibitors of cerebral protein or RNA synthesis and memory. *In* "Macromolecules and Behavior" (J. Gaito, Ed.), 2nd Edn, pp. 61–82. Appleton-Century-Crofts, New York.

Squire, L. R. and Barondes, S. H. (1974). Anisomycin, like other inhibitors of cerebral protein synthesis, impairs "long-term" memory of a discrimination task. *Brain Res.* **66,** 301–308.

Squire, L. R. and Becker, C. L. (1975). Inhibition of cerebral protein synthesis impairs long-term habituation. *Brain Res.* **97,** 367–372.

Squire, L. R. and Davis, H. P. (1975). Cerebral protein synthesis inhibition and discrimination training: Effects of extent and duration of inhibition. *Behav. Biol.* **13,** 49–57.

Squire, L. R., Emanuel, C. A., Davis, H. P. and Deutsch, J. A. (1975). Inhibitors of cerebral protein synthesis: Dissociation of aversive and amnesic effects. *Behav. Biol.* **14,** 335–341.

Stein, B. E., Magalhaes-Castro, B. and Kruger, L. (1976). Relationship between visual and tactile representations in cat superior colliculus. *J. Neurophysiol.* **39,** 401–419.

Stein, G. S., Spelsberg, T. C. and Kleinsmith, L. J. (1974). Nonhistone chromosomal proteins and gene regulation. *Science* **183,** 817–824.

Stephan, H., Bauchot, R. and Andy, O. J. (1970). Data on the size of the brain and of various brain parts in insectivores and primates. *Advanc. Primatol.* **1,** 289–298.

Stern, P. and Potkonjak, D. (1970). The correlation between degeneration of the nigro-neostriatal dopamine neuron system and oxotremorine and guanethidine induced tremor. *Pharmakopsychiatrie Neuro-Psychopharmakologie,* **3,** 127–131.

Stern, W. C., Miller, F. P., Cox, R. H. and Maickel, R. P. (1971). Brain norepinephrine and serotonin levels following REM sleep deprivation in the rat. *Psychopharmacologia* **22,** 50–55.

Stewart, W. J. (1975). Environmental complexity does affect scopolamine-induced changes in activity. *Neurosci. Lett.* **1,** 121–125.

St. Laurent, J., Roizen, M. F., Miliaressis, E. and Jacobowitz, D. M. (1975). The effects of self-stimulation on the catecholamine concentration of discrete areas of the rat brain. *Brain Res.* **99,** 194–200.

Stock, M. B. and Smithe, P. M. (1967). The effect of undernutrition during infancy on subsequent brain growth and intellectual development. *S. Afr. Med. J.* **41,** 1027–1029.

Stöckel, K., Solomon, F., Paravicini, U. and Thoenen, H. (1974). Dissociation between effects of nerve growth factor on tyrosine hydroxylase and tubulin synthesis in sympathetic ganglia. *Nature* (*London.*) **250,** 150–151.

Stolk, J. M., Kaufman, M. E., Hanlon, D. P., Barchas, J. D. and Nisula, B. C. (1975). Brain dopamine-β-hydroxylase activity in vivo: Pharmacological, endocrinological, and psychological implications for adaptation. *In* "Neurobiological Mechanisms of Adaptation and Behavior" (A. J. Mandell, Ed.), pp. 199–245. Raven Press, New York.

Stone, E. A. (1970). Swim-stress-induced inactivity: Relation to body temperature and brain norepinephrine, and effects of *d*-amphetamine. *Psychosomatic Med.* **32,** 51–59.

Sugaya, E., Takato, M. and Noda, Y. (1975). Neuronal and glial activity during spreading depression in cerebral cortex of cat. *J. Neurophysiol.* **38,** 822–841.

Sun, M.-C., McIlwain, H. and Pull, I. (1976). The metabolism of adenine derivatives in different parts of the brain of the rat, and their release from hypothalamic preparations on excitation. *J. Neurobiol.* **7,** 109–122.

Suomi, S. J. and Harlow, H. F. (1972). Depressive behavior in young monkeys subjected to vertical chamber confinement. *J. Comp. Physiol. Psychol.* **180,** 11–18.

Suomi, S. J., Harlow, H. F. and McKinney, W. T., Jr. (1972). Monkey psychiatrists. *Amer. J. Psychiat.* **128,** 927–932.

Suomi, S. J., Harlow, H. F. and Novak, M. A. (1974). Reversal of social deficits produced by isolation rearing in monkeys. *J. Human Evolution* **3,** 527–534.

Susz, J. P., Hof, H. I. and Brunngraber, D. G. (1973). Isolation of concanavalin A-binding glycoproteins from rat brain. *FEBS Letters* **32,** 289–292.

Suzuki, K. (1965). The pattern of mammalian brain gangliosides—II. Evaluation of the extraction procedures, post mortem changes and the effect of formalin preservation. *J. Neurochem.* **12,** 629–638.

Suzuki, K. (1972). Neurochemical aspects of mucopolysaccharidoses. *In* "Handbook of Neurochemistry" (A. Lajtha, Ed.), Vol. 7, pp. 17–32. Plenum Press, New York.

Suzuki, K. and Zagoren, J. C. (1976). Variations of Schmidt-Lanterman incisures in Quaking mouse. *Brain Res.* **106,** 146–151.

Svendgaard, N. A., Björklund, A. and Stenevi, U. (1976). Regeneration of central cholinergic neurones in the adult rat brain. *Brain Res.* **102,** 1–22.

Svennerholm, L. (1970). Gangliosides. *In* "Handbook of Neurochemistry" (A. Lajtha, Ed.), Vol. 3, pp. 425–452. Plenum Press, New York.

Sytinsky, I. A., Guzikov, B. M., Gomanko, M. V., Eremin, V. P. and Konovalova, N. N. (1975). The gamma-aminobutyric acid (GABA) system in brain during acute and chronic ethanol intoxication. *J. Neurochem.* **25,** 43–48.

Szasz, T. S. (1973). Mental illness as a metaphor. *Nature* (*Lond.*) **242,** 305–307.

Sze, P. (1970). Neurochemical factors in auditory stimulation and development of susceptibility to audiogenic seizures. *In* "Physiological Effects of Noise", pp. 259–269. Plenum Press, New York.

Sze, P. Y. and Maxson, S. C. (1976). Involvement of corticosteroids in acoustic induction of audiogenic seizure susceptibility in mice. *Psychopharmacologia* (In press).

Takahashi, T., Konishi, S., Powell, D. and Leeman, S. (1974). Identification of the motoneuron-depolarizing peptide in bovine dorsal root as hypothalamic substance P. *Brain Res.* **73,** 59–69.

Takemori, A. E. (1971). Intermediary and energy metabolism. *In* "Narcotic Drugs: Biochemical Pharmacology" (D. H. Clouet, Ed.), pp. 159–189. Plenum Press, New York.

Tallan, H. H. (1962). A survey of the amino acids and related compounds in nervous tissue. *In* "Amino Acid Pools" (J. T. Holden, Ed.), pp. 471–485. Elsevier, Amsterdam.

Talwar, G. P. and Iqbal, Z. (1972). Macromolecules—functional and biochemical correlates. *In* "Macromolecules and Behavior" (J. Gaito, Ed.), 2nd Edn, pp. 13–38. Appleton-Century-Crofts, New York.

Talwar, G. P. and Singh, V. B. (1971). Excitation. *In* "Handbook of Neurochemistry" (A. Lajtha, Ed.), Vol. 6, pp. 29–61. Plenum Press, New York.

Talwar, G. P., Sadasivudu, B. and Chitre, V. S. (1961). Changes in the pentose-nucleic acid content of sub-cellular fractions of the brain of the rat during "Metrazol" convulsions. *Nature* (*Lond.*) **191,** 1007–1008.

Talwar, G. P., Chopra, S. P., Goel, B. K. and D'Monte, B. (1966). Correlation of the functional activity of the brain with metabolic parameters—III: Protein metabolism of the occipital cortex in relation to light stimulus. *J. Neurochem.* **13,** 109–116.

Tannhauser, M. and Izquierdo, I. (1974). Effect of convulsant and anticonvulsant agents on hippocampal RNA concentration. *Pharmacology* **11,** 139–145.

Tanzi, E. (1893). I fatti e le induzioni nell' odierna istologia del sistema nervoso. *Riv. Sper. Freniat.* **19,** 419–472.

Tapia, R. (1975). Biochemical pharmacology of GABA in CNS. *In* "Handbook of Psychopharmacology" (L. L. Iversen, S. D. Iversen and S. H. Snyder, Eds), Vol. 4, pp. 1–58. Plenum Press, New York.

Tapia, R. and Massieu, G. H. (1969). Estudios sobre algunus vias metabólicas en el sistema nervioso central y su relación con las funciones nerviosas. *In* "Ensayos Bioquimicos" (G. Soberón, Ed.), pp. 185–203. Prensa Médica Mexicana, México.

Tapia, R. and Sandoval, M. E. (1974). Possible participation of γ-aminobutyric acid in the regulation of protein synthesis in brain *in vivo*. *Brain Res.* **69,** 255–263.

Tapp, J. T. and Markowitz, H. (1963). Infant handling: Effects on avoidance learning, brain weight, and cholinesterase activity. *Science* **140,** 486–487.

Tatetsu, S. (1974). On histologic findings in schizophrenia and schizophrenic state. *In* "Biological Mechanisms of Schizophrenia and Schizophrenia-like Psychoses" (H. Mitsuda and T. Fukuda, Eds), p. 288. Igaku Shoin Ltd., Tokyo.

Tatetsu, S., Goto, A. and Fujiwara, T. (1956). "The Methamphetamine Psychosis". Igaku Shoin Ltd., Tokyo.

Tellez-Nagel, I., Johnson, A. B. and Terry, R. D. (1973). Ultrastructural and histochemical study of cerebral biopsies. *In* "Huntington's Chorea: 1872–1972, Advances in Neurology" (A. Barbeau, T. N. Chase and G. W. Paulson, Eds), Vol. 1, pp. 387–398. Raven Press, New York.

Tennyson, V. M., Kremzner, L. T., Dauth, G. W. and Gilman, S. (1975). Chronic cerebellar stimulation in the monkey: Electron microscopic and biochemical observations. *Neurology* **25,** 650–654.

Terenius, L. and Wahlstrom, A. (1975). Morphine-like ligand for opiate receptors in human CSF. *Life Sci.* **16,** 1759–1764.

Terry, R. D. (1963). The fine structure of neurofibrillary tangles in Alzheimer's disease. *J. Neuropathol. Exptl. Neurol.* **22,** 629–642.

Terry, R. D. and Wisniewski, H. M. (1970). The ultrastructure of the neurofibrillary tangle and the senile plaque. *In* "CIBA Foundation Symposium on Alzheimer's Disease and Related Conditions" (G. E. W. Wolstenholme and M. O'Connor, Eds), pp. 145–168. Churchill, London.

Terry, R. D. and Wisniewski, H. M. (1972). Ultrastructure of senile dementia and of experimental analogs. *In* "Aging and the Brain, Adv. Behav. Biol." (C. M. Gaitz, Ed.), Vol. 3, pp. 89–116.

Teschemacher, H., Opheim, K. E., Cox, B. M. and Goldstein, A. (1975). A peptide-like substance from pituitary that acts like morphine. 1. Isolation. *Life Sci.* **16,** 1771–1776.

Tews, J. K. and Stone, W. E. (1965). Free amino acids and related compounds in brain and other tissues: Effects of convulsant drugs. *In* "Progress in Brain Research" (W. A. Himwich and J. P. Schadé, Eds), Vol. 16, pp. 135–163. Elsevier, Amsterdam.

Thesleff, S. (1962). A neurophysiological speculation concerning learning. *Prospect. Biol. Med.* **5,** 293–295.

Thierry, A.-M., Javoy, F., Glowinski, J. and Kety, S. S. (1968). Effects of stress on the metabolism of norepinephrine, dopamine and serotonin in the central nervous system of the rat. 1. Modification of norepinephrine turnover. *J. Pharmacol. Exptl. Therapeut.* **163,** 163–171.

Thierry, A.-M., Blanc, G. and Glowinski, J. (1970). Preferential utilization of newly synthesized norepinephrine in the brain stem of stressed rats. *Eur. J. Pharmacol.* **10,** 139–142.

Thoa, N. B., Eichelman, B. and Ng, L. K. Y. (1972). Shock induced aggression: Effects of 6-hydroxydopamine and other pharmacological agents. *Brain Res.* **43,** 467–475.

Thoenen, H. (1975). Transynaptic regulation of neuronal enzyme synthesis. *In* "Handbook of Psychopharmacology" (L. L. Iversen, S. D. Iversen and S. H. Snyder, Eds), Vol. 3, pp. 443–475. Plenum Press, New York.

Thoenen, H. and Otten, U. (1975). Trans-synaptic enzyme induction in adrenergic neurons as a model for neuronal plasticity. *First European Neurosciences Meeting, Munich.* 9–13 September, Abstr. No. 390.

Thoenen, H. and Stöckel, K. (1975). Ortho- and retrograde axonal transport: Importance for the function of adrenergic neurones. *Clin. Exptl. Pharmacol. Physiol. Suppl. 2,* 1–5.

Thoenen, H., Mueller, R. A. and Axelrod, J. (1970). Phase difference in the induction of tyrosine hydroxylase in cell body and nerve terminals of sympathetic neurones. *Proc. Nat. Acad. Sci., U.S.A.* **65,** 58–62.

Thoenen, H., Angeletti, P. U., Levi-Montalcini, R. and Kettler, R. (1971). Selective induction by nerve growth factor of tyrosine hydroxylase and dopamine-β-hydroxylase in rat superior cervical ganglia. *Proc. Nat. Acad. Sci., U.S.A.* **68,** 1598–1602.

Thoenen, H., Saner, A., Angeletti, P. U. and Levi-Montalcini, R. (1972a). Increased activity of choline acetyltransferase in sympathetic ganglia after prolonged administration of nerve growth factor. *Nature* (*Lond.*) **236,** 26–28.

Thoenen, H., Saner, A., Kettler, R. and Angeletti, P. U. (1972b). Nerve growth factor and preganglionic cholinergic nerves: Their relative importance to the development of the terminal adrenergic neuron. *Brain Res.* **44,** 593–602.

Thompson, W. R. and Heron, W. (1954). The effects of restricting early experience on the problem-solving capacity of dogs. *Can. J. Psychol.* **8,** 17–31.

Thompson, R. and Pryor, R. S. (1956). The effect of anoxia on the retention of a discrimination habit. *J. Comp. Physiol. Psychol.* **49,** 297–300.

Tomasch, J. (1971). Comments on "neuromythology". *Nature* (*Lond.*) **233,** 60.

Tower, D. B. (1958). The effects of 2-deoxy-D-glucose on metabolism of slices of cerebral cortex incubated *in vitro*. *J. Neurochem.* **3,** 185–205.

Trachtenberg, M. C. and Pollen, D. A. (1970). Neuroglia: Biophysical properties and physiologic function. *Science* **167,** 1248–1252.

Treistman, S. N. and Levitan, I. B. (1976). Alteration of electrical activity in molluscan neurones by cyclic nucleotides and peptide factors. *Nature* (*Lond.*) **261,** 62–64.

Tsou, K. and Jang, C. S. (1964). Studies on the site of analgesic action of morphine by intracerebral micro-injection. *Sci. Sinica* **13,** 1099–1109.

Tsukahara, N. and Fujito, Y. (1976). Physiological evidence of formation of new synapses from cerebrum in the red nucleus neurons following cross-union of forelimb nerves. *Brain Res.* **106,** 184–188.

Tsukahara, N., Hultborn, H., Murakami, F. and Fujito, Y. (1975). Electrophysiological study of the formation of new synapses and collateral sprouting in the red nucleus neurons after partial denervation. *J. Neurophysiol.* **38,** 1359–1372.

Tucek, S. and Cheng, S.-C. (1970). Precursors of acetyl groups in acetylcholine in the brain *in vivo*. *Biochim. Biophys. Acta* **208,** 538–540.

Tunnicliff, G., Wimer, C. C. and Wimer, R. E. (1973). Relationships between neurotransmitter metabolism and behaviour in seven inbred strains of mice. *Brain Res.* **61,** 428–434.

Ueda, T., Maeno, H. and Greengard, P. (1973). Regulation of endogenous phosphorylation of specific proteins in synaptic membrane fractions from rat brain by adenosine 3′:5′-monophosphate. *J. Biol. Chem.* **248,** 8295–8305.

Ungar, G. (1968). Molecular mechanisms in learning. *Perspect. Biol. Med.* **11,** 217–232.

Ungar, G. (1973a). Evidence for molecular coding of neural information. *In* "Memory and Transfer of Information" (H. P. Zippel, Ed.), pp. 317–341. Plenum Press, New York.

Ungar, G. (1973b). Evidence for molecular coding of memory. *Proc. 1st Latin Amer. Congr. Psychobiolology, Sao Paulo*, Abstr., pp. 43–44.

Ungar, G. (1975a). Peptides and behavior. *Int. Rev. Neurobiol.* **17,** 37–60.

Ungar, G. (1975b). Is there a chemical memory trace? *Symp. on Chemistry of Memory, IUPAC Congress, Jerusalem*, pp. 1–8.

Ungar, G. and Cohen, M. (1966). Induction of morphine tolerance by material extracted from brain of tolerant animals. *Int. J. Neuropharmacol.* **5,** 183–192.

Ungar, G. and Irwin, L. N. (1967). Transfer of acquired information by brain extracts. *Nature* (*Lond.*) **214,** 453–455.

Ungar, G. and Ocequera-Narvarro, C. (1965). Transfer of habituation by material extracted from brain. *Nature* (*Lond.*) **207,** 301–302.

Ungar, G., Galvan, L. and Clark, R. H. (1968). Chemical transfer of learned fear. *Nature* (*Lond.*) **217,** 1259–1261.

Ungar, G., Desiderio, D. M. and Parr, W. (1972). Isolation, identification, and synthesis of a specific-behavior-inducing brain peptide. *Nature* (*Lond.*) **238,** 198–202.

Ungerstedt, U. (1968). 6-Hydroxydopamine induced degeneration of central monoamine neurons. *Eur. J. Pharmacol.* **5,** 107–110.

Ungerstedt, U. (1971). On the anatomy, pharmacology and function of the nigrostriatal dopamine system. *Acta Physiol. Scand.*, Suppl. 367.

Ungerstedt, U. (1974). Functional dynamics of central monoamine pathways. *In* "The Neurosciences, Third Study Program" (F. O. Schmitt and F. G. Worden, Eds), pp. 979–988. MIT Press, Cambridge, Mass.

Uphouse, L. L., MacInnes, J. W. and Schlesinger, K. (1972). Effects of conditioned avoidance training on polyribosomes of mouse brain. *Physiol. Behav.* **8,** 1013–1018.

Urban, I. and DeWied, D. (1975a). Differences in hippocampal RSA in vasopressin deprived Brattleboro rats: Effect of DG-AVP. *First European Neurosciences Meeting, Munich,* 9–13 September, Abstr. No. 401.

Urban, I. and DeWied, D. (1975b). Inferior quality of RSA during paradoxical sleep in rats with hereditary diabetes insipidus. *Brain Res.* **97,** 362–366.

Urban, I., Lopes da Silva, F. H., Strom van Leeuwen, W. and DeWied, D. (1974). A frequency shift in the hippocampal theta activity: An electrical correlate of central action of ACTH analogues in the dog. *Brain Res.* **69,** 361–365.

Utakoji, T. and Hsu, T. C. (1965). Nucleic acids and protein synthesis of isolated cells from chick embryonic spinal ganglia in culture. *J. Exptl. Zool.* **158,** 181–202.

Utena, H. (1974). On relapse-liability; schizophrenia, amphetamine psychosis and animal model. *In* "Bioloigical Mechanisms of Schizophrenia and Schizophrenia-like Psychoses" (H. Mitsuda and T. Fukuda, Eds), pp. 285–287. Igaku Shoin Ltd., Tokyo.

Valcana, T. and Timiras, P. S. (1969). Effect of hypothyroidism on ionic metabolism and Na-K activated ATP phosphohydrolase activity in the developing rat brain. *J. Neurochem.* **16,** 935–943.

Valtin, H. (1967). Hereditary hypothalamic diabetes insipidus in rats (Brattleboro strain). A useful experimental model. *Amer. J. Med.* **42,** 814–827.

Valtin, H. and Schroeder, H. A. (1964). Familial hypothalamic diabetes insipidus in rats (Brattleboro strain). *Amer. J. Physiol.* **206,** 425–430.

Valverde, F. (1967). Apical dendritic spines of the visual cortex and light deprivation in the mouse. *Exptl. Brain Res.* **3,** 337–352.

Valverde, F. (1968). Structural changes in the area striata of the mouse after enucleation. *Exptl. Brain Res.* **5,** 274–292.

Valverde, F. (1971). Rate and extent of recovery from dark rearing in the visual cortex of the mouse. *Brain Res.* **33,** 1–11.

Valverde, F. and Esteban, M. E. (1968). Peristriate cortex of mouse: Location and the effects of enucleation on the number of dendritic spines. *Brain Res.* **9,** 145–148.

Valverde, F. and Ruiz-Marcos, A. (1967). Light deprivation and the spines of apical dendrites of the visual cortex of the mouse. *Anat. Rec.* **157,** 392.

Valverde, F. and Ruiz-Marcos, A. (1969). Dendritic spines in the visual cortex of the mouse. Introduction to a mathematical model. *Exptl. Brain Res.* **8,** 269–283.

Valzelli, L. (1967). Drugs and aggressiveness. *Adv. Pharmacol.* **5,** 79–108.

Valzelli, L. (1971). Aggressivité chez le rat et la souris: Aspects compartmentaux et biochimiques. *Actual. Pharmacol.* **24,** 133–152.

Valzelli, L. (1973a). The "isolation syndrome" in mice. *Psychopharmacologia* **31,** 305–320.

Valzelli, L. (1973b). Environmental influences upon neurometabolic processes in learning and memory. *In* "Current Biochemical Approaches to Learning and Memory" (W. B. Essman and S. Nakajima, Eds), pp. 29–47. Spectrum Publ., Flushing, New York.

Valzelli, L. and Garattini, S. (1968). Behavioral changes and 5-HT turnover in animals. *Adv. Pharmacol.* **6B,** 249–260.

Van den Berg, C. J. (1970). Glutamate and glutamine. *In* "Handbook of Neurochemistry" (A. Lajtha, Ed.), Vol. 3, pp. 355–379. Plenum Press, New York.

Van Gelder, N. M. (1971). Brain weight and growth of mice fed gamma-amino-butyric acid, glycine or L-glutamic acid diet. *Brain Res.* **33,** 571–577.

Van Harreveld, A. (1958). A change in the diameter of apical dendrites during spreading depression. *Amer. J. Physiol.* **192,** 457–468.

Van Harreveld, A. (1966). "Brain Tissue Electrolytes". Butterworths, London.

Van Harreveld, A. (1972). The extracellular space in the vertebrate central nervous system. *In* "The Structure and Function of Nervous Tissue" (G. H. Bourne, Ed.), Vol. IV, pp. 447–511. Academic Press, New York and London.

Van Harreveld, A. and Fifkova, E. (1973). Mechanism involved in spreading depression. *J. Neurobiol.* **4,** 375–388.

Van Harreveld, A. and Fifkova, E. (1974). Involvement of glutamate in memory formation. *Brain Res.* **81,** 455–467.

Van Harreveld, A. and Fifkova, E. (1975a). A mechanism for potentiation and short term memory. *Koninkl. Nederl. Akad. Van Wetenschappen, Proc. Ser. C.* **78,** 21–24.

Van Harreveld, A. and Fifkova, E. (1975b). Swelling of dendritic spines in the fascia dentata after stimulation of the perforant fibers as a mechanism of post-tetanic potentiation. *Exptl. Neurol.* **49,** 736–749.

Van Hof, M. W. (1967). Visual acuity in the rabbit. *Vision Res.* **7,** 749–751.

Van Hof, M. W. (1974). Retinal input and early development of the visual system. *Documenta Ophthalmologica Proceedings Series, XIth I.S.C.E.R.G. Symposium,* pp. 495–500. Dr. W. Junk B. V. Publishers. The Hague, The Netherlands.

Van Hof, M. W. and Kobayashi, K. (1972). Pattern discrimination in rabbits deprived of light for 7 months after birth. *Exptl. Neurol.* **35,** 551–557.

Van Hof-Van Duin, J. (1976a). Development of visuomotor behavior in normal and dark-reared cats. *Brain Res.* **104,** 233–241.

Van Hof-Van Duin, J. (1976b). Early and permanent effects of monocular deprivation on pattern discrimination and visuomotor behavior in cats. *Brain Res.* **111,** 261–276.

Van Rossum, J. M. (1967). The significance of dopamine-receptor blockade for the action of neuroleptic drugs. *In* "Neuropsychopharmacology" (H. Brill, Ed.), pp. 321–329. Excerpta Medica Foundation, New York.

Van Wimersma Greidanus, Tj. B. and DeWied, D. (1969). Effects of intracerebral implantation of corticosteroids on extinction of an avoidance response in rats. *Physiol. Behav.* **4,** 365–370.

Van Wimersma Greidanus, Tj. B. and DeWied, D. (1971). Effects of systemic and intracerebral administration of two opposite acting ACTH-related peptides on extinction of conditioned avoidance behavior. *Neuroendocrinology* **7,** 291–301.

Van Wimersma Greidanus, Tj. B., Balvers, A., Rapis, M., Wielaard, W. and DeWied, D. (1975a). Limbic midbrain structures and the effect of ACTH analogues and of vasopressin on avoidance behavior. *First European Neurosciences Meeting, Munich,* 9–13 September, Abstr. No. 406.

Van Wimersma Greidanus, Tj. B., Dogterom, J. and DeWied, D. (1975b). Intraventricular administration of anti-vasopressin serum inhibits memory consolidation in rats. *Life Sci.* **16,** 637–644.

Vardaris, R. M. and Schwartz, K. E. (1971). Retrograde amnesia for passive avoidance produced by stimulation of dorsal hippocampus. *Physiol. Behav.* **6,** 131–135.

Varon, S. (1975). Nerve growth factor and its mode of action. *Exptl. Neurol.* **48** (Part 2), 75–92.

Vaughn, J. E. and Skoff, R. P. (1972). Neuroglia in experimentally altered central nervous system. *In* "The Structure and Function of Nervous Tissue" (G. H. Bourne, Ed.), Vol. V, pp. 39–72. Academic Press, New York and London.

Vergnes, M. and Karli, P. (1965). Étude des voies nerveuses d'une influence inhibitrice s'exerçant sur l'aggressivité interspecifique du rat. *C.R. Soc. Biol. de Strasbourg* **159,** 972–975.

Verhave, T. (1958). Effect of methamphetamine on operant level and avoidance behavior. *J. Exptl. Anal. Behav.* **1,** 207–217.

Vernadakis, A. and Woodbury, D. M. (1960). Effects of diphenylhydantoin and adrenocortical steroids on free glutamic acid, glutamine and gamma-aminobutyric acid concentrations of rat cerebral cortex. *In* "Inhibition in the Nervous System and Gamma-Aminobutyric Acid" (E. Roberts, Ed.), pp. 242–248. Macmillan (Pergamon), New York.

Verndakis, A., Valcana, T., Curry, J. J., Maletta, G. J., Hudson, D. and Timiras, P. S. (1967). Alterations in growth of brain and other organs after electroshock in rats. *Exptl. Neurol.* **17,** 505–516.

Verzar, F. (1968). Intrinsic and extrinsic factors of molecular aging. *Exptl. Gerontol.* **3,** 69–75.

Vesco, C. and Giuditta, A. (1968). Disaggregation of brain polysomes induced by electroconvulsive treatment. *J. Neurochem.* **15,** 81–85.

Vierck, C. J. (1974). Tactile movement detection and discrimination following dorsal column lesions in monkeys. *Exptl. Brain Res.* **20,** 331–346.

Virchow, R. (1846). Ueber das granulierte Aussehen der Windungen der Gehirnventrikel. *Allg. Z. Psychiat.* **3,** 242–885.

Vogel, W. H., Orjei, V. and Century, B. (1969). Activities of enzymes involved in the formation and destruction of biogenic amines in various areas of human brain. *J. Pharmacol. Exptl. Therapeut.* **165,** 196–203.

Vogt, M. (1973). Functional aspects of catecholamines in central nervous system. *Brit. Med. Bull.* **29,** 168–172.

Volkmar, F. R. and Greenough, W. T. (1972). Rearing complexity affects branching of dendrites in the visual cortex of the rat. *Science* **176,** 1445–1447.

Von Euler, U. S. and Gaddum, J. H. (1931). An unidentified depressor substance in certain tissue extracts. *J. Physiol.* (*Lond.*) **72,** 74–87.

Vrba, R. (1962). Glucose metabolism in rat brain *in vivo. Nature* (*Lond.*) **195,** 663–665.

Vrensen, G. and DeGroot, D. (1973). Quantitative stereology of synapses: A critical investigation. *Brain Res.* **58,** 23–35.

Vrensen, G. and DeGroot, D. (1974a). The effect of dark rearing and its recovery on synaptic terminals in the visual cortex of rabbits. A quantitative electron microscopic study. *Brain Res.* **78,** 263–278.

Vrensen, G. and DeGroot, D. (1974b). Osmium-zinc iodide staining and the quantitative study of synapses. *Brain Res.* **74,** 131–142.

Vrensen, G. and DeGroot, D. (1975). The effect of monocular deprivation on synaptic terminals in the visual cortex of rabbits. A quantitative electron microscopic study. *Brain Res.* **93,** 15–24.

Vrensen, G. and Van Hof, M. W. (1975). Behavioral and ultrastructural aspects of visual deprivation in the rabbit. *First European Neurosciences Meeting, Munich,* 9–13 September, Abstr. No. 419.

Wade, G. N. and Feder, H. H. (1974). Stimulation of ^{3}H-leucine incorporation into protein by estradiol-17β or progesterone in brain tissues of ovariectomized guinea pigs. *Brain Res.* **73,** 545–549.

Waelsch, H. (Ed.) (1955). "Biochemistry of the Developing Nervous System". Academic Press, New York and London, 537 pp.

Wald, F., Mazzuchelli, A., Lapetina, E. G. and DeRobertis, E. (1968). The effect of antiserum against nerve-ending membranes from cat cerebral cortex on the bioelectrical activity of mollusc neurons. *Exptl. Neurol.* **21,** 336–345.

Waldeyer, W. (1893). Ueber einige neuere Forschung im Gebiete der Anatomie des Centralnervensystems. *Deutsch med. Wochenschr.*, No. **44.**

Waley, S. G. (1957). Acidic peptides of the lens. 2. The use of ion-exchange resins as molecular sieves. *Biochem. J.* (*Lond.*) **67,** 172–176.

Waley, S. G. (1958). Acidic peptides of the lens. *Biochem. J.* (*Lond.*) **68,** 189–192.

Walker, C. A. (1972). Twenty-four hour LD_{50} rhythms of DL-amphetamines in isolated and aggregated mice. *In* "Drug Addiction, Experimental Pharmacology". Vol. 1, pp. 247–253. Futura Publ. Co., New York.

Walker, C. A., Speciale, S. G. and Friedman, A. H. (1971). The influence of drug treatment on norepinephrine levels and ultrastructure of the rat hypothalamus and caudate nucleus during a programmed light-dark cycle. *Neuropharmacology* **10,** 325–334.

Walker, J. E., Lewin, E., Sheppard, J. R. and Cromwell, R. (1973). Enzymatic regulation of adenosine 3′, 5′-monophosphate (cyclic AMP) in the freezing epileptogenic lesion of rat brain and in homologous contralateral cortex. *J. Neurochem.* **21,** 79–85.

Walker, J. E., Lewin, E. and Moffitt, B. (1975). Production of epileptiform discharges by application of agents which increase cyclic AMP levels in rat cortex. *In* "Epilepsy: Proc. of the Hans Berger Centenary Symposium" (P. Harris and C. Mawdsley, Eds), pp. 30–36. Churchill Livingstone, New York.

Wall, P. D. (1960). Cord cells responding to touch, damage, and temperature of skin. *J. Neurophysiol.* **23,** 197–210.

Wall, P. D. (1975). The somatosensory system. *In* "Handbook of Psychobiology" (M. S. Gazzaniga and C. Blakemore, Eds), pp. 373–392. Academic Press, New York and London.

Wallace, E. F., Krantz, M. J. and Lovenberg, W. (1973). Dopamine-β-hydroxylase: A tetrameric glycoprotein. *Proc. Nat. Acad. Sci., U.S.A.* **70,** 2253–2255.

Walsh, R. N. and Cummins, R. A. (1975). Mechanisms mediating the production of environmentally induced brain changes. *Psychol. Bull.* **82,** 986–1000.

Walsh, R. N. and Greenough, W. T. (Eds) (1976). "Environments as Therapy for Brain Dysfunction". Plenum Press, New York (in press).

Walsh, R. N., Budtz-Olsen, O. E., Penny, J. E. and Cummins, R. A. (1969). The effects of environmental complexity on the histology of the rat hippocampus. *J. Comp. Neurol.* **137,** 361–368.

Walsh, R. N., Budtz-Olsen, O. E., Torok, A. and Cummins, R. A. (1971). Environmentally induced changes in the dimensions of the rat cerebrum. *Develop. Psychobiol.* **4,** 115–122.

Walsh, R. N., Cummins, R. A. and Budtz-Olsen, O. E. (1973). Environmentally induced changes in the dimensions of the rat cerebrum: A replication and extension. *Develop. Psychobiol.* **6,** 3–7.

Walvarens, P. and Chase, H. P. (1969). Influence of thyroid on formation of myelin lipids. *J. Neurochem.* **16,** 1477–1484.

Warecka, K., Möller, H. J., Vogel, H.-M. and Tripatzis, I. (1972). Human brain-specific $alpha_2$-glycoprotein: Purification by affinity chromatography and detection of a new component; localization in nervous cells. *J. Neurochem.* **19,** 719–725.

Waser, P. G. (Ed.) (1975). "Cholinergic Mechanisms". Raven Press, New York, 573 pp.

Wasterlain, C. G. (1974). Brain ribosomes in intracranial hypertension. *J. Neurochem.* **23,** 253–259.

Watkins, J. C. (1967). Metabolic derangements and other causative factors in toxic convulsions. *Biochem. J.* (*Lond.*) **106,** 4–7 P.

Way, E. L., Iwamoto, E. T., Bhargava, H. N. and Loh, H. H. (1975). Adaptive cholinergic-dopaminergic responses in morphine dependence. *In* "Neurobiological Mechanisms of Adaptation and Behavior" (A. J. Mandell, Ed.), pp. 169–184. Raven Press, New York.

Weaver, L. C. and Kerley, T. L. (1962). Strain difference in response of mice to *d*-amphetamine. *J. Pharmacol. Exptl. Therapeut.* **135,** 240–244.

Weeks, J. R. (1975). Environmental influences affecting the voluntary intake of drugs: An overview. *Fed. Proc.* **34,** 1755–1758.

Weigert, C. (1895). "Encyclopädie d. mikroskop. Technik". Cited by Hydén (1973b).

Weight, F. F. (1971). Mechanisms of synaptic transmission. *Neurosci. Res.* **4,** 1–27.

Weil-Malherbe, H. (1970). The biochemistry of affective disorders. *In* "Handbook of Neurochemistry" (A. Lajtha, Ed.), Vol. 3, pp. 371–416. Plenum Press, New York.

Weinberger, J., Nicklas, W. J. and Berl, S. (1976). Mechanism of action of anticonvulsants: Role of the differential effects on the active uptake of putative neurotransmitters. *Neurology* **26,** 162–166.

Weiskrantz, L. (1958). Sensory deprivation and the cat's optic nervous system. *Nature* (*Lond.*) **181,** 1047–1050.

Weiss, B. and Laties, V. G. (1962). Enhancement of human performance by caffeine and the amphetamines. *Pharmacol. Rev.* **14,** 1–36.

Weiss, B. F., Roel, L. E., Munro, H. N. and Wurtman, R. (1974). *In* "Aromatic Amino Acids in the Brain", pp. 325–334. Ciba Fdn. Symp. 22 (New Series).

Weiss, B. L. and Aghajanian, G. K. (1971). Activation of brain serotonin metabolism by heat: Role of raphe neurons. *Brain Res.* **26,** 37–48.

Weiss, J. M., Stone, E. A. and Harrell, N. (1970). Coping behavior and brain norepinephrine level in rats. *J. Comp. Physiol. Psychol.* **72,** 153–160.

Weiss, J. M., Glazer, H. I. and Pohorecky, L. A. (1974). Coping behavior and neurochemical changes: An alternative explanation for the original "learned helplessness" experiments. *In* "Relevance of the Psychopathological Animal Model to the Human". Plenum Press, New York (in press).

Weiss, P. and Hiscoe, H. B. (1948). Experiments on the mechanism of nerve growth. *J. Exptl. Zool.* **107,** 315–395.

Weiss, P. and Pillai, A. (1965). Convection and fate of mitochondria in nerve fibers: Axonal flow as vehicle. *Proc. Nat. Acad. Sci., U.S.A.* **54,** 48–56.

Weiss, P. A. and Mayr, R. (1971). Organelles in neuroplasmic ("axonal") flow: Neurofilaments. *Proc. Nat. Acad. Sci., U.S.A.* **68,** 846–850.

Welch, A. S. and Welch, B. L. (1968a). Effect of stress and *para*-chlorophenylalanine upon brain serotonin, 5-hydroxyindoleacetic acid and catecholamines in grouped and isolated mice. *Biochem. Pharmacol.* **17,** 699–708.

Welch, A. S. and Welch, B. L. (1968b). Reduction of norepinephrine in the lower brainstem by a psychological stimulus. *Proc. Nat. Acad. Sci., U.S.A.* **60,** 478–481.

Welch, A. S. and Welch, B. L. (1971). Isolation, reactivity and aggression: Evidence for an involvement of brain catecholamines and serotonin. *In* "The Physiology of Aggression and Defeat" (B. E. Eleftheriou and J. P. Scott, Eds), pp. 91–142. Plenum Press, New York.

Welch, B. L. (1965). Psychophysiological response to the mean level of environmental stimulation: A theory of environmental integration. *In* "Symp. Med. Aspects of Stress in Military Climate" (D. Mck. Rioch, Ed.), pp. 39–96. U.S. Govt. Printing Off., Washington, D.C.

Welch, B. L. (1967). Discussion of aggression, defense and neurohumors by A. B. Rothballer. *In* "Aggression and Defense, Neural Mechanisms and Social Patterns: U.C.L.A. Forum in Medical Sciences" (C. D. Clemente and D. B. Lindsley, Eds), Vol. 7, p. 150. University of California Press, Los Angeles.

Welch, B. L. and Goldberg, A. M. (1973). Adrenal choline acetyl-transferase activity: Sustained effects of chronic intermittent psychological and psychosocial stimulation. *Int. J. Neurosci.* **5,** 95–99.

Welch, B. L. and Welch, A. S. (1966). Graded effect of social stimulation upon *d*-amphetamine toxicity, aggressiveness and heart and adrenal weight. *J. Pharmacol. Exptl. Therapeut.* **151,** 331–338.

Welch, B. L. and Welch, A. S. (1968c). Greater lowering of brain and adrenal catecholamines in group-housed than in individually-housed mice administered DL-α-methyltyrosine. *J. Pharmacol. Exptl. Therapeut.* **20,** 244–246.

Welch, B. L. and Welch, A. S. (1968d). Rapid modification of isolation-induced aggressive behavior and elevation of brain catecholamines and serotonin by the quick-acting monoamine oxidase inhibitor, pargyline. *Comm. Behav. Biol.* **1,** 347–351.

Welch, B. L. and Welch, A. S. (1969a). Aggression and the biogenic amine neurohumors. *In* "Aggressive Behaviour" (S. Garattini and E. B. Sigg, Eds), pp. 188–202. Excerpta Medica Fdn., Amsterdam.

Welch, B. L. and Welch, A. S. (1969b). Sustained effects of brief daily stress (fighting) upon brain and adrenal catecholamines and adrenal, spleen, and heart weights of mice. *Proc. Nat. Acad. Sci., U.S.A.* **64,** 100–107.

Welch, B. L. and Welch, A. S. (1973). Chronic social stimulation and tolerance to amphetamine: Interacting effects of amphetamine and natural nervous stimulation upon brain amines and behavior. *In* "Current Concepts on Amphetamine Abuse" (E. H. Ellinwood, Jr. and S. Cohen, Eds), pp. 107–115. U.S. Govt. Printing Office, Washington, D.C.

Welch, B. L., Brown, D. G., Welch, A. S. and Lin, D. C. (1974a). Isolation, restrictive confinement or crowding of rats for one year. I. Weight, nucleic acids and protein of brain regions. *Brain Res.*, **75,** 71–84.

Welch, B. L., Hendley, E. D. and Turek, I. (1974b). Norepinephrine uptake into cerebral cortical synaptosomes after one fight or electroconvulsive shock. *Science* **183,** 220–221.

Welford, A. T. (1958). "Ageing and Human Skill". Oxford University Press, London.

Wenzel, J., David, H., Pohle, W., Marx, I. and Matthies, H. (1975). Free and membrane-bound ribosomes and polysomes in hippocampal neurons during a learning experiment. *Brain Res.* **84,** 99–109.

Werman, R. (1966). A review—Criteria for identification of a central nervous system transmitter. *Comp. Biochem. Physiol.* **18,** 745–766.

Westman, J. (1971). Quantitative estimation of the proportion of perikaryal surface area covered with boutons—a possibility to distinguish different nerve cell populations. *Brain Res.* **32,** 203–207.

Westrum, L. E., White, L. E. and Ward, A. A. (1964). Morphology of the experimetal epileptic focus. *J. Neurosurg.* **21,** 1033–1044.

Wheeler, D. D., Boyarsky, L. L. and Brooks, W. H. (1966). Release of amino acids from nerve during stimulation. *J. Cell. Physiol.* **67,** 141–148.

Wherrett, J. R. and McIlwain, H. (1962). Gangliosides, phospholipids, protein and ribonucleic acid in subfractions of cerebral microsomal material. *Biochem. J.* (*Lond.*) **84,** 232–237.

White, W. A. (1931). Medical psychology. The mental factor in disease. *Nerv. and Ment. Dis. Monogr. Series*, No. **54,** 141 pp.

Whittaker, V. P. (1965). The application of subcellular fractionation techniques to the study of brain function. *Progr. Biophys. Mol. Biol.* **15,** 41–88.

Whittaker, V. P. (1973). The structural and chemical properties of synaptic vesicles. *In* "Proteins of the Nervous System" (D. J. Schneider, Ed.), pp. 155–169. Raven Press, New York.

Whittaker, V. P., Michaelson, I. A. and Kirkland, R. J. (1964). The separation of synaptic vesicles from nerve-ending particles ("synaptosomes"). *Biochem. J.* (*Lond.*) **90,** 293–303.

Wiegandt, H. (1967). The subcellular localization of gangliosides in the brain. *J. Neurochem.* **14,** 671–674.

Wiesel, T. N. and Hubel, D. H. (1963). Single-cell responses in striate cortex of kittens deprived of vision in one eye. *J. Neurophysiol.* **26,** 1004–1017.

Wiesel, T. N. and Hubel, D. H. (1965a). Comparison of the effects of unilateral and bilateral eye closure on cortical unit responses in kittens. *J. Neurophysiol.* **28,** 1029–1040.

Wiesel, T. N. and Hubel, D. H. (1965b). Extent of recovery from the effects of visual deprivation in kittens. *J. Neurophysiol.* **28,** 1060–1072.

Wigglesworth, J. S. (1964). Experimental growth retardation in the fetal rat. *J. Path. Bact.* **88,** 1–14.

Wiklund, L. and Bjorklund, A. (1975). Regrowth of central serotonin neurons after 5,6-dihydroxytryptamine-induced axotomy in the rat brain stem. *First European Neurosciences Meeting, Munich,* 9–13 September, Abstr. No. 434.

Wilburn, M. W. and Kesner, R. P. (1972). Differential amnestic effects produced by electrical stimulation of the caudate nucleus and nonspecific thalamic system. *Exptl. Neurol.* **34,** 45–50.

Wilburn, M. W. and Kesner, R. P. (1974). Effects of caudate nucleus stimulation upon initiation and performance of a complex motor task. *Exptl. Neurol.* **45,** 61–71.

Will, B. E., Rosenzweig, M. R. and Bennett, E. L. (1976a). Effects of differential environments on recovery from neonatal brain lesions, measured by problem-solving scores and brain dimensions. *Physiol. Behav.* **16,** 603–611.

Will, B. E., Rosenzweig, M. R., Bennett, E. L., Hebert, M. and Morimoto, H. (1976b). Relatively brief environmental enrichment aids recovery of learning capacity and alters brain measures after postweaning brain lesions in rats. *J. Comp. Physiol. Psychol.* (In press).

Williams, R., Little, S. and Ensinck, J. (1969). Adenyl cyclase and phosphodiesterase activities in brain areas of man, monkey and rat. *Amer. J. Med. Sci.* **258,** 190–202.

Wilson, J. (1968). Chemical and metabolic aspects of heredodegenerative neurological disorders. *In* "Applied Neurochemistry" (A. N. Davison and J. Dobbing, Eds), pp. 401–423. F. A. Davis Co., Philadelphia.

Wilson, L. and Meza, I. (1973). The mechanism of action of colchicine. *J. Cell. Biol.* **58,** 709–719.

Wiltse, H. E. and Menkes, J. H. (1972). Brain damage in the amino acidurias. *In* "Handbook of Neurochemistry" (A. Lajtha, Ed.), Vol. 7, pp. 143–167. Plenum Press, New York.

Winer, B. J. (1962). "Statistical Principles in Experimental Design". McGraw-Hill, New York.

Winick, M. (1969). Malnutrition and brain development. *J. Pediat.* **74,** 667–679.

Winick, M. and Noble, A. (1965). Quantitative changes in DNA, RNA, and protein during prenatal and postnatal growth in the rat. *Develop. Biol.* **12,** 451–466.

Winick, M., Rosso, P. and Waterlow, J. (1970). Cellular growth of cerebrum, cerebellum and brain stem in normal and marasmic children. *Exptl. Neurol.* **26,** 393–400.

Winick, M., Meyer, K. K. and Harris, R. C. (1975). Malnutrition and environmental enrichment by early adoption. *Science* **190,** 1173–1175.

Wisniewski, H. M. and Terry, R. D. (1970). An experimental approach to the morphogenesis of neurofibrillary degeneration and the argyrophilic plaque. *In* "Alzheimer's Disease and Related Conditions" (G. E. W. Wolstenholme and M. O'Connor, Eds), pp. 223–248. J. & A. Churchill, Ltd., London.

Wisniewski, H. and Terry, R. (1973). Morphology of the aging brain, human and animal. *In* "Neurological Aspects of Mutation and Aging" (D. H. Ford, Ed.), pp. 167–186. Elsevier, Amsterdam.

Wisniewski, H. M., Terry, R. D. and Hirano, A. (1970). Neurofibrillary pathology. *J. Neuropathol. Exptl. Neurol.* **29,** 163–176.

Wisniewski, H. M., Narang, H. K. and Terry, R. D. (1976). Neurofilbrillary tangles of paired helical filaments. *J. Neurol. Sci.* **27,** 173–181.

Wolburg, H. (1972). Intraaxonaler transport von ethidium-bromidsensitiven RNS- und niedermolekularen ^{3}H-uridin-verbindungen im tractus opticus von teleosteern. *Exptl. Brain Res.* **15,** 348–363.

Wolfe, L. S. (1961). The distribution of gangliosides in subcellular fractions of guinea-pig cerebral cortex. *Biochem. J.* (*Lond.*) **79,** 348–355.

Wolthuis, O. L. (1972). Experiments with UCB 6215, a drug which enhances acquisition in rats: Its effects compared with those of methamphetamine. *Eur. J. Pharmacol.* **16,** 283–297.

Woodburne, L. S. (1967). "The Neural Basis of Behavior". Charles E. Merrill Publ. Co., Columbus, Ohio, 378 pp.

Woodbury, D. M. and Kemp. J. W. (1971). Pharmacology and mechanisms of action of diphenylhydantoin. *Psychiat. Neurol. Neurochir.* **74,** 91–115.

Woodward, D. J., Bickett, D. and Chanda, R. (1975). Purkinje cell dendritic alterations after transient developmental injury of the external granular layer. *Brain Res.* **97,** 195–214.

Woolley, D. W. and Gommi, B. W. (1964). Serotonin receptors: V. Selective destruction by neuroaminidase plus EDTA and reactivation with tissue lipids. *Nature* (*Lond.*) **202,** 1074–1075.

Woolley, D. W. and Gommi, B. W. (1965). Serotonin receptors. VII. Activities of various pure gangliosides as the receptors. *Proc. Nat. Acad. Sci., U.S.A.* **53,** 959–963.

Worden, F. G. (1966). Attention and auditory electrophysiology. *In* "Progress in Physiological Psychology" (E. Stellar and J. Sprague, Eds), Vol. 1, pp. 45–116.

Wurtman, R. J., Axelrod, J. and Kelly, D. E. (1968). "The Pineal". Academic Press, New York and London, 199 pp.

Yahr, M. D. (Ed.) (1974). "Current Concepts in the Treatment of Parkinsonism". Raven Press, New York, 268 pp.

Yamada, K. M. and Wessells, N. K. (1971). Axon elongation: Effect of nerve growth factor on microtubule protein. *Exptl. Cell Res.* **66,** 346–352.

Yamori, Y., Lovenberg, W. and Sjoerdsma, A. (1970). Norepinephrine metabolism in brainstem of spontaneously hypertensive rats. *Science* **170,** 544–546.

Yanai, J. and Ginsburg, B. E. (1976). Increased sensitivity to chronic ethanol in isolated mice. *Psychopharmacologia* **46,** 185–189.

Yen, H. C., Day, C. A. and Sigg, E. B. (1962). Influence of endocrine factors on the development of fighting behavior in rodents. *Pharmacology* **4,** 173.

Yoshida, M. and Precht, W. (1971). Monosynaptic inhibition of neurons of the substantia nigra by caudatonigral fibers. *Brain Res.* **32,** 225–228.

Young, A. B., Oster-Granite, M. L., Herndon, R. M. and Snyder, S. H. (1974). Glutamic acid. Selective depletion by viral induced granule cell loss in hamster cerebellum. *Brain Res.* **73,** 1–13.

Young, J. Z. (1938). *In* "Evolution" (G. R. de Beer, Ed.), pp. 179–204, Oxford.

Young, J. Z. (1965), The Croonian lecture: The organization of a memory system. *Proc. Roy. Soc., B.* **163,** 285–320.

Young, P. T. (1961). "Motivation and Emotion: A Survey of the Determinants of Human and Animal Activity". John Wiley & Sons, Inc., New York.

Yuwiler, A. (1971). Problems in assessing biochemical ontogeny. *In* "Brain Development and Behavior", pp. 43–57. Academic Press, New York and London.

Yuwiler, A. and Geller, E. (1965). Serotonin depletion by dietary leucine. *Nature* (*Lond.*) **208,** 83–84.

Yuwiler, A. and Geller, E. (1969). Brain serotonin changes in phenylalanine-fed rats: Synthesis, storage and degradation. *J. Neurochem.* **16,** 999–1005.

Yuwiler, A. and Louttit, R. T. (1961). Effects of phenylalanine diet on brain serotonin in the rat. *Science* **134,** 831–832.

Yuwiler, A., Geller, E. and Slater, G. G. (1965). On the mechanism of the brain serotonin depletion in experimental phenylketonuria. *J. Biol. Chem.* **240,** 1170–1174.

Yuwiler, A., Geller, E. and Schapiro, S. (1973). Effect of neonatal corticoids on tryptophan pyrrolase and brain serotonin. *In* "Hormones, Metabolism and Stress; Recent Progress and Perspectives" (Š. Németh, Ed.), pp. 215–225, Publ. House of the Slovak Academy of Sciences, Bratislava.

Yuzepchuk, S. A., Rozina, E. E., Kaptsova, T. I. and Kulish, E. A. (1975). Morphological differences in the central nervous system and organs of monkeys inoculated intracerebrally with virulent and attenuated strains of Mumps virus. *Acta Virol.* **19,** 305–310.

Zanetta, J. P., Morgan, I. G. and Gombos, G. (1975). Synaptosomal plasma membrane glycoproteins: Fractionation by affinity chromatography on concanavalin A. *Brain Res.* **83,** 337–348.

Zangwill, O. I. (1961). Lashley's concept of cerebral mass action. *In* "Current Problems in Animal Behavior" (W. H. Thorpe and O. G. Zangwill, Eds). Cambridge University Press, Cambridge.

Zemp, J. W., Wilson, J. E., Schlesinger, K., Boggan, W. O. and Glassman, E. (1966). Brain function and macromolecules. I. Incorporation of uridine into RNA of mouse brain during short-term training experience. *Proc. Nat. Acad. Sci., U.S.A.* **55,** 1423–1431.

Zemp, J. W., Wilson, J. E. and Glassman, E. (1967). Brain function and macromolecules. II. Site of increased labelling of RNA in brains of mice during a short-term training experience. *Proc. Nat. Acad. Sci., U.S.A.* **58,** 1120–1125.

Zerbolio, D. J. (1969). The proactive effect of electroconvulsive shock on memory storage: With and without convulsions. *Commun. Behav. Biol.* **4,** 23–27.

Ziesenis, J. S., Davis, D. E. and Smith, D. E. (1975). Diel variations in the aggressive behaviour of the mouse, *Mus musculus*. *Anim. Behav.* **23,** 941–948.

Zimmerman, E. and Critchlow, V. (1973). Inhibition of morphine-induced pituitary-adrenal activation by dexamethasone in the female rat. *Proc. Soc. Exptl. Biol. Med.* **143,** 1224–1226.

Zimmerman, R. R., Geist, C. R. and Wise, L. A. (1974). Behavioral development, environmental deprivation, and malnutrition. *Adv. Psychobiol.* **2,** 133–192.

Zimmerman, R. R., Geist, C. R. and Ackles, P. K. (1975). Changes in the social behavior of Rhesus monkeys during rehabilitation from prolonged protein-calorie malnutrition. *Behav. Biol.* **14,** 325–333.

Zinkin, S. and Miller, A. J. (1967). Recovery of memory after amnesia induced by electroconvulsive shock. *Science* **155,** 102–104.

Zomzely-Neurath, C., York, C. and Moore, B. W. (1973). In vitro synthesis of two brain-specific proteins (S 100 and 14-3-2) by polyribosomes from rat brain. I. Site of synthesis and programming by polysome-derived messenger RNA. *Arch. Biochem. Biophys.* **155,** 58–69.

Zornetzer, S. F., Chronister, R. B. and Ross, B. (1973). The hippocampus and retrograde amnesia: Localization of some positive and negative memory disruptive sites. *Behav. Biol.* **8,** 507–518.

Zubek, J. P. (Ed.) (1969). "Sensory Deprivation: Fifteen Years of Research". Appleton-Century-Crofts, New York.

Subject Index

I

Author Index

The numbers in italic are those pages where references are listed in full

H

Q

R

S